CW01499022

Vocabulary for Sustainable Consumption and Lifestyles

Vocabulary for Sustainable Consumption and Lifestyles: A Language for Our Common Future curates a shared vocabulary of concepts that enables a society-wide conversation about sustainable consumption and lifestyles, the future of consumer society, and ways to transcend it.

Since the United Nations (UN) Earth Summit in Rio de Janeiro in 1992, the global environmental and social consequences of mass consumption have been well documented, yet progress is slow. Overconsumption and extractive practices continue to drive ecological overshoot. Set against this backdrop, each of the 87 essays in this book imparts a meaning to a concept, highlights its history, and offers different perspectives, interpretations, and applications for social change. The two premises of this book are that we need to transition to a society in which the *well-being* and *dignity* of people are achieved with a much smaller footprint and that technological solutions are inadequate for the challenge. Policies and actions are greatly lagging behind the growing understanding of the system of production-consumption because social change is often slow, and sustainable consumption does not have a clear political champion. The book addresses tensions that also interfere with progress, such as science versus politics, economic winners versus losers, traditions versus an uncertain future, and present needs versus future costs.

This innovative volume is an important resource for students, scholars, policymakers, grassroots activists, and agents of change interested in sustainable consumption and sustainable living more broadly.

Lewis Akenji is the executive director of the Hot or Cool Institute in Berlin, a public-interest think tank that explores the intersection between society and sustainability. Lewis has served as the executive director of SEED, founded as a United Nations partnership to promote entrepreneurship for sustainable development. He has consulted with multilateral institutions, including the UN, the Asian and African Development Banks, the European Commission, the Association of Southeast Asian Nations (ASEAN), and has served as technical or policy adviser to several national governments. He serves on several boards and international committees, including as a Full Member of the Club of Rome and Commissioner on the Transformational Economics Commission of Earth4All.

Philip J. Vergragt is a climate activist, professor emeritus of technology assessment at TU Delft, the Netherlands, and a research professor at Clark University, USA. He is one of the co-founders and a current board member of SCORAI. He co-chairs the Electric Vehicles Task Force and is an advisory member of the Energy Commission at Newton, MA. His current research interests are sustainable consumption, sustainable cities, and systemic change. He is the co-author of more than 100 scientific publications and five books. Philip holds a PhD in physical chemistry at the University of Leiden, the Netherlands (1976).

Halina Szejnwald Brown is professor emerita of environmental science and policy at Clark University. Her recent academic research has focused on the interface between culture, technology, and policy in facilitating a transition beyond the current consumer society. She is a co-founder and board member of Sustainable Consumption Research and Action Initiative and chairs Citizens Commission on Energy in her home city of Newton, Massachusetts. She is a fellow of the American Association for the Advancement of Science, fellow of the International Society for Risk Analysis, and fellow of Tellus Institute in Boston. Brown holds a doctoral degree in chemistry from New York University.

Thomas S.J. Smith is a researcher, writer, and editor based in the north of Spain. He received his PhD in geography and sustainable development at the University of St Andrews and has since held numerous roles including postdoctoral researcher in environmental studies at Masaryk University, Brno, and Marie Skłodowska-Curie postdoctoral fellow in geography at Ludwig Maximilian University (LMU), Munich. He is a member of the Community Economies Institute (CEI) and on the board of the Sustainable Consumption Research and Action Initiative (SCORAI). His research interests relate to social ecological transformations, economic localization, and degrowth.

Laura Maria Wallnöfer is a postdoctoral research and teaching associate at the Institute of Marketing and Innovation, Department of Economics and Social Sciences at BOKU University of Natural Resources and Life Sciences, Vienna. She has an interdisciplinary background in energy and transport management and sustainable development and did her PhD on the Integration of Perspectives and Concepts about Individuals as Change Agents at the Doctoral School for Transitions to Sustainability at BOKU University. Her current research focuses on the intersections of different transition actors' influence spheres and how the multi-actor process required for a sustainable transformation can be better coordinated if those intersections are known.

Routledge – SCORAI Studies in Sustainable Consumption

This series aims to advance conceptual and empirical contributions to this new and important field of study. For more information about The Sustainable Consumption Research and Action Initiative (SCORAI) and its activities please visit http://scorai.net.

Series Editors:

Halina Szejnwald Brown, *Professor Emerita at Clark University, USA.*
Philip J. Vergragt, *Emeritus Professor at TU Delft, The Netherlands;*
Research Professor at Clark University, USA.
Soumyajit Bhar, *Assistant Professor and Assistant Dean of Admissions and Outreach,*
School of Liberal Studies, BML Munjal University, India.

Subsistence Agriculture in the US
Reconnecting to Work, Nature and Community
Ashley Colby

Sustainable Lifestyles after Covid-19
Fabián Echegaray, Valerie Brachya, Philip J. Vergragt and Lei Zhang

Sustainable Products in the Circular Economy
Impact on Business and Society
Edited by Magdalena Wojnarowska, Marek Ćwiklicki and Carlo Ingrao

Narrating Sustainability through Storytelling
Edited by Daniel Fischer, Sonja Fücker, Hanna Selm and Anna Sundermann

The Low-Carbon Good Life
Jules Pretty

Teaching and Learning Sustainable Consumption
A Guidebook
Edited by Jen Dyer, Daniel Fischer, Jordan King, Marlyne Sahakian, and Gill Seyfang

Consuming the Environment
Edited by Myra J. Hird

Vocabulary for Sustainable Consumption and Lifestyles
A Language for Our Common Future
Edited by Lewis Akenji, Philip J. Vergragt, Halina Szejnwald Brown, Thomas S.J. Smith and Laura Maria Wallnöfer

For more information about this series, please visit: www.routledge.com/Routledge-SCORAI-Studies-in-Sustainable-Consumption/book-series/RSSC

Vocabulary for Sustainable Consumption and Lifestyles

A Language for Our Common Future

Edited by Lewis Akenji, Philip J. Vergragt, Halina Szejnwald Brown, Thomas S.J. Smith and Laura Maria Wallnöfer

Routledge
Taylor & Francis Group

LONDON AND NEW YORK

Designed cover image: Halina Szejnwald Brown and Philip J. Vergragt

First published 2026
by Routledge
4 Park Square, Milton Park, Abingdon, Oxon OX14 4RN

and by Routledge
605 Third Avenue, New York, NY 10158

Routledge is an imprint of the Taylor & Francis Group, an informa business

British Library Cataloguing-in-Publication Data
A catalogue record for this book is available from the British Library

ISBN: 978-1-032-95274-1 (hbk)
ISBN: 978-1-032-95248-2 (pbk)
ISBN: 978-1-003-58405-6 (ebk)

DOI: 10.4324/9781003584056

Typeset in Times New Roman
by Apex CoVantage, LLC

Contents

Preface

This book seeks to create a shared language. It aims to spearhead a conversation about the future of consumer society in a world threatened by interlinked ecological and social tensions. Its premise is that we need to transition away from mass consumption as the organizing principle of societal life to a society in which the well-being and dignity of people are achieved with a much smaller impact on life-supporting earth systems. This book is especially aimed at researchers, teachers, policymakers, activists, businesspeople, professional communicators, and, crucially, members of the general public who recognize that the current trajectory for addressing the ecological crises is inadequate. This trajectory is largely based on technological solutions and economic quick-fixes to what is essentially a social problem.

Since the United Nations (UN) Earth Summit in Rio de Janeiro in 1992, consumption has been recognized as an essential dimension of sustainable development. However, the concepts of sustainable consumption and lifestyles have remained poorly understood and articulated and, thus, underdeveloped for policy or consistent practical interventions. Since 1992, the few scholars working on sustainable consumption did so in isolation from each other and often from others, even within their own disciplinary silos. They tended to be inexperienced in policy processes and had weak links with grassroots advocacy and activism.

In 2008, a group of scholars in the United States created a forum for interconnecting these researchers and bridging their work with practitioners. It became the Sustainable Consumption Research and Action Initiative (SCORAI) – following a European Union-funded (EU) project a few years earlier under the name SCORE! The premise was that addressing the ecological problem by increasing energy efficiency and replacing fossil- with non-fossil energy sources addressed the supply side but ignored the demand side of the energy balance sheet and that this approach was woefully inadequate for reducing impacts associated with energy use. A better understanding was needed of why affluent societies consume so much and how that system of consumption functions.

Today, with approximately 1,500 members and activities across all continents and many countries, SCORAI is one of several nodes of research, policy analysis, and practice in the field of sustainable consumption and lifestyles. The Hot or Cool Institute, SCORAI's partner in creating this book, is prominent among them. The work of Hot or Cool Institute is predicated on the understanding that the magnitude, urgency, and scale of the ecological challenge require a rethink of our systems and how we organize ourselves as a society to meet our needs.

The interdisciplinary understanding of how this complex system that we call consumer society functions and reproduces itself has made huge progress. Various branches of the United Nations and European governments, including the EU, have adopted official proclamations about the need to reduce consumption. Among the most recent examples, in 2024 the UN Environment Assembly adopted a dedicated resolution on "promoting sustainable lifestyles". Even the Intergovernmental Panel on Climate Change (IPCC), which since its inception in 1988 has primarily focused on assessments of impacts and interventions, devoted an entire chapter of its Sixth Assessment Report in 2022 to consumption. It concluded that there is the untapped potential to reduce greenhouse gas emissions by 40–70% by 2050 through changes in consumption and lifestyles. Consumption also appears in the media with growing frequency.

But policies and actions greatly lag behind this growing awareness, knowledge, and understanding. To some extent, this is not surprising, as social change is often slow. Internal tensions also interfere with progress: between the interests and capabilities of the Global North and Global South; between science and politics; between the familiarity of established culture and the uncertainty of rapid change; between the huge cost of immediate action and the even greater future costs of inaction; between the need for a steady-state economy and concern for the well-being of people; between individual behavioral change and system change; between eroded social trust and cohesion in highly unequal societies and the need for collective action. In addition, sustainable consumption, unlike other urgent environmental problems and potential solutions, does not have a clear political champion.

This book has two objectives. One is to curate a common language – a shared vocabulary of concepts – that will enable people from very diverse walks of life to understand and talk about the roots of the current ecological crisis and potential solutions. By assembling and cross-connecting the elements of that language in one place, we seek to create a conversation about consumption and lifestyles. In the digital age, anyone can find some kind of explanation online for each of the concepts included in this book. But we seek to impart a specific meaning to these concepts, to interpret them, their history, different perspectives, and applications in the context of consumption and social change. This vocabulary will hopefully result in a more robust and productive discourse and new insights on how to transition to a post-consumer society.

The second objective is to strengthen the multidisciplinary community or network of researchers, practitioners, and activists, and to create a common understanding of what we mean by "sustainable consumption and lifestyles". As we will see from the various contributions, the understandings, framings, focal points, problem definitions, and even language are quite different from each other, although there are obvious overlaps and similarities that we have tried to highlight. A common language is vital for shared and cohesive social change. We hope that ultimately this book will open new doors for action and mobilize the change makers, be they academics, activists, citizens, or policymakers.

We thank the following colleagues for reviewing and commenting on an earlier version of the introduction which follows: Prof. Julia Steinberger, University of Lausanne; Dr. Elias T. Ayuk, International Resource Panel; Dr Yasuhiko Hotta, Institute for Global Environmental Strategies, Dr. Kathleen Rest, Union of Concerned Scientists; and Cory Alperstein, climate activist and independent scholar. We also want to thank all authors for patiently bearing with the editorial team and their many demands.

Figures

Tables

Boxes

Introduction

Consumption Officially Enters the Sustainability Debate

As world leaders prepared to gather in Rio de Janeiro in 1992 on the eve of the United Nations (UN) Conference on Environment and Development (also known as the "Earth Summit"), the complexities of unsustainable development were just beginning to dawn on them in uncomfortable ways. Until then, in the "Global North", the widely held view was that population growth and poverty were the main unaddressed drivers of environmental harm. Following this framing, negotiations at the Summit would need to focus on the "Global South", given its high rates of fertility and poverty.

But scientific research was beginning to tell a richer, unacknowledged story: in the Global South, negative impacts of activities such as cutting down forests, digging out minerals, and growing bananas and coffee in unsustainable ways, arose primarily from activities undertaken to satisfy the ever-growing appetites of a minority global population in the Global North. In a globalized economy, *overproduction* in the Global South was the flip side of *overconsumption* in the Global North.

This reframing of the problem would threaten to derail negotiations at the Summit – it would position countries from the Global North versus those from the Global South, Big Agriculture versus small farmers, foreign aid versus fair compensation for labor and resources, and accusations of neo-colonialism versus corrupt local governance. It would also be a major shift in how environmental protection would be perceived by industrial countries; since the 1960s, these countries had invested primarily in controlling pollution generated within their own borders.

In the end, the wording of the final resolution was a balancing act. It stated: "inappropriate development resulting in overconsumption, coupled with an expanding world population" are the cause of environmental degradation (UN, 1992, para 6.1). An entire chapter of the action document was dedicated to "changing consumption patterns": calling on relevant parties to "develop a better understanding of the role of consumption" and to develop "policies and strategies to encourage changes in unsustainable consumption patterns".

The resulting *Report of the United Nations Conference on Environment and Development* (UN, 1992, para 4.5), better known as Agenda 21, declares:

Although consumption patterns are very high in certain parts of the world, the basic consumer needs of a large section of humanity are not being met. This results in excessive demands and unsustainable lifestyles among the richer segments, which place immense stress on the environment. The poorer segments, meanwhile, are unable to meet food, health care, shelter and educational needs. Changing consumption patterns will require a multipronged strategy focusing on demand, meeting the basic needs of the poor, and reducing wastage and the use of finite resources in the production process.

DOI: 10.4324/9781003584056-1

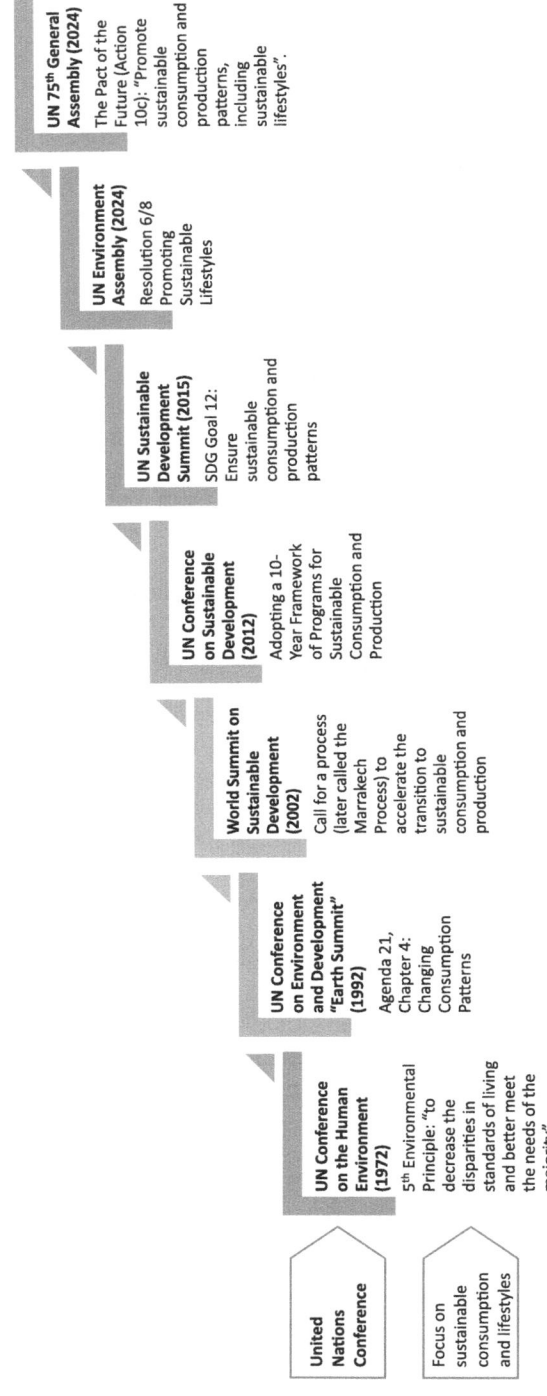

Figure 0.1 United Nations sustainability conferences, their outcome documents, and their focus on consumption and lifestyles from 1972 to 2024

This quote provides an understanding that would shape research, policy and actions on sustainable consumption and lifestyles to this day. Versions of the declaration have percolated through every UN sustainability summit since Agenda 21 (see Figure 0.1), albeit in watered-down interpretations, including the current framework for the sustainable development goals (SDGs).

Global efforts to address unsustainable development from this consumption-population perspective – by world leaders at the UN negotiations, research scientists, businesses, civil society organizations, and activists – resulted in near-constant tensions and little apparent success. This demonstrates the complexity unveiled by the new problem acknowledgment at the Earth Summit. The continuing rise in global temperatures, biodiversity loss, an increasing number of natural disasters, and rising ecosocial tensions, all demonstrate the failure to bring these efforts together effectively. It further suggests the urgent need to revisit what and why we consume, how we produce, distribute, and discard the things we do consume, and how our complex production-consumption system is linked to social and ecological tipping points.

This introductory chapter begins by sketching out the origins of consumer society (Section 2), the history and meaning of the term "sustainable consumption" in the global sustainability discourse (Section 3), and the evolution of the understanding of how the modern societal system of consumption functions (Section 4). Section 5 discusses some of the major challenges standing in the way of a transition to a different organizing principle of societal life, and Section 6 makes the case for creating a vocabulary in which these issues can be widely discussed. The 87 entries in this book represent a non-exhaustive list of established and emerging concepts at the core of language being used in these discussions. Like all languages, this one will evolve over time through use; hopefully, it will result in mobilizing powerful actors to set priorities and affect social change.

Engineering a Consumer Society: From the United States to the World

More than half a century before the 1992 Earth Summit, the British economist John Maynard Keynes published his well-known essay *Economic Possibilities for our Grandchildren*. After observing in his lifetime the huge increase in workers' productivity in the manufacturing and energy sectors – owing to technological and process innovations, and the harnessing of fossil energy – Keynes predicted that a hundred years from then people will be able to drastically reduce their working hours to about 15 hours per week in order to satisfy their needs. The "economic problem" would be solved. A new challenge would be *"for the ordinary person, with no special talents, to occupy himself, especially if he no longer has roots in the soil or in custom or in the beloved conventions of a traditional society"*.

Keynes' predictions have not come to pass. In Europe and North America, working hours have indeed decreased since then, though not nearly in step with continuing productivity growth, as Keynes anticipated. Where he was most mistaken, perhaps, was the assumption that the motivation for people to work is mostly to satisfy their basic needs. He did not consider that, once the "economic problem" of basic subsistence was solved, cultivated and seemingly insatiable wants would take their place as the motivator for work and the pursuit of well-being.

In Keynes' days, it was impossible to foresee that a complex societal system would be developed to perpetuate the human drive to satisfy wants through the acquisition and maintenance of goods and services. Nor could it be envisioned that this behavior would become the organizing principle of societal life, including culture, institutions, politics, and the economy. Today, we call this complex system the *consumer society*.

In her magisterial work *A Consumer Republic* (2003), historian Lisabeth Cohen describes how, during the three decades after the end of World War II (WWII), consumer society was constructed in the United States. It is instructive to retell this story because the US model soon became a prototype to be exported and replicated throughout the world, overriding alternative models of development (see Box 0.1).

Box 0.1   Construction of consumer society in the United States

After the deprivations of the Great Depression and the sacrifices of WWII, the United States in 1945 was ready for a victorious shopping spree. The government – fearful that the return of war veterans would lead to widespread unemployment – looked to industry to shift its wartime production capacity toward the civilian sector. The Employment Act of 1946 stated that *"federal government's responsibility . . . [is to] . . . promote maximum employment, production, and purchasing power"*. Industry happily complied by deploying aggressive and sophisticated modern marketing and advertising methods to increase demand for their products. The labor unions too were willing partners in the effort. As early as 1944, the American Federation of Labor (AFL) wrote, *"Without adequate purchasing power in the form of wages we cannot get full employment"*.

The 1944 "GI Bill" helped returning World War II veterans to get free college education, as well as down payments and government-guaranteed loans for purchasing homes and other goods (Black Americans were excluded). The mortgage interest deductions, government-financed infrastructure, and local zoning laws facilitated the growth of sprawling suburbs, with their endless procession of appliances and furnishings, and high dependence on car-based mobility. The private suburban shopping mall became a public space – stratified by race and income – replacing the previously more egalitarian public spaces of city streets, cafes, and places of commerce. The proliferation of credit cards allowed people to buy now and pay later. The earlier creation of Social Security in 1935 facilitated the transition to mass consumption by relieving Americans from the need to save for old age.

The results were astonishing. National output of goods and services doubled between 1946 and 1956 and doubled again by 1970, driven by private consumption expenditures. From then on, economic growth became a measure of general prosperity driven by consumption. By 1960, 62% of Americans owned their homes, compared to 44% in 1940.

At the peak of the Cold War, American lifestyles also served as an important symbol of the superiority of the capitalist system over Soviet-style socialism. In the famous "kitchen" debate between Vice President Nixon and Soviet Premier Khrushchev at the American Exhibition in Moscow in 1959, Nixon boasted: "The United States [has] come closest to the ideal of prosperity for all in a classless society".

In short, a major cultural and economic transition took place in the United States in the span of a single generation. The transition not only changed the lifestyles of most Americans in profound ways but also fostered a cultural shift: mass consumption and the suburban lifestyle became almost a national civic religion, conflated with such fundamental human aspirations as well-being and democracy.

Source: Adapted from Cohen (2003)

By the 1970s, 20 years before at the Earth Summit consumerism was recognized as a global ecological problem, all key pillars of the growth-oriented consumer society were firmly in place: the debt-financed economy – both household and government – that demanded growth to survive; pension funds dependent on growth to deliver on their commitments; the financial system happy to fuel growth, both real and imagined; and the ever-rising aspirations of households. By the 1970s, consumers as a distinct class, transcending the traditional sociological groupings by age, race, wealth, education, or political leanings, became a political force to reckon with. Through **boycotts, buycotts**, and other public campaigns, consumer movements punished and rewarded companies for their social and ecological impacts. Consumer protection laws were adopted to protect that distinct class.

The neoliberal ideology – which proclaimed that greed and wealth accumulation are good, markets know best, and government is a drag on the economy – came to the fore in the late 1970s and extended its influence to the present day. Its policies emphasize free trade, deregulation, lower taxes, increasing consumption to grow the economy, and promoting the ideology of consumer sovereignty. It intentionally conflates a consumer with a **citizen** and creates a transactional relationship between citizens and the government: the purpose of public policy would increasingly not be *to serve the public good* but rather to satisfy the consumer-voter.

In the next logical stage of maturation, by the mid- to late 20th century, many elements of neoliberalism and the US-style consumer society spread to other parts of the world, first in the industrial economies of North America and Western Europe, then in Asia and post-socialist Europe, and finally in capitals and major cities of the rest of the world.

Defining Sustainable Consumption, Sustainable Lifestyles – and Sustainable Living[1]

Despite its growing use in research and policy circles, the operational meaning of the term sustainable consumption is quite fluid. Some may associate it with daily household decisions such as minimal shopping, leisure flying, and taking cruises, and with green and ethical shopping. For social activists, sustainable consumption may bring to mind changing personal and community **value systems**, **norms**, **social practices**, personal priorities, diet, or adopting a minimalist way of life. From a policy perspective, sustainable consumption may signify a **personal carbon budget**, shorter workweek, limits on advertising, and more efficient buildings and mobility infrastructure; while the **political economy** perspective points toward a diminished role of the financial sector, **degrowth**, a **steady-state economy**, taxation reform, and reigning in corporate power. Each perspective has its own language and shared understandings among its adherents, but communication between them is inadequate.

Most definitions of sustainable consumption and lifestyles are modifications of the now-classic definition of sustainable development offered by the World Commission on Environment and Development (known as the Brundtland Report, titled *Our Common Future*): "*sustainable development is development that meets the needs of the present without compromising the ability of future generations to meet their own needs*" (WCED, 1987). Although the root concepts "lifestyle" and "consumption" have been previously defined by academics, the modifier "sustainable" brings additional complexity to the definitions.

Sustainable consumption is both a concept and a practice, and research on it sets out to understand and promote the types of consumption behaviors that are conducive to a sustainable society. This is reflected in one of its earliest and most widely used definitions (known as the "Oslo Declaration"):

Sustainable consumption is "*the use of goods and services that respond to basic needs and bring a better quality of life, while minimizing the use of natural resources, toxic materials and emissions of waste and pollutants over the life cycle, so as not to jeopardize the needs of future generations*".

(Ofstad et al., 1994)

The concept of sustainable lifestyles goes beyond consumption, however. It speaks to more complex and intangible aspects of human life, including habits, social practices, traditions, aspirations, and the search for meaning, all embedded in societal structures: physical, economic, cultural, and political. A 2016 definition captures this complexity:

> *A sustainable lifestyle minimizes ecological impacts while enabling a flourishing life for individuals, households, communities, and beyond.*
>
> (Vergragt et al., 2016)

In policy and academic discourse, the two terms – sustainable consumption and sustainable lifestyles – are often used interchangeably, and confusingly so (Gabriel & Lang, 2006; Miles, 1998). One helps to define the other and vice versa: sustainable consumption research emphasizes the material aspect of sustainable lifestyles, and conversely, lifestyles determine one's consumption patterns. While sustainable consumption tends to address decisions and actions linked to the purchase, use, and disposal of material products and services, the term sustainable lifestyle incorporates other actions in the context of values, education, community, and infrastructure.

One term that easily connects sustainable consumption and sustainable lifestyles is "sustainable living". While encompassing both terms, it does so without taking on the notion of "lifestyle" as advertised by corporate marketing.

> *Sustainable living is equitable consumption and lifestyles that contribute to wellbeing of individuals and society within ecological limits.*
>
> (Akenji, 2019)

This definition has at its core three key elements: ecological limits as a basis and boundary for providing individual and societal needs; equity and justice in how we organize ourselves as a society and pursue our needs; and well-being as a shared objective. It also recognizes that consumption and lifestyles are embedded in a societal context, including institutions, norms, and infrastructures that frame and shape individual and collective choices. Implicit in the definition is the recognition that there is more than one way of living sustainably; there are different approaches by different individuals and societies that could be described as sustainable.

The concept of sustainable living also accommodates a broader set of relevant concepts, such as **voluntary simplicity**, minimalism, **sufficiency,** and healthy living – often combining physical, emotional, and spiritual health with a more limited role for materialism. In public health parlance, sustainable living promotes health and well-being for individuals, communities, and the planet by balancing ecological, economic, and social systems; in business, it would be reflected in **work-life balance** and employment conditions; and in marketing, promotion of "green" or healthy products and services.

Because of their overlaps and variations, the above three terms and definitions are used across different entries in this book. The emergence of these concepts can be seen in the evolving understanding of consumer society elaborated in the following section.

Evolving Understanding of Consumer Society as a Complex System

Over the millennia, scholars, religious leaders, and social commentators have expressed criticism of unrestrained wealth accumulation, consumption, and their underpinnings – greed, gluttony, acquisitiveness, profit-seeking, excessive luxury, and positional consumption. In the 20th century, writers such as Veblen (*The Theory of the Leisure Class*, 1899), Galbraith (*The Affluent Society*, 1958), Marcuse (*One-Dimensional Man,* 1964), Elgin (*Voluntary Simplicity*, 1981), and others focused largely on the moral, class, and existential dimensions of mass consumption and its impact on the quality of modern life.

That changed in 1992 when at the Earth Summit, **consumerism** was officially declared a *global ecological problem*. Since then, sustainable consumption and consumerism have attracted increasing attention from scientists, community activists, and policymakers. Much of the early research on the topic of consumption focused on consumer behavior. After all, the consumer is the most visible end-user of market products and is often assumed to have free will and rational expression of choice. Economic, social, and psychological theories have been used to explain the role of consumers as economic actors and to theorize about their motivations and drivers (Ajzen, 1991; Bourdieu, 1984; Thaler, 1980, 2016).

But people do not go about as rational, fully informed, and fully autonomous actors in the formal economy, as often attributed to them in economic theories. And while their pursuit of meaning and well-being must include the acquisition of goods and services, people do not solely seek these things through material means. *"People also love, generate ideas, express their value systems, take care of family, create art, cherish silence, pray, fast, in ways that material flows and market economics cannot account for"* (Akenji, 2019). They are also influenced by traditions and **social norms**, constrained by physical and regulatory infrastructure, and manipulated by sophisticated **advertising**. Part of the challenge of research for sustainable consumption and lifestyles is that these mixtures of material and psychological, rational and emotional, biological and cultural, all come together in vastly differing configurations across billions of people.

Steven Miles (1998) has noted that *"How we consume, why we consume, and the parameters laid down for us within which we consume have become increasingly significant influences on how we construct our everyday lives"*. The statement recognizes that lifestyles occur within, or are railroaded by, broader social and physical contexts; in approaching sustainable living, it is important to differentiate between factors at the individual or the household level, and those that are beyond (direct) individual control (Wallnoefer et al., 2024). Scholars thus began to argue that focusing on the individual consumer is problematic because it fails to recognize the historical, political, and social conditions that shape everyday life, including our consumption patterns. The frame thus expanded from consuming individuals to a *system of consumption*.

Books such as Schor's *The Overworked American* (1993) and *The Overspent American* (1998), De Graaf's *Affluenza* (2001), Maniates's and Princen's *Confronting Consumption* (2002), and works of several other authors featured in Jackson's *Earthscan Reader on Sustainable Consumption* (2006) recognized that consumption is a manifestation of both basic human psychology and intentionally designed societal systems, including the ways employment and economy are organized. Their work was further extended by Shove et al. (2012) who used **social practice theory** to emphasize that consumption is a collective act transmitted through widely accepted social norms and practices.

The psychological perspective on consumption – widely used until then and accused by many of "blaming the victim" or **consumer scapegoatism** – gave way to institutional theories, cultural analyses, macroeconomics, and theories of social change and collective action.

Scholars have emphasized the deliberate construction of the system of consumption in the United States and linked it specifically to the **economic growth** paradigm and, importantly, to its ideological underpinnings created during the Cold War period (see Section 2). While the idea of a deliberate construction implies that, at least in principle, the system of consumption could be dismantled, the ideological dimension also highlights the difficulty of doing so. This was illustrated by US President Bush's famous declaration in 1992 that *"the American way of life is not up for negotiation. Period"*.

Research on consumption also began to increasingly draw on more quantitative methodologies like natural resource accounting. Using methods such as Environmentally-Extended Multi-Regional Input-Output analysis enabled researchers to estimate the material and **carbon "footprint"** of a given consumption pattern and to show a clear correlation between the **level of income**, resource consumption, and environmental impacts. This focus on materials and energy, economic behaviors, and the desire for modeling and quantification partly explains why process and product optimization and technological innovations have become prevalent policy recommendations for climate mitigation (Akenji, 2019).

The landmark report by Jackson (2009) *Prosperity Without Growth* delved deeper into the systemic nature of mass consumption. It specifically highlighted the links between mass consumption and economic growth, the associated role of the financial sector, and the impact of economic globalization, which made mass production cheaper and the useful life of products shorter. The report challenged the notion that absolute decoupling of economic growth from the use of energy is achievable through technological solutions. Jackson followed in the footsteps of Herman Daly's (1993) theory of the **steady-state economy** and Peter Victor's (2008) macroeconomic models of such an economy in Canada.

From then on, questioning the powerful dominance of technological solutions to ecological crisis (also referred to as "techno-optimism", "weak sustainability approach", and "green technology approach") and of absolute decoupling of economic growth from energy and material use – once the position of a tiny minority – would gain more currency among academics (Alfredsson et al., 2018; Parrique et al., 2019). We clearly need both green technology and demand reduction through lifestyle changes (see Box 0.2). Nonetheless, in the very real world of politics and policymaking, the dominance of technology-based over lifestyle-based approaches to climate mitigation continues.

Box 0.2   Limitations of the green technology approach

There are a number of key limitations to an approach that pins its hopes on green technology and efficiency improvements. These include the following:

- Efficiency is blind to the upper limits of emissions, and so we can keep improving it even as we transgress the planetary boundaries.
- Owing to the rebound effect, the sheer increase in the scale of consumption during the past several decades has canceled out the efficiency gains. No country in the world has succeeded in "absolute decoupling" of economic growth from environmental impacts. Not even (or, especially not) the Scandinavian countries that appear at the top of most indexes of progress – but with per capita footprints that would need multiple Earths.
- The technology-based approach focuses on the symptoms, not the causes of the unsustainability. At the fundamental level, we need to address overproduction in relation to overconsumption.
- The promises of the technology-based approach have failed to deliver over the last several decades (see carbon-sucking machines, geoengineering, and electric vehicles, for example). There is yet no plausible scenario for replacing our entire car stock or providing every family in the world with an efficient electric vehicle without great ecological damage.
- The technology-based approach widens the wealth gap because it is the rich who tend to own the patents or invest in these technologies.

- According to the International Energy Agency, global growth of energy demand in 2024 and 2025 is expected to be 4% annually relative to the preceding year. At this rate of increase, the supply of renewable energy does not keep up with the growth of demand, despite its own exponential growth rate (IEA, 2024). This means that fossil-fuel-generated electricity makes up the difference.
- Increasing renewable energy generation brings about other ecological problems owing to the increased demand for minerals and rare earth metals as well as political instability in mining regions (see **Energy Overshoot**).

Source: Adapted from *"The (Technology) Efficiency Paradox"*, blog post by Lewis Akenji, available at *www. hotorcool.org*

The understanding of the attraction of consumer society and a potential transition beyond it has been greatly enriched by so-called happiness and subjective well-being research (Smith & Reid, 2018). Numerous scholars, among them Layard (2006), Kahneman and Deaton (2010), Tsurumi et al. (2021), Graham (2011), and Skidelsky and Skidelsky (2012), to name a few, interrogated such questions as: Does wealth accumulation and consumption make people happy? Is there a saturation point? How much is enough?

The answers are not straightforward. On the one hand, deprivation is a source of unhappiness in life; and friends, family, and a sense of belonging are the true long-lasting sources of life satisfaction. Research findings also suggested a point of saturation; beyond a moderate per capita income, additional income does not lead to greater happiness or satisfaction with life. On the other hand, satisfaction with material life is a relative concept. It is deeply grounded in social comparisons and how much one has *relative* to others (Kahneman & Deaton, 2010; Killingsworth et al., 2023).

That leads to a kind of arms race: while the top earners strive to distance themselves from the rest (including their peers) by accumulating and spending more, those in the lower economic classes strive to emulate them and distance themselves from those below them. The phenomenon applies to all economic strata and brings a lot of stress to people's lives and little happiness to most.

At the time of this writing, complexities in the system of mass consumption – its components and mutual interdependencies in the context of a global economy – are generally recognized, but points of intervention are less clear. While the case for transitioning toward a different organizing principle of societal life is strong, conversations about the point of departure and destination lack common ground. There are other major challenges to contend with. Three among those – the socio-economic impact of reducing consumption, the inequality between the Global North and Global South, and wealth inequality within countries – especially stand out.

Challenges to System Transition

Economic and Political Implications of Diminished Consumption

In recent years, various branches of the United Nations, as well as European governments and the EU, have adopted official proclamations about the need to reduce consumption.

For example, the European Commission's New Energy Efficiency Directive (EU/2023/1791) established a weak, politically tainted but legally binding target to reduce the EU's final energy consumption by 11.7% by 2030, based on 2020 scenarios. In the UN Sustainable Development Goals (SDGs) framework, Goal 12 is on Responsible Consumption, Goal 10 is on Reducing Inequalities, and Goal 3 is on Wellbeing. The adoption of the UN 10-Year Framework of Programmes for Sustainable Consumption and Production (10YFP) in 2012 at the Rio+20 Summit so far stands out as the most ambitious approach to the issue by the UN system, but the mandate is poorly under-resourced and has no functional implementation mechanism. In 2024, the UN Environment Assembly adopted a resolution (6/8) titled "Promoting Sustainable Lifestyles". The same year, at the UN's 75th General Assembly, the 193-member organization adopted "The Pact for the Future", in which it commits (Action 10c) to "Promote sustainable consumption and production patterns, including sustainable lifestyles".

Even the Intergovernmental Panel on Climate Change (IPCC), which since its inception in 1988 has primarily focused on technical assessments and mitigation, devoted an entire chapter of its Sixth Report (2022) to sustainable consumption and **sufficiency.** It estimates that changes in systems of consumption and lifestyles by 2050 can potentially reduce greenhouse gas emissions by 40 to 70% (IPCC, 2023). A study published in *Nature* went beyond carbon emissions, finding that reducing consumption among just the top 10% or 20% of the world's consumers would go a long way to reducing overshoot of most planetary boundaries, decreasing environmental pressure by 25 to 53% (Tian et al., 2024). The media are also writing about consumption with growing frequency. The detailed and accessible analysis in the 1.5-degree lifestyles reports by Hot or Cool Institute (Akenji et al., 2021) has been repeatedly featured in mainstream media such as the BBC, Financial Times, The New York Times, Bloomberg News, Forbes, and multiple international and local mainstream media outlets.

Governments are nonetheless averse to abandoning the economic growth paradigm. And for good reason. What remains unresolved is the potential socio-economic impact of rapid reductions in consumption: recession, unemployment, and massive economic dislocations, both among the consumers in the Global North and among producers in the Global South. Some scholars have attempted to develop scenarios and macroeconomic models for such a transition (Victor, 2008), but many questions remain open. For instance, all private and public pension plans depend on future growth to deliver on their commitments; in the United States, meanwhile, much of the public service sector, such as public radio and television, depends on **advertising** revenues for their operating budgets. These examples are just the tip of the iceberg.

After decades of following the neoliberal economic model, national governments are also deeply constrained by the enormous power of multinational corporations, which demand growth and mass consumption (Slobodian, 2018). Governments thus cling to the idea that economic growth can continue as long as it is decoupled from resources. Green consumerism and green growth are the operative words. This framing has allowed governments to pay lip service to sustainable consumption while still tacitly or explicitly encouraging mass consumption. They look to technological and market solutions for what is essentially a social and political problem.

Inequality

As noted earlier, inequality within countries drives consumption because people strive to raise their social status by emulating those "above them". The groundbreaking 2009 and 2019 studies by British epidemiologists Wilkinson and Pickett illuminate other corrosive effects of inequality by showing that it is a powerful social stressor that is increasingly rendering societies dysfunctional. For example, bigger gaps between the rich and the poor are accompanied by higher rates of homicide and imprisonment, more infant mortality, obesity, drug abuse, and COVID-19 deaths, as well as higher rates of teenage pregnancy and lower levels of child well-being and social mobility.

These findings imply an alternative to economic growth as a way to solve many social ills: reducing inequality. But approaching sustainability from this inequality perspective conjures the specter of wealth redistribution: a political third rail. This is another great challenge of our times.

Inequality *within societies* has another corrosive effect on the politics of system transition. Reaching broad societal support for action on that scale requires social cohesion and a feeling that everybody contributes their share. But in highly unequal societies, social trust, cohesion, and solidarity are greatly eroded. It is hard to build support for collective action if people feel that the burden is not being shared fairly (Wilkinson & Pickett, 2024).

Inequality *between countries* is another unresolved issue. The tension between the rich and poor countries of the Global North and Global South was already evident at the 1992 Earth Summit and continues to clog up global negotiations on who should pay for loss and damage due to impacts of climate change, and the costs of climate mitigation and adaptation. Although climate change is a consequence of environmental destruction, overconsumption, and historical emissions by the rich countries of the Global North, countries of the Global South are facing its worst impacts.

On top of that, poorer countries are trapped in a neo-colonial and extractive global economic system that is forcing them to use their limited financial and natural resources to continue to supply the rich countries of the Global North, instead of developing their own economies or building their own resilience. Analysis by the International Resources Panel (IRP, 2024) shows that, through global trade, rich countries displace environmental impacts onto others, and that rich countries use six times more materials per capita and are responsible for ten times more climate impacts per capita than low-income countries.

In addition to the issue of historical responsibility, the shrinking size of the global carbon budget and the power disparities in international negotiations make changes in energy consumption a zero-sum game: an increase in the Global South requires a decrease in the Global North. In 2017, Hubacek et al. estimated that bringing the 837 million people living in extreme poverty to the level of consumption that is referred to as poor would have a minimal impact on the global carbon budget; but bringing the poor of the world (half of its population) to a more dignified state of existence might raise the global temperature by 0.6 degrees by the end of the century. More recent estimates put the impact of eradicating poverty somewhat lower, but not negligible (Oswald et al., 2021; Baltruszewicz et al., 2021, 2023).

Even as climate change forces the need to reduce emissions and as resource stocks dwindle, the International Resources Panel warns that

> *without urgent and concerted action to change the way resources are used, material resource extraction could increase by almost 60 per cent from 2020 levels by 2060. . . far exceeding what is required to meet essential human needs for all in line with the SDGs.*

Without addressing these tensions between resource needs and availability, and the asymmetries in political and economic power, countries of the Global South are unlikely to meet their material needs, and the Global North would pull the rest of the world into a deeper climate overshoot.

Envisioning Sustainable Consumption and Lifestyles

Aiming for global carbon equality would mean a radical change in lifestyles in the Global North. What would such a low-impact life look like? In this volume, we include some metaphors that attempt to define it by adopting the idea of minima and maxima: earth systems boundaries (Röckström et al., 2023), **doughnut economics**, **consumption corridors**, and **fair consumption space**. These differ in respective emphases on physical and social factors, equity, justice, and the degree of quantification, but all aim to define boundaries: the lower boundary ensures dignified, equitable living below which no one should fall; and the upper one defines the biophysical limits that should not be exceeded.

We also include examples of visions – some descriptive, others using quantitative models – of an economy capable of delivering such a life within boundaries: **steady-state economy, sharing economy, circular economy, and society** and **foundational economy**. Some papers highlight principles and policies that could form the basis of a society with sustainable lifestyles: **sufficiency** (Princen, 2003), for instance, or changing provisioning systems by adopting **universal basic services** for fundamental needs (Gough, 2019).

Using theoretical models, some researchers have sharpened the picture of a life within boundaries and under the scenario of equality by producing specific numbers to describe it. In modeling so-called **1.5-degree lifestyles,** they consider factors such as the size of living space, access to basic amenities, sufficient nutrition, basic mobility, and others (Akenji et al., 2021; Oswald et al., 2021; Baltruszewicz et al., 2021, 2023). The results all demonstrate a large gap between the current average lifestyles of citizens in affluent societies and lifestyles of sufficiency.

The irony of envisioning and calling for sufficiency lifestyles in wealthy societies is that low-income people in these countries already provide elements of a living model of it (see, for instance, examples from Norway, reported by Korsnes et al., 2024; see also Pungas et al., 2024). By necessity, they develop procedures and understandings that support lower consumption levels, like sharing, volunteering, **repairing**, negotiating needs, and calculating costs. Many more examples of sustainable lifestyles, especially in the Global South, can be found in a compendium similar to this one, called *Pluriverse: A post-development dictionary* (Kothari et al., 2019).

Notably, the frugality, simplicity, and sharing that are practiced by low-income people are often the same ones that the abundant literature on sustainable lifestyles presents as a model for achieving more life satisfaction through community participation and more leisure time. But this kind of life also often comes with stigmas, a sense of social exclusion, and low social standing – hardly a situation to emulate. This in turn undermines social cohesion and solidarity, which are necessary for social change.

Thus the challenge for policymakers and advocates is to create the conditions under which low consumption lifestyles are the norm, fair to everyone, and easiest to practice. Efficiency-oriented policies, such as subsidies for heat pumps and electric cars, which work best for economically strong groups, are not up to the task and still have their own environmental impacts. Social activism to bring about value shifts plays a relevant role in sustainability transitions for mainstreaming sustainable lifestyles out of their current niche. Several concepts explored in this book are relevant for considering how to create such conditions, among them **choice editing, carbon budgets, grassroots innovation, buen vivir and buenos convivires, ubuntu, living labs, eco-communities, and community-supported agriculture**.

The Need for a Common Language

A major barrier to making sustainable consumption and lifestyles a high-profile issue is that it does not have a political champion; it is a political orphan. At the time of this writing, the websites of leading global and national environmental organizations do not mention unsustainable consumption or unrestrained economic growth, although inequality is highlighted. This should not come as a surprise. The research roots of the modern environmental movement are in natural sciences and technology; technological and supply-side solutions to ecological overshoot are therefore their natural choices. Furthermore, it is much easier to mobilize their constituency by targeting the business world as villains and human health as under threat – as was the case with environmental pollution from the 1960s on – than by challenging dominant lifestyles (see Box 0.3). Neither is there much explicit discussion of sustainable lifestyles among advocates for social justice or public health.

Box 0.3 Environmental movement coalition in the twentieth century

In the United States and Europe, great reforms were introduced in controlling air, water, and soil pollution during the early 20th century, largely owing to the political advocacy of the public health community and, by the 1960s, also environmental organizations. The community of epidemiologists, medical professionals, and environmental scientists that emerged shared a professional language and worldviews. They performed an essential role in generating scientific data about the adverse effects of pollution and chemical contamination on health, disseminated that body of knowledge in scientific publications and mainstream media, and vigorously advocated for government policies.

In the 1960s, they were instrumental in building a broad-based coalition of scientists, environmental activists, and the alarmed public in affecting social change: building regulatory institutions, enacting laws, and allocating public funds (Brown et al., 1997). This same type of coalition was responsible for banning indoor tobacco smoking in the 1990s and early 2000s.

The problem of overconsumption is of course more complex than that of toxic pollution. And consumers who are concerned about their future quality of life are slow to accept the idea of sustainable consumption and lifestyles.

Conceptual barriers to collective action accompany the political ones. The transdisciplinary nature of sustainable consumption and lifestyles research hampers the emergence of a coherent community of scholar-advocates who share a language and worldviews. Suggested points of intervention are often based on fragmented findings and siloed perspectives rather than a unified perspective (Wallnoefer, 2022). The result is that people and institutions who should communicate with each other do so poorly or not at all and graduates from sustainability programs who are eager to engage in social activism are often only familiar with a sliver of the big picture but without a good understanding of how all the pieces fit together.

These obstacles do not, however, justify inaction. Indeed, they are a clarion call for action. It is necessary to build a *coalition* of multidisciplinary academics, activists, policymakers, and business leaders who understand the urgency of shifting to a different organizing principle of societal life, and who can see opportunities in doing so. Creating a *broad-based discourse* that draws on the many concepts included in this book is the first step.

We are encouraged by the signs of a growing interest in consumption reduction policies among citizens in high-consuming countries. A recent study in ten European countries, for instance, showed that citizens are more committed to sufficiency policies than the authors of the National Energy and Climate Plans (Lage et al., 2023). In France, several recent reports on sufficiency have received considerable public attention (Bourliaguet, 2025). Cities like The Hague are banning fossil fuel-related advertisements, which are a relevant driver of unsustainable consumption. Conversations around degrowth and post-growth approaches have also blossomed, fundamentally questioning what has been called the "imperial mode of living" (Brand & Wissen, 2021). This has brought such debates into prominent conferences in the European Union and national parliaments.

There is also a long tradition and a more recent surge of civil experimentation and activism, quite often not under the banner of sustainable consumption, but very relevant for its development. Many of these are in the lower-consuming countries of the Global South (Kothari et al., 2019). Various promising civil society developments are tackled in the book under topics such as **subvertising, grassroots innovation, alternative hedonism, green parenting, and community-supported agriculture**.

Structure of the Book

This book seeks to curate and organize a shared vocabulary for the broad, transdisciplinary communities working on a transition toward sustainable consumption and lifestyles, especially in the high-consuming countries. It is also aimed at people who want to better understand those ideas and how they relate to each other. It strives to assess how far the discipline has matured and to highlight some of the emerging concepts that could become relevant for sustainable futures. Together, the entries in this book attempt to reclaim some of the language from appropriation and greenwashing, and to challenge some myths and misconceptions about sustainable consumption and lifestyles. Ultimately, we hope that it provides inspiration for researchers, practitioners, activists, innovators, and observers on how to collectively move toward a more sustainable society.

The 87 essays are organized into five clusters with overlapping boundaries (see Box 0.4). Cluster I includes entries focusing mainly on actions by individuals, households, and social groups; Cluster II is more theoretical, including abstract concepts, frameworks, and applied theories. Cluster III takes a political economy perspective; Cluster IV focuses on social activism and value shifts; and Cluster V addresses governance, policy, and choice architecture.

Each entry provides a definition, a brief history of the concept, and a reflection on the various perspectives on the topic, as well as its applications and implications for a transition to a sustainable consumption system. To facilitate cross-referencing between entries, we have marked the connections with other entries in bold. The result is a mosaic of concepts, most of them interconnected, which together form the "Vocabulary" that can be used to have a dialogue on sustainable living, consumption and lifestyles. This mosaic is important not only for a collective conversation but also for identifying leverage points for systemic change. Without a shared understanding of the issues at stake and of the various concepts that matter, it is hard to imagine purposeful social action.

In the entries, we avoided large bibliographies, as would be the case in a review article. This is because we seek to offer the reader a generally understood and self-contained description of each concept in the context of a transition to a system of sustainable consumption and lifestyles. Readers who want to know more about a topic can follow up with further research of their own, starting with the list of five additional readings.

Box 0.4   Clusters of concepts

Cluster I: *Daily Household Decisions and Lifestyles* takes the perspective of the consumer. How and when do consumers make decisions on where to live, how they get from point A to B, how they spend their free time, and what they eat and drink? It also explores examples of sustainable lifestyles, emerging ways of living, and how communities are experimenting, as well as some barriers to change.

Cluster II: *Concepts, Frameworks, and Applied Theories* brings us to the conceptual underpinnings of sustainable and unsustainable consumption. It explores how lifestyles and consumption patterns are determined by behavioral and structural drivers and barriers, points of intervention to trigger change, and potential outcomes of transformation processes.

Cluster III: *Political Economy* takes a systemic view of consumer society, including economic structures, the role of finance and money, power relations, and inequality. It reflects on potential types of different economic paradigms, and what principles could be relevant to follow.

Cluster IV: *Social Activism and Value Shifts* looks at the myriad of social experiments and actions to enhance well-being with a smaller footprint. This cluster also reflects on the interpretations of collective well-being in different cultures and social contexts, as well as the role of education. It addresses how individuals can take action within and beyond their role as consumers, and what alternative ways of living could serve as orientation for that.

Cluster V: *Governance, Policy, and Choice Architecture* focuses on institutional arrangements and government policies to facilitate sustainable consumption and lifestyles. This explores how sustainable lifestyles can be enabled through structural, legislative, cultural, and technological changes that default toward mainstreamed sustainable choice options and behavioral patterns.

Note

1 This section draws extensively from discussions and definitions in the publication *Avoiding Consumer Scapegoatism: Towards a Political Economy of Sustainable Living* (Akenji, 2019).

References

Ajzen, I. (1991). The theory of planned behavior. *Organizational Behavior and Human Decision Processes*, 50(2), 179–211. https://doi.org/10.1016/0749-5978(91)90020-T.

Akenji, L. (2019). *Avoiding consumer scapegoatism: Towards a political economy of sustainable living*. University of Helsinki.

Akenji, L., Bengtsson, M., Toivio, V., Lettenmeier, M., Fawcett, T., Parag, T., Saheb, Y., Andersen, R., Hoff, H., Nissinen, K., & Rees, W. (2021). *1.5-degree lifestyles: Towards a fair consumption space for all*. Hot or Cool Institute. Available at: https://hotorcool.org/1-5-degree-lifestyles

Alfredsson, E., Bengtsson, M., Brown, H.S., Isenhour, C., Lorek, S., Stevis, D., & Vergragt, P. (2018). Why achieving the Paris Agreement requires reduced overall consumption and production. *Sustainability: Science, Practice and Policy*, 14(1), 1–5. https://doi.org/10.1080/15487733.2018.1458815.

Auden, S. (2021). Worrying about your carbon footprint is exactly what big oil wants you to do. *The New York Times*.

Baltruszewicz, M., Steinberger, J.K., Ivanova, D., Brand-Correa, L.I., Paavola, J., & Owen, A. (2021). Household final energy footprints in Nepal, Vietnam and Zambia: Composition, inequality and links to well-being. *Environmental Research Letters*, 16(2), 025011. https://doi.org/10.1088/1748-9326/abd588.

Baltruszewicz, M., Steinberger, J.K., Paavola, J., Ivanova, D., Brand-Correa, L.I., & Owen, A. (2023). Social outcomes of energy use in the United Kingdom: Household energy footprints and their links to well-being. *Ecological Economics*, 205. https://doi.org/10.1016/j.ecolecon.2022.107686.

Bauman, Z., & Miles, S. (1999). Consumerism as a way of life. *Social Forces*, 78. https://doi.org/10.2307/3005817.

Bourdieu, P. (1984). *Distinction: A social critique of the judgement of taste*. Cambridge, MA: Harvard University Press.

Bourliaguet, B. (2025). Rethinking the energy transition: Sufficiency and the French strategy. *Energy Research & Social Science*, 124, 104055. https://doi.org.10.1016/j.erss.2025.104055.

Brand, U., & Wissen, M. (2021). *The imperial mode of living*. New York: Verso.

Brown, H.S., Cook, B., Shatkin, J.A., Krueger, J.R. (1997). Reassessing the History of Hazardous Waste Disposal Policy in the United States: Problem Definition, Expert Knowledge, and Agenda-Setting. *Risk, Issues and Health, Safety and Environment*, 249–272.

Cohen, L. (2003). *A consumers' republic: The politics of mass consumption in postwar America*. 1st Paperback ed. Vintage Books.

Daly, H.E. (1993). Steady-state economics: A new paradigm. *New Literary History*, 24. https://doi.org/10.2307/469394.

De Graaf, J., Wann, D., & Naylor, T.H. (2001). *Affluenza: The all-consuming epidemic*. Berrett-Koehler Publishers.

Demand, Services and Social Aspects of Mitigation. (2023). *Climate change 2022 – Mitigation of climate change*. https://doi.org/10.1017/9781009157926.007.

Dubois, G., Sovacool, B., Aall, C., Nilsson, M., Barbier, C., Herrmann, A., Bruyère, S., Andersson, C., Skold, B., Nadaud, F., Dorner, F., Moberg, K.R., Ceron, J.P., Fischer, H., Amelung, D., Baltruszewicz, M., Fischer, J., Benevise, F., Louis, V.R., & Sauerborn, R. (2019). It starts at home? Climate policies targeting household consumption and behavioral decisions are key to low-carbon futures. *Energy Research & Social Science*, 52. https://doi.org/10.1016/j.erss.2019.02.001.

Elgin, D. (1981). *Voluntary simplicity: Toward a way of life that is outwardly simple, inwardly rich*. New York: William Morrow.

Gabriel, Y., & Lang, T. (2006). *Unmanageable consumer*. 2nd ed. London: Sage Publications. https://doi.org/10.1017/CBO9781107415324.004.

Graham, C. (2011). *The pursuit of happiness: An economy of well-being*. Brookings Institution.

Gough, I. (2019). Universal basic services: A theoretical and moral framework. *Political Quarterly*, 90(3), 534–542. ISSN: 0032-3179.

Gupta, J., Liverman, D., Prodani, K., Aldunce, P., Bai, X., Broadgate, W., Ciobanu, D., Gifford, L., Gordon, C., Hurlbert, M., Inoue, C.Y.A., Jacobson, L., Kanie, N., Lade, S.J., Lenton, T.M., Obura, D., Okereke, C., Otto, I.M., Pereira, L., Rockström, J., Scholtens, J., Rocha, J., Stewart-Koster, B., David Tàbara, J., Rammelt, C., & Verburg, P.H. (2023). Earth system justice needed to identify and live within Earth system boundaries. *Nature Sustainability*, 6, 630–638. https://doi.org/10.1038/s41893-023-01064-1.

Hirao, M., Tasaki, T., Hotta, Y., & Kanie, N. (2021). Policy development for reconfiguring consumption and production patterns in the Asian region. *Global Environmental Research*, 25(1&2).

IEA. (2024). *Global electricity demand set to rise strongly this year and next, reflecting its expanding role in energy systems around the world*. International Energy Agency. Available at: https://www.iea.org/news/global-electricity-demand-set-to-rise-strongly-this-year-and-next-reflecting-its-expanding-role-in-energy-systems-around-the-world

IPCC. (2023). IPCC report 2023. In *Climate change 2023: Synthesis report*. International Panel for Climate Change. Available at: https://www.ipcc.ch/report/sixth-assessment-report-cycle/.

IRP. (2024). *Global resources outlook 2024: Bend the trend – pathways to a liveable planet as resource use spikes*. International Resource Panel. Available at: https://www.resourcepanel.org/reports/global-resources-outlook-2024.

Jackson, T. (2006). *The earthscan reader on sustainable consumption*. Routledge.

Jackson, T. (2009). *Prosperity without growth: Economics for a finite planet*. 1st ed. London: Routledge. https://doi.org/0.4324/9781849774338.

Jackson, T., Victor, P.A., & Naqvi, A. (2016). *Towards a Stock-Flow Consistent Ecological Macroeconomics*. PASSAGE Prosperity and Sustainability in the Green Economy.

Jungell-Michelsson, J., & Heikkurinen, P. (2022). Sufficiency: A systematic literature review. *Ecological Economics*, 195. https://doi.org/10.1016/j.ecolecon.2022.107380

Kahneman, D., & Deaton, A. (2010). High income improves evaluation of life but not emotional well-being. *Proceedings of the National Academy of Sciences of the United States of America*, 107(38), 16489–15493. https://doi.org/10.1073/pnas.1011492107

Keynes, J.M. (1930). *Economic possibilities for our grandchildren* (1930). In *Essays in persuasion*. New York: W.W. Norton & Co., 1963.

Killingsworth, M.A., Kahneman, D., & Mellers, B. (2023). Income and emotional well-being: A conflict resolved. *Proceedings of the National Academy of Sciences of the United States of America*, 120(1). 1–8. https://doi.org/10.1073/pnas.2208661120

Korsnes, M., & Solbu, G. (2024). Can sufficiency become the new normal? Exploring consumption patterns of low-income groups in Norway. *Consumption and Society*. https://doi.org/10.1332/27528499y2024d000000009.

Kothari, A., Salleh, A., Escobar, A., Demaria, F., & Acosta, A. (2019). *Pluriverse – a post-development dictionary*. New Delhi: Tulika Books.

Lage, J., Thema, J., Zell-Ziegler, C., Best, B., Cordroch, L., & Wiese, F. (2023). Citizens call for sufficiency and regulation – A comparison of European citizen assemblies and National Energy and Climate Plans. *Energy Research & Social Science*, 104, 103254. Available at: https://doi.org/10.1016/j.erss.2023.103254.

Layard, R. (2006). *Happiness: Lessons from a new science*. London: Penguin Group.

Marcuse, H. (1964). *One-dimensional man: Studies in the ideology of advanced industrial society*. Boston: Beacon Press.

Miles, S. (1998). *Consumerism: As a way of life*. London: Sage Publications.

Millward-Hopkins, J., Steinberger, J.K., Rao, N.D., & Oswald, Y. (2020). Providing decent living with minimum energy: A global scenario. *Global Environmental Change*, 65. https://doi.org/10.1016/j.gloenvcha.2020.102168

Ofstad, S., Westly, L., & Bratelli, T. (1994). *Symposium: Sustainable consumption: 19–20 January 1994: Oslo, Norway*. Oslo, Norway: Ministry of Environment.

Olmsted, M.S., & Galbraith, J.K. (1958). The affluent society. *American Sociological Review*, 23. https://doi.org/10.2307/2089073

Oswald, Y., Steinberger, J.K., Ivanova, D., & Millward-Hopkins, J. (2021). Global redistribution of income and household energy footprints: A computational thought experiment. *Global Sustainability*, 4, e4. https://doi.org/10.1017/sus.2021.1

Parrique, T., Barth, J., Briens, F., Kerschner, C., Kraus-Polk, A., Kuokkaren, A., & Spangenberg, J.H. (2019). *Decoupling debunked: Evidence and arguments against green growth as a sole strategy for sustainability*. European Environmental Bureau.

Princen, T., Miniates, M., & Conca, K. (2002). *Confronting consumption*. Cambridge, MA: MIT Press.

Princen, T. (2003). Principles for sustainability: From cooperation and efficiency to sufficiency. *Global Environmental Politics*, 3(1), 33–50. https://doi.org/10.1162/152638003763336374

Pungas, L., Kolínský, O., Smith, T.S.J., Cima, O., Fraňková, E., Gagyi, A., Sattler, M., & Sovová, L. (2024). Degrowth from the East – between quietness and contention. Collaborative learnings from the Zagreb Degrowth Conference. *Czech Journal of International Relations*, 59(2), 79–113. https://doi.org/10.32422/cjir.838

Riefler, P., Baar, C., Büttner, O.B., & Flachs, S. (2024). What to gain, what to lose? A taxonomy of individual-level gains and losses associated with consumption reduction. *Ecological Economics*, 224, 108301. https://doi.org/10.1016/j.ecolecon.2024.108301

Rockström, J., Gupta, J., Qin, D. et al. (2023). Safe and just earth system boundaries. *Nature*, 619, 102–111. https://doi.org/10.1038/s41586-023-06083-8.

Rockström, J., Steffen, W., Noone, K., Persson, Å., Chapin, F.S., Lambin, E., Lenton, T.M., Scheffer, M., Folke, C., Schellnhuber, H.J., Nykvist, B., de Wit, C.A., Hughes, T., van der Leeuw, S., Rodhe, H., Sörlin, S., Snyder, P.K., Costanza, R., Svedin, U., Falkenmark, M., Karlberg, L., Corell, R.W., Fabry, V.J., Hansen, J., Walker, B., Liverman, D., Richardson, K., Crutzen, P., & Foley, J. (2009). Planetary boundaries: Exploring the safe operating space for humanity. *Ecology and Society*, 14. https://doi.org/10.5751/ES-03180–140232

Schor, J.B. (1993). *The overworked American: The unexpected decline of leisure*. New York: Basic Books.

Schor, J.B. (1999). *The overspent American: Why we want what we don't need*. New York: Harper Perennial.

Senate and House of Representatives of the United States of America. (1946). *Employment Act of 1946*.

Shove, E., Pantzar, M., & Watson, M. (2012). *The dynamics of social practice: Everyday life and how it changes*. London: Sage Publications. https://doi.org/10.4135/9781446250655

Skidelsky, R., & Skidelsky, E. (2012). *How much is enough? Money and the good life*. Revised ed. New York: Other Press.

Slobodian, Q. (2018). *Globalists: The end of empire and the birth of neoliberalism*. Cambridge, MA: Harvard University Press. Available at: http://www.hup.harvard.edu/catalog.php?isbn=9780674979529.

Smith, T.S.J., & Reid, L. (2018). Which being in 'wellbeing'? Ontology, wellness and the geographies of happiness. *Progress in Human Geography*, 42(6). https://doi.org/10.1177/0309132517717100

Thaler, R.H. (1980). Toward a positive theory of consumer choice. *Journal of Economic Behavior & Organization*, 1(1), 39–60.

Thaler, R.H. (2016). From cashews to nudges: The evolution of behavioral economics. *American Economic Review*, 106(7), 1577–1600.

Thaler, R.H., & Sunstein, C.R. (2008). *Nudge: Improving decisions about health, wealth, and happiness*. New Haven, CT: Yale University Press.

Tian, P., Zhong, H., Chen, X., Feng, K., Sun, L., Zhang, N., Shao, X., Liu, Y., Hubacek, K., 2024. Keeping the global consumption within the planetary boundaries. *Nature*, 635, 625–630. https://doi.org/10.1038/s41586-024-08154-w

Tsurumi, T., Yamaguchi, R., Kagohashi, K. et al. (2021). Are cognitive, affective, and eudaimonic dimensions of subjective well-being differently related to consumption? Evidence from Japan. *Journal of Happiness Studies*, 22, 2499–2522. https://doi.org/10.1007/s10902-020-00327-4

United Nations Conference on Environment & Development (UNCED). (1992). *Agenda 21: The United Nations Programme of Action from Rio*. Rio de Janeiro.

UN Secretary-General, World Commission on Environment and Development. (1987). *Report of the World Commission on Environment and Development.* Note by the Secretary-General.

Veblen, T. (1994). *The theory of the leisure class*. 1899. Mineola, NY: Dover.

Vergragt, P.J., Dendler, L., de Jong, M., & Matus, K. (2016). Transitions to sustainable consumption and production in cities. *Journal of Cleaner Production*, 134, 491–501. https://doi.org/10.1016/j.jclepro.2016.05.050.

Victor, P.A. (2008). *Managing without growth: Slower by design, not disaster*. Cheltenham: Edward Elgar. https://doi.org/10.4337/9781848442993.

Wackernagel, M., & Rees, W. (1996). *Our ecological footprint: Reducing human impact on the earth*. Gabriola Island, BC: New Society Publishers.

Wallnoefer, L.M. (2022). *Individual-level environmental sustainability – integrating perspectives and concepts across Research Fields*. PhD dissertation, University of Natural Resources and Life Sciences, Vienna. https://doi.org/10.13140/RG.2.2.17115.75045.

Wallnoefer, L.M., Svensson-Hoglund, S., Bhar, S., & Upham, P. (2024). At the intersections of influence: Exploring the structure–agency nexus across sufficiency goals and time frames. *Sustainability Science*, 19(3), 683–686. https://doi.org/10.1007/s11625-024-01467-9.

Wilkinson, R.G., & Pickett, K.W. (2010). *The spirit level: Why equality is better for everyone*. London: Penguin UK.

Wilkinson, R.G., & Pickett, K.E. (2024). Why the world cannot afford the rich. *Nature*. https://doi.org/10.1038/d41586-024-00723-3

Cluster I

Daily Household Decisions and Lifestyles

1 Consumerism

Erik Assadourian

Definition

Consumerism is a cultural pattern that orients people's meaning, contentment, and acceptance primarily around the consumption of goods and services. As cultural beings, much of what humans take for being "natural" is actually cultural in nature. Dietary choices (whether we prefer tea or soda, insects or meat); what we live in (a yurt vs. a multi-thousand-square-foot house); how we dress (**fast fashion** or durable well-crafted clothing); how we spend our leisure time (gardening or shopping); even which animals we perceive as companions rather than food – all of these are primarily shaped by the cultures we are born and live in.

In consumer cultures (i.e., those where consumerism is the dominant cultural paradigm), consumerism takes a central role in shaping how we live, typically encouraging those activities that increase access to stuff and the immaterial benefits that come with it. Moreover, consumption is typically seen as the primary solution to most problems: if you are overweight, you take medicines or buy expensive diet aids or gym memberships. If you are lonely, you get a pet. If you are bored, you stream entertainment or surf social media. If you want to express love, you buy a gift. If you have so much stuff you cannot fit it in your home, you rent a storage unit to put your extra belongings in. In reality, many problems stem directly from overconsumption (e.g., of food, media, or goods), and the best solution is often consuming less, even if in consumer cultures that can be overlooked or even seen as culturally taboo. This reality applies to environmental problems as much as to individual and social ones. For example, the most effective solution to reduce fossil fuels is not to replace them with renewables but to reduce overall consumption levels (see **Sufficiency, Energy Overshoot**).

History

Some argue that consumerism started in the 1950s, as the United States scrambled to keep production going after the end of World War II, cultivating a "Consumer Republic" to replace the wartime economy (Cohen, 2003). Economist Victor Lebow, writing in 1955, even suggested that "our enormously productive economy demands that we make consumption our way of life, that we convert the buying and use of goods into rituals, that we seek our spiritual satisfactions, our ego satisfactions, in consumption".

However, the development of consumerism was a longer, more organic process. It evolved over several hundred years, in ways subtle and overt. For example, an 18th-century British pottery manufacturer, Josiah Wedgwood, had salespeople go door-to-door drumming up excitement for new pottery designs, creating demand for newer lines of products, even from customers who already had perfectly good, but now seemingly outdated, pottery sets (Stearns, 2001). Global trade

DOI: 10.4324/9781003584056-3

and colonial expansion also played a role, opening up new markets, and creating demand for new products, from exotic new fruits and vegetables to porcelain, sugar, and tobacco. The development of the corporation also created structures to enable more business risk, as well as debt structures that drove growth and cultivation of demand.

Marketing, in simple forms like store displays and newspaper ads, informed potential customers about new products, and over time evolved to encourage and drive consumption, particularly as mass-media technologies and sophisticated psychologically manipulative **advertising** developed. The global growth curve for consumerism has certainly been exponential, with the last 75 years accelerating in the overall reach of this pattern. For example, one measure of consumerism – global advertising expenditures – grew from $79 billion in 1950 to $1.63 trillion in 2023 (2023 dollars).

Human nature also played a part in cementing the consumerism paradigm. As social animals, we compare ourselves to others in our social group. As global interconnection grew (telegraph, radio, cinema, TV, internet), this led to comparing ourselves to the very richest around the world and finding our own consumption patterns inadequate. Even "keeping up with the Joneses" (i.e., coveting new things your neighbors bought and striving to buy these yourself) more locally can lead people to upgrade cars, homes, clothing, and televisions, in a perpetual hamster wheel of consumption (see **Conspicuous/Positional Consumption**).

This is further exacerbated by the **hedonic treadmill**, in which humans habituate to the new (good or bad) and thus require new experiences to get another thrill, which is most readily visible in the consumer practice of travel, exploring new restaurants, and regularly upgrading their vehicles. The Diderot Effect, named after the French philosopher who got a new dressing gown that made the rest of his home look shabby, can aggravate this further as people keep consuming more to replace old items, to align with the aesthetic of their newer ones. This has all been intensified further by the cultivation of obsolescence – both physical (where products are intentionally designed to break) and psychological, where products no longer feel satisfying (often driven by marketing for newer versions) – thus driving individuals to replace these goods (see **Product Returns and Right of Withdrawal**).

Along with ill-health, social inequality, and environmental devastation, consumerism can lock people into this system via indebtedness (see **Sustainable Finance, Money**). The average per capita credit card debt in the United States in 2023 was $6,501, adding up to $1.12 trillion in total. Adding in mortgages and other debts, the average American household carried $104,215 of debt in 2023. Individual, business, and government debt further lock societies into consumer cultural patterns, requiring individuals to work more, businesses to sell more, and governments to encourage growth to pay back their debts.

Different Perspectives

Recognizing that consumerism is embedded in and comes out of culture makes it clearer that to address the phenomenon, we have to shift the dominant cultural paradigm. All cultures are shaped by (and shape) key societal institutions, including business, education, media, government, traditions, and **social movements** (Assadourian, 2010) (see **Political Economy of Consumerism, The Role of Business, Education for Sustainable Consumption, Values and Consumption, Social Norms**). Many of these institutions perpetuate consumerism – such as schools that provide curricula sponsored by junk food companies or media that market to children, socializing the next generation of consumers. However, they can also be harnessed to cultivate an alternative (e.g., a sustainability paradigm).

There have been many attempts to make sustainability feel as natural as consumerism in classrooms, through policy, by greening religious traditions (see **Spiritual Consumption**), with ecocentric film and media, and through movements like Slow Food (see **Community Supported Agriculture, Food Miles**). But these all add up to a drop in the bucket compared to global advertising and the broader marketing and lobbying efforts of companies and interests promoting facets of consumerism (as well as the sheer momentum of several centuries of consumerism).

Applications

At this point, we are immersed in a world where consumerism is natural, and where in the ten most spoken languages in the world, consumer is synonymous with person (Assadourian, 2010). Breaking out of this pattern willingly (as opposed to waiting until ecological breakdown makes consumerism impossible for the majority of people) will require bold – and potentially unpopular – interventions (see Box 1.1).

Box 1.1   Three provocative intervention points to address consumerism

Ending advertising as a driver of media: Taking away tax breaks for advertising is a good starting point, as is taxing certain forms of advertising (such as junk food) and using that revenue to subsidize advertising for healthy foods and lifestyles.

Denormalizing pet ownership: Over the past several decades, pet ownership has skyrocketed (with possibly 900 million dogs and 600 million cats worldwide today) and has spread to cultures that never owned pets – in the process creating an ecological burden that cannot be sustained. Now the pet industry is normalizing pet parenting and encouraging spending on gourmet food, healthcare, flights, and even doggy hotels (even as one-tenth of people do not have enough to eat) (Vale & Vale, 2009).

Normalizing Low-consumption Rites of Passage: The average American spends $30,000 on a wedding and $10,000 on a funeral. Embedded in that are huge ecological costs and cultural reminders that purchases translate to expressions of love. Simplifying these traditions could serve as a powerful model of sustainability and help reveal that love does not equate to how much one spends.

Of course, there are hundreds of other potential intervention points as well: developing ecocentric schools and curricula; denormalizing regular meat consumption, and shifting instead to "celebratory" meat consumption (on festive days only, as was the way in many cultures historically) (see **Protein Shift, Social Tipping Points**); even shifting the culture of travel so that flying becomes normal as a "once-in-a-life-phase" experience, rather than a routine luxury.

The key to this, however, is to see consumerism through a cultural lens and recognize that changing consumption patterns requires a transformation in cultures, which in turn requires strategically utilizing societal institutions as levers of cultural change.

Further Reading

Assadourian, Erik. (2010). The rise and fall of consumer cultures. In Worldwatch Institute (Ed.), *State of the world 2010: Transforming cultures: From consumerism to sustainability*, pp. 3–20. New York: W. W. Norton & Company.

Cohen, Lizabeth. (2003). *A consumer's republic: The politics of mass consumption in postwar America*. New York: Alfred A. Knopf.

Lebow, Victor. (1955). Price competition in 1955. *Journal of Retailing* (Spring), 7. Available at: https://mron-line.org/wp-content/uploads/2019/07/Lebow.pdf (accessed: 13 January 2025).

Stearns, Peter N. (2001). *Consumerism in world history: The global transformation of desire*. New York: Routledge.

Vale, Robert, & Vale, B. (2009). *Time to eat the dog? The real guide to sustainable living*. London: Thames & Hudson.

2 Household Income Versus Carbon Footprint

Andrew M. M. Reeves and Jared Starr

Definition

A carbon footprint measures the total carbon dioxide (CO_2) emissions generated by specific processes or products and can encompass emissions by individuals, households, collectives, institutions, or countries. The approach of analyzing carbon footprints against income examines how carbon emissions correlate with income levels, be they individual, household, or national.

In addition to CO_2, total greenhouse gas emissions are often quantified in terms of equivalent CO_2 needed to cause a given amount of global warming. A household or entity's carbon footprint is often based on emissions associated with its consumption (see **Consumption-Based Accounting**). Alternatively, carbon footprints can be based on emissions associated with income or wealth. For the wealthiest ~1% of the population in most countries, emissions related to wealth or investments form a significant proportion of carbon footprints.

The motivation for the study and understanding of how carbon footprints relate to income emerges from well-established links between inequality and negative impacts on both sustainability and human well-being (see **Carbon Inequality, Climate Justice**). Practical considerations, such as pinpointing responsibility for carbon emissions and therefore targeting ways to reduce them are also important motivators.

History

The footprint term itself emerged in the academic literature as a subset of the "ecological footprint" concept, coined by William E. Rees in 1992. Originally, in this usage, carbon footprint represented the land area needed to sequester the CO_2 produced by human activity (minus that absorbed by the oceans) for the whole world or a nation's population. Rather than equivalent land area, "carbon footprint" is now typically measured directly in terms of mass of CO_2 emissions.

Income inequality's relationship to CO_2 emissions became an important part of discussion and research in the 1990s, with an initial focus primarily on per capita footprint versus per capita Gross Domestic Product (GDP). GDP per capita remains a common proxy for average national income, although it is worth noting that it does not include income from offshore holdings, like measures such as Gross National Income (GNI). Higher per capita income is strongly correlated with an increased per capita carbon footprint (see Figure 2.1 for illustration).

In the 2000s, the study of carbon emissions gradually shifted from large global or national scales to individual corporations, products, households, and individuals. This came from a need to understand the sources of carbon emissions in depth, to develop detailed policies and specific actions to decrease them (see **Co-Benefits of Climate Policy**). These footprints are calculated using environmentally extended input-output tables and household survey data on consumption

DOI: 10.4324/9781003584056-4

CO₂ emissions per capita vs. GDP per capita, 2022

This measures CO₂ emissions from fossil fuels and industry¹ only – land-use change is not included. GDP per capita is adjusted for inflation and differences in living costs between countries.

Data source: Global Carbon Budget (2024); Population based on various sources (2024); Bolt and van Zanden - Maddison Project Database 2023
Note: GDP per capita is expressed in international-$² at 2011 prices.
OurWorldinData.org/co2-and-greenhouse-gas-emissions | CC BY

1. **Fossil emissions:** Fossil emissions measure the quantity of carbon dioxide (CO₂) emitted from the burning of fossil fuels, and directly from industrial processes such as cement and steel production. Fossil CO₂ includes emissions from coal, oil, gas, flaring, cement, steel, and other industrial processes. Fossil emissions do not include land use change, deforestation, soils, or vegetation.

2. **International dollars:** International dollars are a hypothetical currency that is used to make meaningful comparisons of monetary indicators of living standards. Figures expressed in international dollars are adjusted for inflation within countries over time, and for differences in the cost of living between countries. The goal of such adjustments is to provide a unit whose purchasing power is held fixed over time and across countries, such that one international dollar can buy the same quantity and quality of goods and services no matter where or when it is spent. Read more in our article: What are Purchasing Power Parity adjustments and why do we need them?

Figure 2.1   Per capita carbon footprint (tons of CO₂-equivalent emissions) versus GDP per capita (adjusted for purchasing power parity) for countries of the world. The relative size of the circle indicates population size

Source: Our World in Data (2022)

and incomes. Much popularization of the term at an individual or household level was made, particularly in the form of easy-to-use household carbon footprint calculators. These revealed to the public the high carbon emissions associated with things like driving, airline travel, and meat consumption (see **Sustainable Mobility**, **Protein Shift**).

Additionally, thanks to improved income and wealth data within countries, and more recently on a global level, a growing body of work focuses on how carbon footprints relate to household or individual income. The highest-income (per year) or wealthiest (in terms of net worth) individuals have the largest and increasing carbon footprints and a disproportionate share of the global carbon footprint. To make this concrete, the top 10% of the global income distribution is responsible for about 50% of all global emissions (see Figure 2.2). Not only this, but carbon footprint inequality within countries surpassed between-country inequality around 2010. High carbon footprints in high-income households can be attributed to the large emissions associated with investment and consumption factors like (i) larger detached or multiple homes (see **Sustainable Housing**, **Urban**

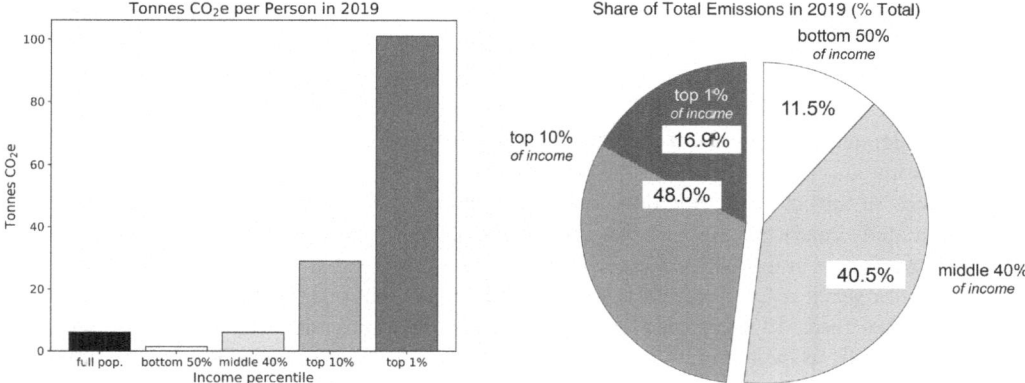

Figure 2.2 Breakdown of consumption-based global carbon emissions by income percentiles. Own compilation, based on results of Chancel (2022). In the left plot *(a)*, average carbon footprints (emissions) for 2019 are shown for a breakdown of income groups. Calculated carbon footprint includes all domestic consumption and income from investments, and imports and exports embedded in goods and services traded with the world. The right plot *(b)*, using the same income breakdown and colors, shows each income group's overall share of the total global carbon footprint (emissions). The highest-income groups have a disproportionate share of the global carbon footprint

Source: Chancel (2022)

Planning and Spatial Allocation) and (ii) high levels of luxury and service consumption (see **Conspicuous/Positional Consumption**), highly individualized and energy-intensive transportation (e.g., private vehicles, private jet travel, motorized yachts), and frequent global travel. The huge emissions coming from this group's investment and lifestyle choices are wholly incompatible with keeping global warming impacts below 1.5 degrees (see **1.5-Degree Lifestyles**).

Different Perspectives

An issue with a focus on carbon footprints versus individual or household income is that individuals are often limited in the realistic choices available to them (see **Freedom of Choice, Choice Editing**). For example, an individual cannot choose to take public transit to work if there is no transit available to take them there in a timely fashion. Additionally, although individual measures of carbon footprint include the corresponding portion of production, because of concentrated private ownership of production, at least in capitalist countries, an individual or household generally does not have agency over the production process itself.

An abundance of household and individual carbon footprint calculators is now available, which include socioeconomic factors. While useful, this focus tends to suggest households can and will choose to make small tweaks in their consumption patterns rather than requiring changes to collective consumption or re-examination of the metabolism of the economic system (see **Consumer Scapegoatism, Stocks Versus Flows, Steady-State Economy**). Such an individual focus tends to yield very minimal changes in carbon footprint at a country level. Additionally, it has been widely reported that carbon footprint – in an individual/household consumption framing – was originally popularized in the early 2000s by media campaigns of large fossil fuel companies, particularly the oil giant British Petroleum, to delay systemic action to transition away from fossil fuels.

Increasing work on attributing carbon footprint to different income brackets has provided a partial antidote to the limitations of individual carbon footprints by putting such work into a larger context. Excess unsustainable final consumption, and therefore additional fossil fuel consumption, is partly driven by the distribution of incomes in an economy (see **Money**).

For the very wealthiest individuals, in the top 0.1% or so globally, most income comes passively from investments. Work remains to be done to collect sufficiently complete data globally to accurately understand the impact of income generated from investment assets and capital gains on carbon footprint.

Differences in carbon footprint definition, accounting, and modeling can also significantly impact estimated emissions. Accounting for the embodied carbon footprint in globally traded goods and services is important, as are choices in which processes are considered as part of a given product or service. Land use changes for shifts in production are often ignored (e.g., a shift from grassland or forest to industrial farming), which can lead to underestimates of actual carbon footprint changes from increased or changing consumption. The choice of time horizon for warming potential equivalent can complicate the comparison of studies since different greenhouse gases have different warming impacts over time. Characterizing these and other choices in methodology in different contexts remains a highly active area of research.

Applications

Some research indicates that previous consumption-based approaches like carbon taxes can be regressive, impacting low- and medium-income households hardest. Tax rebates can offset this impact (see, e.g., Canada's carbon tax system). However, public support remains poor, especially in car-dependent areas and the highest-income households may have enough savings to protect them from environmentally necessary lifestyle changes.

Instead, a strong but progressive income as well as wealth tax could be more effective, ideally implemented both nationally and globally, as suggested by Thomas Piketty and others. This would serve the additional policy objective of reducing the harms of growing income inequality, such as growing political polarization. If simply redistributed, this would likely result in a net increase in overall consumption and therefore in carbon footprint, since lower-income households spend a higher proportion of their income than high-income households (see **Personal Carbon Allowance**). Instead, re-investment into solutions that reduce average consumption while increasing the quality of life for low- to middle-income people will be essential, such as high-quality shared rather than private services where possible (e.g., public transit, sharing systems) and de-carbonized energy production (see **Foundational Economy**, **Universal Basic Services, Product-Service Systems, Sharing Economy**). This broad policy combination is consistent with that advocated by proponents of **Degrowth**, who argue for overall consumption reduction, particularly in high-income countries, by replacing highly individualistic ways of living and consuming with more efficient collective modes of service provision and shared resources.

Further Reading

Chancel, L. (2022). Global carbon inequality over 1990–2019. *Nature Sustainability*, 5, 931–938. https://doi.org/10.1038/s41893-022-00955-z.

Household carbon footprint calculator. (2024). US EPA. Available at: https://www.epa.gov/ghgemissions/household-carbon-footprint-calculator (accessed: 3 June 2024).

Liobikienė, G. (2020). The revised approaches to income inequality impact on production-based and consumption-based carbon dioxide emissions: Literature review. *Environmental Science and Pollution Research*, 27(9), 8980–8990. https://doi.org/10.1007/s11356-020-08005-x.

Scrucca, F., Barberio, G., Fantin, V., Porta, P.L., & Barbanera, M. (2021). Carbon footprint: Concept, methodology and calculation. In *Carbon footprint case studies: Municipal solid waste management, sustainable road transport and carbon sequestration*. https://doi.org/10.1007/978-981-15-9577-6_1.

Starr, J., Nicolson, C., Ash, M., Markowitz, E.M., & Moran, D. (2023). Income-based US household carbon footprints (1990–2019) offer new insights on emissions inequality and climate finance. *PLoS Climate*, 2(8), e0000190. https://doi.org/10.1371/journal.pclm.0000190.

3 Conspicuous/Positional Consumption

Anna Horodecka

Definition

The term conspicuous consumption (CC) is used to describe extravagant spending on luxuries or leisure activities that display one's wealth to enhance prestige, economic power, and indicate membership of a superior class, rather than to meet basic needs. The term originated in Veblen's (1899) *The Theory of the Leisure Class*, in which he argued that "the conspicuous consumption of valuable goods is a means of securing the respectability of the gentleman of leisure" and included, for example, tapestries, high-quality clothing (such as high heels), expensive entertainment, feasts, jewelry, and other non-functional items that were used to signal status. Many of these items also characterize contemporary consumption.

The term is close to positional consumption (PC) which involves the acquisition of goods and services that confer social status and distinguish the owner from others. These goods are valued not only for their intrinsic qualities but also for their ability to signify social status.

Conspicuous consumption is becoming a widely addressed issue given the increasingly pressing climate crisis, resource degradation, and overconsumption. This concern extends beyond the wealthy, who represent a small percentage of the population with a high carbon footprint. However, through social emulation, where individuals imitate the (extravagant and wasteful) consumption patterns of higher social classes to enhance their own status, CC has a negative environmental impact that affects everyone (see **Carbon Inequality, Household Income Versus Carbon Footprint**). This process is accelerated by growing income and wealth inequality, which leads to a growing demand for positional goods as individuals seek to signal their relative social position through their consumption choices, also known as the "keeping-up-with-the-Joneses" mentality.

Both conspicuous consumption and positional consumption contribute to unsustainable consumption patterns, not only among the wealthy but also among other "emulating" strata of society. Over time, different aspects of this problem have evolved, broadening its meaning to treat these goods more as symbolic goods, which are not necessarily costly from a production perspective, but become costly because they are purchased to differentiate (social distinction) certain groups or individuals. For example, luxury brands respond to conspicuous consumption by continuously releasing limited-edition collections or exclusive collaborations. These goods are not inherently expensive due to production costs but gain value because they signal status and exclusivity.

History

The meaning of conspicuous consumption has shifted and expanded over time, influenced by different perspectives across disciplines. While the sociological perspective was primary (Veblen was considered not only an economist but also a sociologist), with the birth of psychology in the

DOI: 10.4324/9781003584056-5

late 19th century, this discipline also became interested in providing a different individual-focused explanation (e.g., survival, personal traits). The inclusion of ethical thinking in some schools of economics after the 1980s has influenced the moral-indifferent economic attitude, explored in the next section.

Among sociologists, Pierre Bourdieu, in his classic work *Distinction: A Social Critique of the Judgment of Taste*, emphasized that we consume to differentiate ourselves from others. Our consumption patterns reflect our place in the social field, known as habitus (dispositions that guide consumption within a social field), by seeking to consume in accordance with one's social class. He extended Veblen's ideas, which focused on income and class position, by introducing the concept of status competition through symbolic capital. This includes different types of resources: financial (economic capital), social networks and relationships (social capital), and distinctive "tastes", skills, knowledge, and practices accumulated through upbringing and education (cultural capital). The latter is shaped by private and elite schools, where individuals learn what are considered appropriate or desirable forms of consumption ("tastes") to differentiate themselves from others through specific consumption patterns. This cultural capital also has a significant impact on consumer choice.

Psychology has viewed CC/PC behavior as an expression of motivational structures and survival mechanisms, linking it to basic needs such as survival and sexual selection, where displaying wealth attracts mates (De Fraja, 2009). Some personal traits, such as narcissism, may reinforce this mechanism.

Ethical studies have identified the "wasteful" and "immoral" nature of CC, particularly in the context of individuals unable to satisfy their basic needs. This common intuition regards excessive luxury spending amid suffering as unethical. In times of crisis or war, regulations have often restricted luxury products that are considered wasteful (e.g., clothing regulations during World War II in the United States).

Economics has played a crucial role in shaping policy. By analyzing the impact of conspicuous consumption on resource use and sustainability, economists have influenced policies aimed at promoting sustainable consumption patterns. As we will see in the next section, however, there is no single view among economists.

Different Perspectives

Policy decisions, especially those affecting sustainability goals, are often influenced by economic theories. Some of the policies based on mainstream theories stimulated CC/PC instead of reducing it.

Neoclassical economics, which views choices as derivatives of preferences and budget constraints, initially showed little interest in conspicuous consumption. However, Keynesian economics, which focuses on the demand side of the economy, recognized its importance. Keynesian theory, inspired by thinkers such as Mandeville who suggested that "private vices" create "public benefits", saw such spending as beneficial during economic downturns (see **Consumerism**). Keynes believed that increased spending could boost aggregate demand, support employment, and stimulate economic growth.

This concept was further discussed in conjunction with the "trickle-down effect", which is a rationale employed to justify neoliberal policies, by suggesting that benefits given to the wealthy or corporations, such as tax cuts or subsidies, would eventually trickle down to society at large through increased investment and spending. Creating greater wealth at the top of the economic pyramid was justified as stimulating economic growth and job creation, thereby indirectly benefiting all segments of society. Unfortunately, exactly the opposite effect has occurred – increasing inequalities and CC contribute to unnecessary consumption levels, exceeding **consumption corridors** and basic needs-based consumption levels (see **Fair Consumption Space**).

Institutional economists such as Wisman (2014) emphasize that CC, especially when combined with inequality, can push individuals into debt as they strive to meet societal standards, creating a vicious cycle of consumption and debt (see **Money**). This exacerbates economic inequalities and puts additional pressure on resources. CC also has an impact on the erosion of public goods and services, as individuals demonstrate their status through private consumption, leading to reduced support for common resources (self-exclusion from public goods).

Ethically driven economic literature (e.g., humanistic economics such as Schumacher's *Small is Beautiful*) and **ecological economics** (to be distinguished from environmental economics) send strong signals of the necessity of absolute reduction of resource use and social equality (see **Steady-State Economics**). As high levels of CC/PC can have severe environmental impacts if emulated globally, wealthier classes and countries need to reduce CC. Knowing that even the lowest standard of basic income exceeds our environmental limits, the policies of wealthier countries must focus on reducing consumption in general and among the wealthier classes in particular to mitigate environmental damage (see **Carbon Inequality**, **Climate Justice**).

Application

Research on Conspicuous Consumption (CC) and Positional Consumption (PC) has led to several politically interesting and applicable ideas dedicated to remedying this phenomenon (see also **Behavior Change**, **Attitude-Behavior Gap**), with the scope to minimize CC:

1. Fiscal policy:

   - Implement redistributive mechanisms that improve access to public goods and services and reduce debt incurred to maintain social status (see **Foundational Economy**, **Universal Basic Services**).
   - Introduce taxes on luxury goods, especially those that are harmful to the environment, to discourage unnecessary consumption (see also **Personal Carbon Allowance**).
   - Ensure the affordability of sustainable goods to encourage wider uptake.

2. Education and media policies:

   - Address the role of private schools in inculcating "tastes" (as discussed by Bourdieu) that lead to CC/PC by expanding quality public education (**Education for Sustainable Consumption**).
   - Design policies that avoid creating divisions in society that can exacerbate CC/PC patterns (e.g., housing policies).
   - Implement social media literacy programs to highlight the dangers of CC/PC.
   - Promoting anti-consumerist messages through social media.
   - Promote **mindfulness** and the pursuit of inner values.

3. Motivational interventions and other regulations:

   - Encourage prosumption, self-consumption, and shared consumption to reduce dependence on traditional consumption models (see **Prosumerism**, **Sharing Economy**, **Circular Economy and Society**, **Product-Service Systems**).

- Overcome the impact of social media on extensive consumption styles through education and regulations (e.g., regulating the usage of social media, their content, and access) (see **Advertising**).
- Adapt government regulations that impose limits on consumption levels as suggested in the concept of **consumption corridors**.

Further Reading

De Fraja, Gianni. (2009). The origin of utility: Sexual selection and conspicuous consumption. *Journal of Economic Behavior & Organization*, 72(1), 51–69. https://doi.org/10.1016/j.jebo.2009.05.019.

Trigg, Andrew B. (2001). Veblen, Bourdieu, and conspicuous consumption. *Journal of Economic Issues*, 35(1), 99–115. https://doi.org/10.1080/00213624.2001.11506342.

Veblen, Thorstein. (1899). *The theory of the leisure class*. New York, NY: Macmillan.

Watkins, John P. (2019). Veblen's system of conspicuous waste. *Journal of Economic Issues*, 53(4), 914–927. https://doi.org/10.1080/00213624.2019.1657745.

Wisman, Jon D. (2014). Inequality, social respectability, political power, and environmental devastation. *Journal of Economic Issues*, 45(4), 877–900. https://doi.org/10.2753/JEI0021-3624450407.

4 Hedonic Treadmill

Katarzyna Stasiuk

Definition

The hedonic treadmill is a metaphor for a psychological phenomenon where individuals attempt to create various positive experiences to increase their happiness. However, these efforts are typically successful only for a short period, as individuals eventually return to their baseline level of subjective well-being. This return to equilibrium is due to hedonic adaptation, a process in which people become accustomed to positive (and negative) stimuli, leading to a diminished emotional impact over time. Hedonic adaptation to positive circumstances is significant at the individual level, presenting a substantial obstacle to enhancing and sustaining long-term happiness.

The consequences of this continual pursuit of enhanced well-being are evident at both social and economic levels. Many individuals perceive material possessions as a means to enhance satisfaction, making the hedonic treadmill a critical mechanism driving consumer behavior. In an attempt to boost their happiness, individuals purchase material goods, experiencing a temporary increase in satisfaction. When this satisfaction wanes, they make subsequent purchases, mistakenly believing that their long-term well-being can be significantly influenced by individual purchasing decisions. The hedonic treadmill is further amplified by marketing strategies that encourage consumption, thus creating a feedback loop where consumer desires not only drive the market but are also driven by it.

History

The term "hedonic treadmill" was first introduced in the 1970s by psychologist Philip Brickman and colleagues, whose research has been instrumental in shaping contemporary understandings of happiness (Brickman et al., 1978). In their study, they assessed the happiness levels of individuals who had, for example, recently won large sums in the lottery, or experienced life-altering accidents resulting in paraplegia. Although notable changes in happiness levels were expected for the lottery winners and accident victims (an increase and decrease, respectively), all groups reported similar moderate levels of life satisfaction. The authors explained this unexpected result through the theory of habituation to sensory stimuli, which posits that responses to repeated stimuli diminish over time. This led them to the rather pessimistic conclusion that "the nature of [adaptation] condemns men to live on a hedonic treadmill, to seek new levels of stimulation merely to maintain old levels of subjective pleasure, to never achieve any kind of permanent happiness".

The concept of the hedonic treadmill, and happiness more broadly, has long been studied primarily by psychologists. Toward the end of the 20th century, it began to garner interest from other disciplines. Economists believe that it could explain the "Easterlin Paradox", a phenomenon observed in wealthy countries where, beyond a certain point, increased income does not necessarily

DOI: 10.4324/9781003584056-6

correlate with greater happiness. In the early 20th century, sociologist and philosopher Zygmunt Bauman (2001) also addressed the hedonic treadmill, describing it as one of the main mechanisms sustaining consumer society, which "thrives as long as it manages to render the non-satisfaction of its members".

These findings on the hedonic treadmill have recently raised concerns among policymakers that efforts to increase income in affluent countries may not enhance psychological well-being. Therefore, in some views, policy efforts should focus rather on strengthening factors with a more significant impact on happiness (such as enhancing social relationships; see **Well-being Economy**).

Different Perspectives

Since the introduction of hedonic treadmill theory, numerous studies have suggested that it requires significant modifications (mainly related to the drivers of the treadmill, its valence, and the possibility of controlling this process). One such modification pertains to the mechanism underlying the hedonic treadmill. Critics point out that hedonic adaptation encompasses a wide range of processes beyond mere sensory adaptation. According to Daniel Kahneman (2000), aspiration levels may provide an alternative explanation for the hedonic treadmill. As individuals become accustomed to positive circumstances, they begin to perceive them as a standard. In other words, as our standard of living rises, so do our expectations. Social comparisons may also influence this process (see **Social Norms**). When determining aspiration levels, individuals often make upward social comparisons, assessing their circumstances against those they perceive to be better off. As economic well-being increases population-wide, not only do individual situations improve but so do those of comparison groups, perpetually elevating aspiration levels. Hedonic adaptation may also be linked to boredom – a sense of dissatisfaction, lack of interest, and low arousal, which drive people to look for new stimuli.

It is also essential to consider the asymmetry between reactions to negative and positive events when discussing the hedonic treadmill. Although hedonic adaptation to negative circumstances can be beneficial, it is relatively slow, and individuals may never fully return to baseline happiness levels after life events such as unemployment, divorce, or disability. In contrast, hedonic adaptation to positive events occurs quickly and is more likely to be complete.

Sonja Lyubomirsky (2011) offers another significant modification to the concept of the hedonic treadmill. She posits that individuals can exert control over this process through intentional and effortful activities, such as redefining the goals of consumption (e.g., using it for intangible life experiences, enhancing social relationships, or skill development).

Applications

The market drivers and consequences of the hedonic treadmill: The hedonic treadmill persists not only due to the human pursuit of happiness but also because it is accelerated by market forces utilizing the four Ps of marketing (product, price, place, and promotion) (see **Advertising**). For example, the clothing industry has expedited production to offer new styles rapidly, moving from **fast fashion** to even ultra-fast fashion, thereby stimulating consumer boredom with existing garments. The rapid advancement of new technologies leads to electronic products becoming perceived as obsolete shortly after purchase (see **Product Returns and Right of Withdrawal**). Marketing messages use the term "new" even for old products in updated packaging. As the hedonic treadmill may be limited by budget constraints, consumers are frequently offered installment or deferred payment options to facilitate their purchasing decisions.

These marketing strategies boost the hedonic treadmill, which has resulted in tremendous consumption growth, with its destructive impacts on the global environment (see **Consumerism**). On the individual and societal levels, it also led to the focus on materialistic aims which is proven to be negatively correlated with well-being (see Box 4.1).

Box 4.1   Materialists and the consumer culture

Materialism is a concept that applies both at the individual level and at societal and cultural levels. On an individual level, materialism refers to the importance people place on the acquisition and possession of goods, coupled with a tendency to view the acquisition as essential for achieving important goals and desired outcomes. A highly materialistic person believes that acquiring material goods is the primary goal in life, the main determinant of success, and the key to happiness and self-identity. For materialists, "to be" is essentially "to have".

Although materialists often believe that material goods will enhance their happiness, research suggests the opposite is true (see **Well-being** and **Life Satisfaction Versus Income**). Numerous studies on materialism (see Belk, 2015) have shown that it is associated with dissatisfaction – whether with one's standard of living, the amount of enjoyment in life, or with their life as a whole. Instead of feeling content with what they have, materialists tend to focus on what they lack, making them less likely to appreciate the positive aspects of their lives. Furthermore, materialism is linked to lower levels of interpersonal trust and shorter, lower-quality relationships.

Materialism also operates at a societal and cultural level, where it is referred to as consumer culture or **consumerism**. Consumer culture can be broadly defined as a culture in which goals and activities are centered around the purchase and consumption of goods (Belk, 2015). In such a culture, the possession of goods becomes a measure of success and self-worth (see **Conspicuous/Positional Consumption**). People are judged by what they own, not by who they are. Those who own branded and luxury goods are seen as more successful, while those unable to keep up with the latest consumer trends are often looked down upon. As a result, consumer society is caught up in a perpetual cycle of "keeping up with the Joneses" – constant comparisons with others and endless competition for possessions.

Slowing Down the Hedonic Treadmill: Slowing down the hedonic treadmill requires attention not only at the individual (micro) level but also at the community (meso) and policy (macro) levels. Community-based initiatives can effectively facilitate social practices (see **Social Practice Theory**) that offer consumers benefits alternative to the pleasure derived from new purchases. For instance, repair cafes are community-driven workshops where people learn to fix goods rather than replace them (see **Repair**). However, these workshops also provide valuable opportunities to build social relationships and foster local integration. Such cooperative consumption can begin to create **social norms** that will have the effect of modifying individual behaviors (see **Prosumerism** and **Alternative Consumer Cooperatives**).

At the macro level, the hedonic treadmill can be challenged by policies supporting the circular economy, particularly its "Slowing" principle. This principle refers to strategies like designing products to extend their useful lifespans (e.g., with easily repairable components; see **Ecodesign**) and economic models that encourage a **sharing economy** (renting and exchanging goods instead of buying the new ones, Kennedy & Linneluecke, 2022). In 2024, the EU adopted two new regulations addressing these issues: the Ecodesign for Sustainable Product Development and the Right-to-Repair Directive. These regulations aim to extend the producer's responsibility for a product (see **Extended Producer Responsibility**), but they also facilitate changes in consumption patterns.

Further Reading

Bauman, Z. (2001). Consuming life. *Journal of Consumer Culture*, 1, 9–29. https://doi.org/10.1177/1469 54050100100102.

Belk, R. (2015). Culture and materialism. In S. Ng & A. Y. Lee (Eds.), *Handbook of culture and consumer behavior*, pp. 299–323. Oxford University Press

Brickman, P., Coates, D., & Janoff-Bulman, R. (1978). Lottery winners and accident victims: Is happiness relative? *Journal of Personality and Social Psychology*, 36, 917–927. https://doi.org/10.1037/0022-3514.36.8.917.

Kahneman, D. (2000). Experienced utility and objective happiness: A moment-based approach. In D. Kahneman & A. Tversky (Eds.), *Choices, values and frames*, pp. 673–692. New York: Cambridge University Press. https://doi.org/10.1017/CBO9780511803475.038.

Kennedy, S., & Linnenluecke, M.K. (2022). Circular economy and resilience: A research agenda. *Business Strategy and the Environment*, 31, 2754–2765. https://doi.org/10.1002/bse.3004.

Lyubomirsky, S. (2011). Hedonic adaptation to positive and negative experiences. In S. Folkman (Ed.), *The Oxford handbook of stress, health, and coping*, pp. 200–224. Oxford University Press.

5   Choice Paralysis

Soumyajit Bhar

Definition

Choice paralysis occurs when a consumer, overwhelmed by an excessive number of choices, struggles to make a decision or avoids making one altogether. Psychologically, this paralysis is linked to the fear of regret and a desire to make the optimal choice. When too many options are present, this drive for perfection can become incapacitating. This phenomenon manifests across both mundane and significant decisions, from choosing a meal to selecting a career path.

While the prevailing notion is that more choices enhance the quality of life, research shows that beyond a certain threshold, having more options actually leads to indecision, dissatisfaction, and delays. This assumption, deeply embedded in consumer-centric, market-driven economies, equates freedom with consumer choice. However, this belief often overlooks the psychological toll on decision-making, revealing the inherent limitations of consumer freedom when it is overstretched.

History

The concept of choice paralysis gained empirical support through several early studies. One notable experiment showed that consumers presented with a smaller selection of jams were far more likely to make a purchase than those confronted with a larger assortment. Similarly, another study involving college students and gourmet chocolates found that those given fewer choices were more satisfied with their experience.

These early findings laid the groundwork for understanding how too many choices can demotivate consumers, a phenomenon that has since been observed across various domains. Over time, the idea of choice paralysis has gained traction beyond academic research. Books like *The Paradox of Choice* (Schwartz, 2004) brought the concept into mainstream discourse, highlighting the adverse effects that extensive choice options can have on well-being. Media, advocacy groups, and policymakers now use the term to describe how choice overload hinders decision-making in sectors such as healthcare and digital media.

Different Perspectives

This phenomenon, however, might reflect a deeper, systemic issue embedded within our consumer-driven economic model. In a world where liberty is increasingly equated to **freedom of choice**, particularly consumer choice, individuals are bombarded with excessive options, not solely for their benefit but to fuel a neoliberal system dependent on constant economic growth (see **Advertising, Greenwashing**). This model, which thrives on growthism, necessitates insatiable human wants as its oxygen, with each new product, service, or experience positioned as the next essential choice. As such, consumer choice becomes less about meaningful freedom and more

DOI: 10.4324/9781003584056-7

about sustaining an economy that prioritizes growth above all (see **Degrowth**). This pursuit leads not only to choice paralysis but also to the reinforcement of a consumption cycle that is challenging to escape (see **Consumerism, Hedonic Treadmill**).

Research shows that unlimited consumer choice does not foster long-term well-being. Instead, it contributes to frustration, delay, and dissatisfaction, challenging the idea that more choice always leads to better outcomes. To counteract this, we need to rethink what prosperity means, shifting away from the relentless pursuit of limitless consumption toward models that prioritize well-being (see **Well-being Economy**). Policies that simplify decision-making – such as default options in healthcare or retirement – could help alleviate choice paralysis and promote more balanced consumption (see **Green Nudging, Choice Editing**).

Several socio-psychological factors contribute to choice paralysis:

- *Regret*: When a decision leads to less-than-perfect outcomes, individuals may experience regret, believing another option could have been better. The more choices available, the easier it becomes to imagine how an alternative might have led to a superior result, amplifying dissatisfaction (see **Product Returns and Right of Withdrawal**).
- *Missed Opportunities*: Large choice sets make individuals acutely aware of the opportunities they are passing up. This phenomenon, known as FOMO (fear of missing out), exacerbates feelings of loss, as consumers feel they are missing out on better alternatives.
- *High Expectations*: As access to more options increases, so do expectations. With more choices, consumers expect perfection, and even good options may feel inadequate if they do not meet heightened expectations. This "curse of discernment" can lead to dissatisfaction even when choices are objectively better.
- *Self-Blame*: When outcomes are disappointing, individuals are more likely to blame themselves when they have had many options. With limited choices, external factors are more easily blamed, but with a large selection, individuals internalize the blame for not making the "right" choice, further exacerbating decision paralysis.

Applications

As consumer choices grow increasingly complex, choice paralysis is becoming recognized as a significant factor that affects both individual well-being and broader market dynamics. The detrimental effects of choice paralysis have broad applications beyond individual consumers. Public policymakers are designing systems to reduce the cognitive burden in areas such as healthcare and retirement planning. Default choices, like auto-enrolment in pension plans, are increasingly employed to simplify decision-making.

In the digital space, recommendation algorithms employed by e-commerce platforms and streaming services help streamline decision-making by offering curated choices based on individual preferences. These AI-driven systems are instrumental in preventing choice overload in sectors where the volume of options can easily overwhelm users (see **Information and Communication Technology**).

Interestingly, choice paralysis is not entirely negative. In industries plagued by overconsumption – such as **fast fashion** or consumer electronics – the overwhelming number of options can lead to decision avoidance. This hesitancy may result in fewer, more thoughtful purchases or even reduced consumption, offering a potential environmental benefit. The challenge, however, lies in harnessing this phenomenon to promote more sustainable consumption patterns. Ultimately, managing the balance between consumer freedom and decision overload is crucial for ensuring that individuals can make confident, informed decisions without feeling overwhelmed.

Further Reading

Buckner, M.M., & Strawser, M.G. (2016). "Me" llennials and the paralysis of choice: Reigniting the purpose of higher education. *Communication Education*, 65(3), 361–363. https://doi.org/10.1080/03634523.2016. 1177845.

Chernev, A., Bockenholt, U., & Goodman, J. (2010). Choice overload: A conceptual review and meta-analysis. *Journal of Consumer Research*, 37(2), 344–362. https://doi.org/10.1086/651235

Schwartz, B. (2004). *The paradox of choice: Why more is less*. HarperCollins.

Schwartz, B. (2012). Choice, freedom, and autonomy. In P.R. Shaver & M. Mikulincer (Eds.), *Meaning, mortality, and choice: The social psychology of existential concerns*, pp. 271–288. Washington, DC: American Psychological Association. https://doi.org/10.1037/13748-015.

Sunstein, C.R. (2015). *Choosing not to choose: Understanding the value of choice*. Oxford University Press, USA.

6 Generational Consumption Differences (in China)

Wenling Liu, Yulin Zhu, Shahzad Khan Durrani, and Lei Zhang

Definition

Household consumption varies, sometimes vastly, with different generations, shaped by diverse experiences such as economic conditions, technology, and cultural values (Owen & Büchs, 2024). This is particularly evident in China, due to the vast economic and technological developments which have taken place in the past half-century (Hu et al., 2020).

In the 21st century, China's household consumption expenditures have increased significantly, driven by rapid economic growth. The internet and the rise of social media had a profound impact on consumption patterns and lifestyles, accelerating the shift to mobile payments and making online shopping more convenient (see **Information and Communication Technology**, **Money**). As shown in Figure 6.1, between 2003 and 2022 urban household expenditures increased about 2.5 times, and rural household expenditures grew by approximately 4.5 times. However, this also raises concerns about overconsumption and environmental sustainability (Liu et al., 2016).

Over the past 70 years, China's demographic structure was reshaped by rising purchasing power, the one-child policy, and rapid urbanization. The structure shifted from a population pyramid to an "olive" shape, characterized by a substantial middle-aged segment and smaller proportions at the young and old ends, indicative of an aging society (Figure 6.2). Amid this demographic shift and changing family structures, residents' living habits, consumption behaviors, and social characteristics have changed in generation-specific ways, beyond traditional influences of family status and income. For example, (i) elderly households prioritize health care, medical services, and traditional products, (ii) three-generation households exhibit more diverse consumption patterns, and (iii) the DINK (Double Income, No Kids) families (the number of which has been rapidly growing) spend more on lifestyle and leisure activities.

These evolving demographics and consumption patterns have significant implications for businesses and policymakers aiming to address the diverse needs of China's population while considering demands for sustainable consumption and lifestyles.

History

Before the 1978 economic reforms, consumption in China was shaped by a planned economy, with the government controlling the distribution and availability of goods. Chinese households prioritized necessities like food, clothing, and shelter, as consumer goods were scarce and personal income limited.

During the Reform Era (1978–1992), China began shifting to a market-oriented economy. Baby Boomers (born in 1946–1964), who were adults during this period, focused on asset accumulation and financial stability (for international comparability, we have chosen to use a familiar

DOI: 10.4324/9781003584056-8

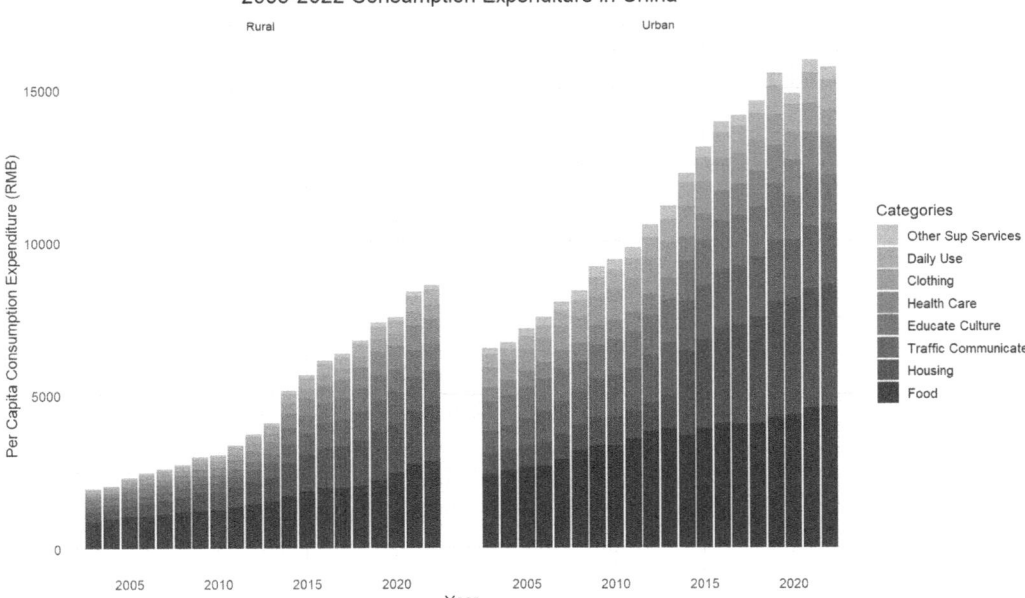

Figure 6.1 2003–2022 household per capita consumption expenditure in China

Data Source: China Statistical Yearbook

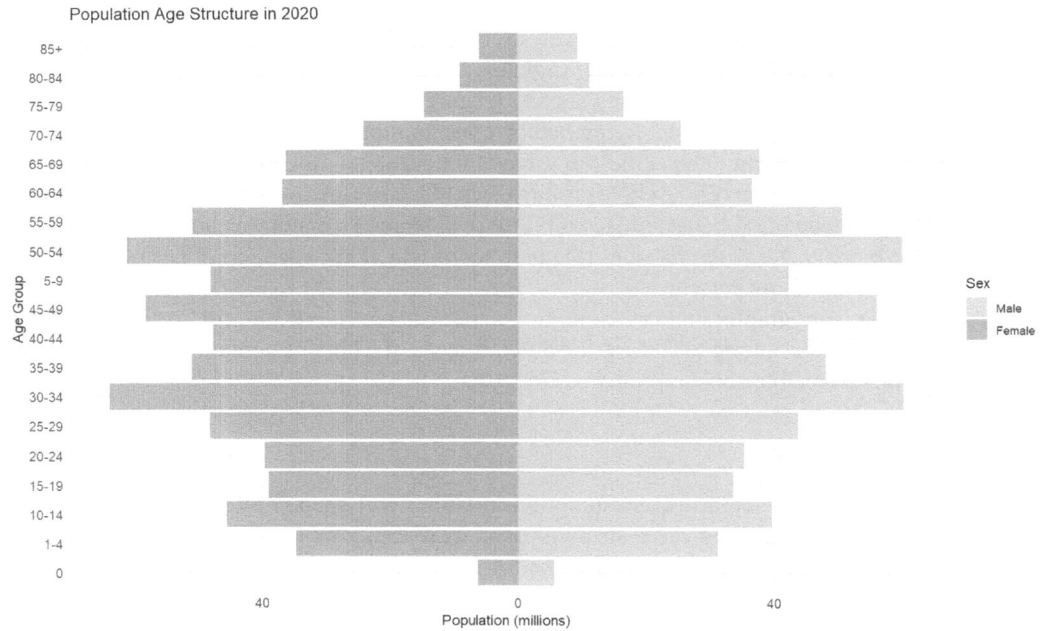

Figure 6.2 Population age structure in 2020

Data Source: Seventh National Census

categorization of generations such as Baby Boomers, Millennials, Generation X, Y, Z, etc.). Limited resources led to conservative consumption practices, such as relying on coal for energy, maintaining simple diets, and having minimal household appliances. Generation X (born in 1964–1980), coming of age during this era, also practiced resource conservation and prioritized savings, gradually increasing spending on travel, education, and household goods as reforms progressed.

Both Baby Boomers and Generation X tend to prioritize practicality and cost over sustainability, often showing a cautious approach to resource use. This generational cohort is typically less receptive to adopting new technologies and sustainable practices due to established habits and skepticism toward new trends. However, their inherent thriftiness aligns very well with sustainable consumption principles (see **Quiet Sustainability**). For instance, older generations are more likely to repair and reuse household items instead of replacing them and thus reduce waste (see **Repair**).

The Market Economy Era (1993–2008) brought increased economic liberalization and access to global goods. The rise of e-commerce, including Alibaba's launch in 1999, began to influence shopping habits. Millennials (born in 1981–1996), who grew up as beneficiaries of the "socialist market economy" and the one-child policy, became China's first digital-savvy generation. With more disposable income and access to digital platforms, they increasingly purchased luxury goods, global brands, and travel experiences. Generation X adapted to this more open economy and also increased spending on education, health, and imported products while maintaining a level of savings.

The Digital Era (2009 to present) signifies China's full integration into the global digital economy, heavily shaping the consumption patterns of Generation Z (born in 1997–2012). As digital natives, they engage extensively with online platforms and are inclined toward sustainable and ethical consumption, influenced by both Western trends and domestic regulations promoting eco-friendly products. Millennials, now fully adapted to digital consumption, also emphasize mobile payments, personalized services, and environmental awareness in their consumer behavior.

Both Millennials and Generation Z in China are more open to sustainable practices and new technologies. They are generally more environmentally conscious, influenced by global trends, and have greater exposure to environmental education and the media (Liang et al., 2024). This demographic is more likely to prioritize ecofriendly products, participate in recycling programs, and embrace digital technologies that promote sustainability. For example, younger Chinese consumers are driving the market for electric vehicles (EVs) and renewable energy solutions (see **Energy Consumption Behavior**, **Energy Overshoot**, **Sustainable Mobility**). They are also more inclined to support brands that emphasize sustainability and ethical practices. This generational cohort often leads the way in integrating sustainable practices into their daily lives, from using energy-efficient appliances to participating in community clean-up events.

Different Perspectives

While global labels like "Millennials" and "Baby Boomers" are recognized in China, unique local terms such as "Post-80s" (born in the 1980s), "Post-90s" (born in the 1990s), and "Post-00s" (born in the 2000s) are popular. The Post-80s and Post-90s, often called the "only-child generation" due to the one-child policy, benefit from concentrated family resources but face pressures like supporting aging parents. The Post-80s are generally more cautious, focusing on practical and quality-oriented spending. In contrast, the Post-90s, influenced by Western culture and digital media, embraced diverse consumption patterns, and spending on hobbies and experiences. The Post-00s, fully immersed in digital technology, value self-identity and favor brands that emphasize individuality and social appeal, with a preference for experiential purchases like livestream shopping and influencer recommendations.

Application

Initially, China's economic growth was driven by policies focused on mass production and exports, transforming the country into a global manufacturing hub. In recent years, the government has shifted its focus toward boosting domestic consumption to create a more resilient economy, less dependent on international markets. The "dual circulation" policy officially emphasizes strengthening the domestic market while remaining engaged in global trade.

Environmental sustainability has recently also become central to China's policy agenda, with ambitious goals for carbon peaking by 2030 and carbon neutrality by 2060. Although sustainable consumption is primarily promoted by government and academic institutions, public interest is gradually growing. Initiatives like the Carbon Generalized System of Preferences (CGSP), which rewards carbon-reducing actions by individuals and small businesses, encourage low-carbon lifestyles (Chen et al., 2023). Though public awareness is still evolving, these efforts can foster a culture of ecoconscious consumption.

The concept of sustainable consumption is also consistent with China's traditional culture, where principles from Confucianism and Taoism foster practical, quality-focused purchases and discourage overt displays of wealth (see **Conspicuous/Positional Consumption**) to maintain collective harmony. Traditional Chinese values such as thrift, family-centered living, and collectivism are often in contrast to the individualism and materialism associated with consumerism. While younger generations are more inclined to embrace consumer culture, there remains an underlying cultural tension between consumerism and Confucian values of moderation and frugality.

Furthermore, economic slowdowns after COVID-19 have dampened consumption, as concerns over job stability, wage growth, and rising living costs make households more cautious with spending. The convergence of strong underlying cultural factors with economic challenges has moderated the expected rise in domestic consumption. This may present a challenge for policymakers aiming to transition from an export-led to a consumption-driven economic growth mode, but it is encouraging from a sustainability perspective.

Further Reading

Chen, F., Chen, Q., Hou, J., & Li, S. (2023). Effects of China's carbon generalized system of preferences on low-carbon action: A synthetic control analysis based on text mining. *Energy Economics*, 124, 106867. https://doi.org/10.1016/j.eneco.2023.106867.

Hu, Z., Wang, M., Cheng, Z., & Yang, Z. (2020). Impact of marginal and intergenerational effects on carbon emissions from household energy consumption in China. *Journal of Cleaner Production*, 273, 123022. https://doi.org/10.1016/j.jclepro.2020.123022.

Liang, J., Li, J., Cao, X., & Zhang, Z. (2024). Generational differences in sustainable consumption behavior among Chinese residents: Implications based on perceptions of sustainable consumption and lifestyle. *Sustainability*, 16, 3976. https://doi.org/10.3390/su16103976.

Liu, W., Oosterveer, P., & Spaargaren, G. (2016). Promoting sustainable consumption in China: A conceptual framework and research review. *Journal of Cleaner Production*, Special Volume: Transitions to Sustainable Consumption and Production in Cities, 134, 13–21. https://doi.org/10.1016/j.jclepro.2015.10.124.

Owen, A., & Büchs, M. (2024). Examining changes in household carbon footprints across generations in the UK using decomposition analysis. *Journal of Industrial Ecology*, 28(6), 1786–1800. https://doi.org/10.1111/jiec.13567.

7 Gender

Stephan Wallaschkowski and Mariëlle Feenstra

Definition

Gender, distinct from sex, refers to social dimensions of being male or female, encompassing traits, behaviors, and roles deemed typical and/or appropriate for men and women (e.g., long vs. short hair). Although partly rooted in biological differences, they are largely socially constructed and vary across societies and times (Wood & Eagly, 2012). People learn the gendered **social norms** of their socio-environment during their socialization from parents, peers, and other role models. Typically, they are internalized without active deliberation, which is why they are often erroneously considered "natural". Because of their pervasiveness and ubiquity in everyday life, gender norms thus significantly influence our identity formation and subsequent self-concept. Consequently, common views of "masculinity" and "femininity" strongly shape our actions, interactions with others, and social (self-)positioning, including consumption patterns and attitudes toward sustainability.

This happens through an internal and an external route. Internally, we regulate our emotions, cognitions, and behaviors to conform to internalized gender norms so that what we feel, think, or do is consistent with our gender identity. Deviations would lead to cognitive dissonance (mental discomfort due to contradictory beliefs and actions), which we typically want to avoid. Externally, these norms affect us via social expectations that others place on us. This enacts gendered emotions, cognitions, and behaviors because not feeling, thinking, or acting in gender-appropriate ways can lead to social sanctions ranging from small cues of disapproval to serious physical or psychological violence.

Both pathways operate in parallel, constantly reinforcing and stabilizing each other. Consequently, people continuously reproduce the gender norms of their socio-environment by adhering to them in their daily social practices – a process called doing gender – and simultaneously expecting the same from others; both usually implicitly, without even realizing it (West & Zimmerman, 1987; see **Social Practice Theory**). In this way, they perpetuate the gendered structure of a society, including the behavioral differences, power dynamics, and hierarchies associated with gender.

In today's consumerist societies, where acquiring goods and services beyond one's basic needs plays a central role in people's lives (see **Consumerism**), gender also has important implications for sustainable consumption. At the individual level, gender norms can hinder or promote sustainable consumption, depending on how the two are interrelated (Figure 7.1). A contradiction between what's considered gender-appropriate and what's required to consume sustainably will create gender-specific barriers to sustainable consumption. The opposite is true, if aspects of sustainable consumption are part of the appropriate "doing gender" for men or women. In this regard, many papers assert that sustainable consumption is more compatible with "femininity", deterring men from it. However, these studies mostly adopt a so-called weak perspective on sustainable consumption (focused primarily on buying "green" products). Looking at strong sustainable consumption, Wallaschkowski (2023) shows that gender differences in **sufficiency** depend on the consumption field: Men tend to overconsume "masculine" and women "feminine" products (see Box 7.1 for examples).

DOI: 10.4324/9781003584056-9

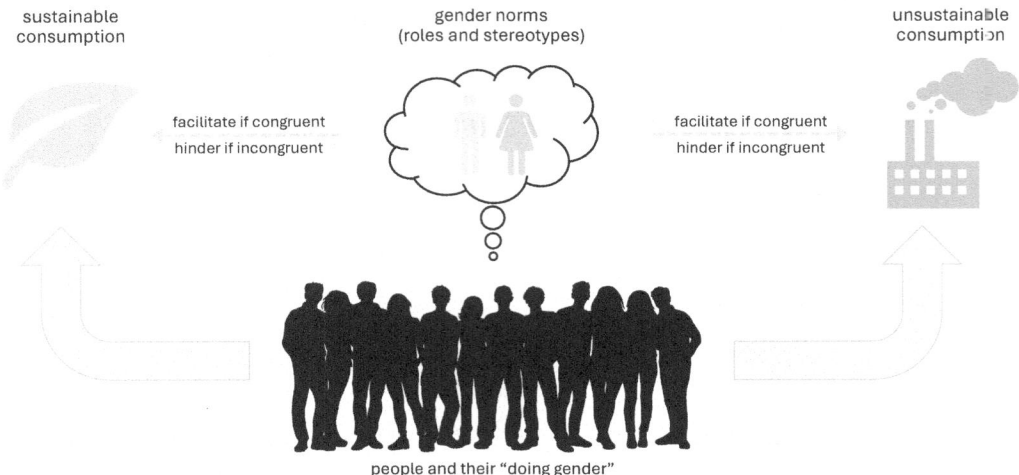

Figure 7.1 Possible interrelations between gender norms and (un-)sustainable consumption at the individual level

Source: By authors

Box 7.1   Examples of gendered barriers to sustainable consumption

Men, meat, and masculinity

In many cultures, there is a strong link between meat consumption and "masculinity". Meat is seen as a symbol of strength and virility – qualities typically attributed to and expected of men. This association is reinforced by media, **advertising**, and common narratives, leading many men toward unsustainable and unhealthy meat-heavy diets (see **Protein Shift**). Conversely, vegetarian men are more likely than women to receive judgmental remarks.

The "female fashionista"

Women's clothing consumption far exceeds men's. This can be attributed to the widespread stereotype of the "female fashionista", associating a fashionable look with "femininity". Reinforced by the common reduction of women to their appearance, it creates a strong pressure to constantly buy new clothes, which men don't face with the same intensity (Wallaschkowski, 2023) (see **Fast Fashion**).

The gendered face of energy poverty

The UN sustainable development goals envision a future with affordable, reliable, and need-adequate access to modern decarbonized energy services for all. However, the persistent *gender pay gap* and gendered division of care responsibilities leave women more vulnerable to energy poverty. Consequently, they are overrepresented in social housing, where investments in energy efficiency to reduce their energy bill hinge on their landlords or are impeded by their limited financial capacity (Feenstra et al., 2024).

The linkage between sustainable consumption policy and gender-responsiveness at the systemic level can be demonstrated in the energy transition. Given that energy access, consumption, and services are gendered globally, the transition toward 100% renewables requires attention to prevailing gender differences and gender relations to be just and successful. In many countries in the Global South, for instance, women cook on biomass, causing more women to die from indoor air pollution than from malaria and AIDS. However, clean-cooking programs entail high up-front investments and challenge long-standing food traditions. It is, thus, necessary to ensure social acceptance and financial feasibility for these women to avoid detrimental consequences and stimulate the consumption of clean energy for cooking.

History

Historically, "sex" and "gender" were seen as the same thing, with people being categorized as male or female based on biological differences – assuming that anatomy determines social roles and identity. In the mid-20th century, however, psychologists, sociologists, and feminist writers began distinguishing "sex" (biological attributes) from "gender" (roles, behaviors, and identities), arguing that many differences between men and women, which often served as the ideological basis for paternalistic suppression, are socially constructed rather than biologically determined. They pushed for greater equality and questioned prevailing gender roles, paving the way for more diverse gender expressions and "queer" identities beyond the traditional binary, seeking gender diversity and justice. Nevertheless, the traditional separation between "masculine" and "feminine" traits, behaviors, and roles still heavily shapes today's social reality.

Today, social scientists point to the heterogeneity also within gender groups, recognizing that human diversity entails many characteristics (e.g., gender, age, education, ethnicity, socio-cultural background, (dis)ability, etc.), which combined constitute complex multidimensional identities. Behaviors like energy consumption are influenced by belonging to several groups simultaneously that together determine one's self-concept and social positioning. The concept of "intersectionality" is therefore increasingly used to amplify when people's multifaceted identity is overlooked or ignored in data collected, analyzed, and reported to inform sustainable consumption strategies and policies.

Different Perspectives

The importance of gender influencing sustainable consumption can be approached from three different perspectives (Weller, 2017), which correspond to the various dimensions in which gender norms and/or their consequences can be observed.

1. *Dimension of gender differences*: This approach examines differences between men and women in sustainable consumption behaviors. Such studies have revealed, for example, that women are more likely to consider ecolabels when selecting products (see **Ecolabeling**). Others have compared average carbon emissions or distinct mobility patterns. Typically, results suggest a higher tendency toward consuming sustainably in women; however, this is usually operationalized in its weak form (Wallaschkowski, 2023).
2. *Structural dimension*: This approach assesses the implications of gender inequalities (e.g., in income, access to resources, distribution of care work, etc.) and power relations for sustainable consumption. However, the interplay of numerous intersecting characteristics is too often overlooked. Differentiating consumers beyond the gender binary enables policies that respond more specifically to individual needs and behaviors (Feenstra et al., 2024).

3. *Symbolic dimension*: This approach explores how common notions of "masculinity" and "femininity" regarding sustainable consumption are socially constructed. What traits, behaviors, and roles are ascribed to men and women, and how do they relate to consuming (un-)sustainably? How are these gender norms formed and maintained in our daily social practices? How are they reinforced and perpetuated by (social-)media? How do they foster or hinder sustainable consumption?

Applications

Applying a gender lens in designing sustainable consumption policies is essential to ensure their effectiveness and widespread adoption. First, this requires sufficiently disaggregated data regarding consumption differences between genders (or more complex intersectional identities). We propose five foci for data collection and analysis: physiological, economic, health, socio-structural, and socio-cultural (see Box 7.2).

Box 7.2   Proposed foci for gender-disaggregated data

Five foci for gender-disaggregated data collection and analysis in the example of energy consumption (see also **Energy Consumption Behavior**):

Physiological: Women are more sensitive to ambient temperature than men, increasing their energy demand for heating and cooling; however, very young or old age significantly reduces the ability of both to cope with heat or cold stress.

Economic: Women with low incomes are disproportionately found as heads of households, either as single-parent families or, due to their greater longevity, living alone at pensionable age. This exposes them to a higher risk of energy poverty.

Health: Living in inadequately heated or cooled homes has gender-specific detrimental implications on respiratory and cardiovascular systems, and mental health and well-being.

Socio-structural: Energy consumption patterns depend on factors like marital and employment status, which differ between but also among genders.

Socio-cultural: Stereotypical ascriptions make men, for example, prefer more fuel-hungry cars.

Second, it is important to get an in-depth understanding of the structural and symbolic basis for these differences by gathering both quantitative and qualitative data. Otherwise, there is a risk of their unintentional reproduction and reinforcement. For instance, hiding the ecofriendliness of male-targeted products – to not threaten customers' "masculinity" – feeds the "green feminine stereotype" underneath this problem. A better option would be marketing communication that deconstructs the notion of green products as "unmanly". This relates to our third point: Marketers, policymakers, activists, etc., should always scrutinize the gendered implications of their measures. If done right, both can go hand in hand: promoting sustainable consumption and a diverse and just society.

Further Reading

Feenstra, M., Laryea, C., & Stojilovska, A. (2024). *Gender impact of the rising costs of living and the energy crisis*. Study requested by the FEMM Committee of the European Parliament, PE 754.488. Available at:

https://www.europarl.europa.eu/RegData/etudes/STUD/2024/754488/IPOL_STU(2024)754488_EN.pdf (accessed: 15 January 2025).

Wallaschkowski, S. (2023). *Gender differences in (un-)sustainable clothing consumption. Implications for fostering strong sustainable consumption.* Paper presented at the joint 5th International Conference of the Sustainable Consumption Research and Action Initiative (SCORAI) and 21st European Roundtable on Sustainable Consumption and Production (ERSCP), Wageningen University and Research, July 5–8, Wageningen, Netherlands.

Weller, I. (2017). Gender dimensions of sustainable consumption. In S. MacGregor (Ed.), *Handbook of gender and environment*, pp. 331–344. Routledge. https://doi.org/10.4324/9781315886572.

West, C., & Zimmerman, D.H. (1987). Doing gender. *Gender & Society*, 1(2), 125–151. https://doi.org/10.1177/0891243287001002002.

Wood, W., & Eagly, A.H. (2012). Biosocial construction of sex differences and similarities in behavior. *Advances in Experimental Social Psychology*, 46, 55–123. https://doi.org/10.1016/B978-0-12-394281-4.00002-7.

8   Attitude-Behavior Gap

Aleksandra Burgiel-Szewc and Laura Maria Wallnöfer

Definition

The "attitude-behavior gap" refers to the discrepancy between people's stated attitudes and their behavioral reactions. Attitudes are enduring positive or negative feelings toward objects, persons, or issues. They can influence a person's behavioral intentions, which in turn drive their actual behaviors. One may assume that recognizing someone's attitudes allows the prediction of their behavior. However, it turns out that what a person expresses as their attitude does not always translate into a corresponding behavior. This points to the phenomenon of actions not being aligned with personal dispositions – also referred to as the "value-action", "intention-behavior", "knowledge-attitudes-practice", or "belief-behavior" gap.

The incompatibility between "words and deeds" is well known across the social and behavioral sciences, where people's self-reported attitudes are not always linked to subsequent actions. This is frequently observed in behavioral domains such as diet, exercising, volunteering, or political participation. It seems to be particularly salient in ethical and green consumption, where pro-environmental attitudes seem to have a limited impact on pro-environmental behavior uptake. Accordingly, numerous studies demonstrate high numbers of consumers concerned about climate change and expressing favorable attitudes toward pro-environmental behaviors, such as buying organic foods, saving energy, or recycling. These behaviors are, however, disproportionately rare in relation to reported concerns, attitudes, and intentions. The term "30:3 syndrome/paradox" was thus coined to describe that 30% of consumers intend to buy ethical goods, while only 3% actually do.

The urgency of widespread adoption of pro-environmental and pro-social behaviors, despite increasingly reported environmental attitudes, thus points to relevant motivational and structural barriers inhibiting **behavior change** toward sustainable lifestyles.

History

In 1935, the central role of attitudes in understanding human behavior, and the dynamic nature of their influence upon actions, was highlighted by social psychologist Gordon Allport. With the emerging consumer society in the 1940s and 1950s, consumer preferences were often assessed to predict purchase decisions, but interferences from situational, economic, or social factors were also already considered. Festinger's Cognitive Dissonance Theory (published 1957) suggested that consumers realizing a disparity between their attitudes and behaviors might experience cognitive dissonance (i.e., psychological discomfort resulting from holding conflicting beliefs,

DOI: 10.4324/9781003584056-10

attitudes, or behaviors, prompting a motivation to reduce this inconsistency). To reduce the dissonance, individuals might use various justifications and neutralization techniques, including **moral
licensing**. To stimulate actual behavior change, Katz in 1960 proposed a functional approach to
studying attitudes and interventions targeting and highlighting the utilitarian or social benefits of
change.

With growing public environmental concern and the rise of **social movements** in the 1960s and
1970s, the absence of respective actions in energy conservation, recycling, or ethical consumption
became increasingly evident. In the 1980s and 1990s, the relationship between attitudes and behaviors was conceptualized in behavioral theories (e.g., Theory of Reasoned Action (TRA), Planned
Behavior (TPB), the Norm Activation- and Value-Belief-Norm Theory) and later empirically validated. Still, certain behaviors, particularly sustainable ones, remained inconsistent with predictions
based on attitudes. Works that initially described the green gap as an unexpected deviation were
soon followed by studies exploring its reasons (e.g., Kollmuss & Agyeman, 2002) and identifying
methods to narrow or close it.

Different Perspectives

There are different positions regarding what causes the attitude-behavior gap and how to effectively close it. The most frequently discussed causes (summarized in Figure 8.1) include external barriers (primarily structural), internal factors (mainly motivational), and methodological and
theoretical issues.

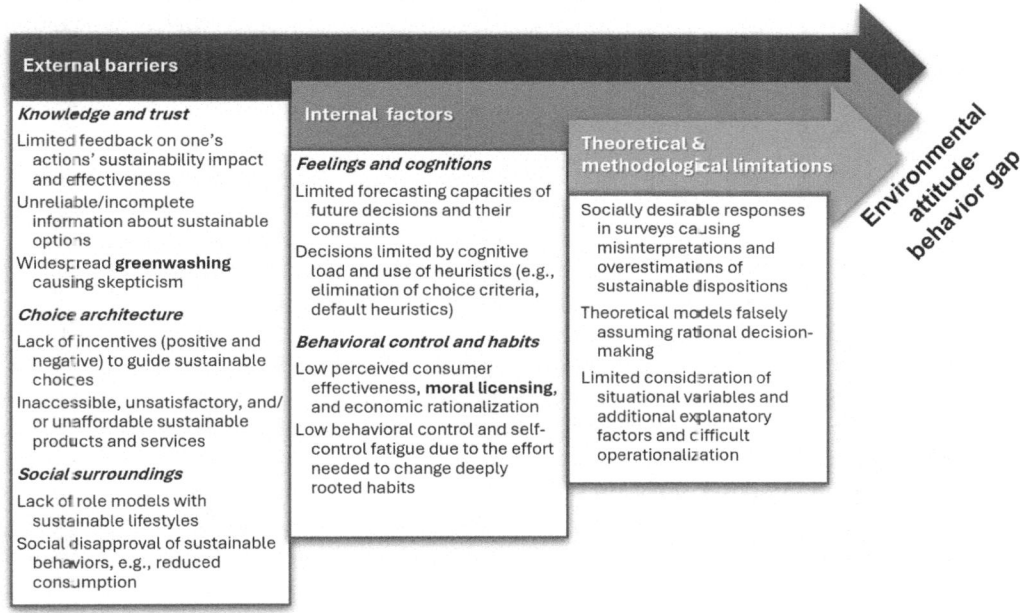

Figure 8.1 Summary of the factors causing the attitude-behavior gap in the sustainability context

Source: Burgiel-Szewc & Wallnoefer (2024)

Methodological sources of the gap lie in the deficiencies of research methods, that is, in the use of self-reported measures (based on respondents' declarations; see Lange, 2022). Meanwhile, a social desirability bias is increasingly likely since environmental concerns have become something of a trendy "must-feel", leading respondents to overestimate the importance of ethical or ecological considerations in their decisions.

Other experts argue that many people genuinely intend to make more pro-environmental decisions and are sincere in their declarations but often overlook constraints that may distort the implementation of their original intentions. Florian Kaiser and colleagues interpret such verbal expressions of attitudes as people's commitment to a goal and thus low-cost behaviors, while subsequent actions represent high-cost behaviors. This perspective shifts the focus from the gap between attitudes and behaviors to the balance of an environmental attitude's strength and behavior's costs, which can facilitate or hamper actions. A further stream of research argues that TPB-based models are incomplete because they neglect the influence of situational factors encountered by people as they move from attitude and intention to action.

In parallel, developments in behavioral economics suggested that theories assuming human rationality (e.g., TPB) cannot effectively explain all forms of behavior. This stimulated a deeper exploration of the psychological causes of the gap, including heuristics (mental shortcuts that simplify decision-making) and cognitive biases (recurring errors in thinking that distort perceptions and interpretations of information and lead to undesirable or illogical decisions).

These observations have two key implications: firstly, they have motivated a search for other factors and variables (i.e., situational, contextual, motivational, structural) that should be included in the previously used models, and secondly, they have shifted attention to alternative theories (e.g., Social Cognitive Theory, Cognitive Dissonance Theory, Dual Action Model) to study the attitude-behavior gap.

Application

In research, the attitude-behavior gap might be narrowed by addressing key methodological challenges. However, the critical remedy for bridging the gap involves **behavioral change** interventions that should enable and motivate consumers to align their actions with stated pro-environmental attitudes. Table 8.1 provides brief insights into recommended actions for the main actor groups to address the key causes of the gap while considering conceptual challenges.

The implementation and impact of the above measures depend on the collaborative effort of individuals in their various roles (as consumers, citizens, or role models), businesses, and policymakers. Businesses may encounter tensions between consumers' expressed preferences for sustainable consumption and actual demand, which could hinder the development of sustainable product offerings.

The broad set of proposed measures reflects the multiple entry points to sustainable consumption. It may offer different actors more flexibility in selecting the most promising actions and studying their effectiveness and appropriateness in transdisciplinary research projects.

Table 8.1 Linking causes with recommended actions to address the attitude-behavior gap

Causes of a gap	Conceptual considerations	Exemplary actions recommended for individual, corporate, institutional, and governmental actors
Feelings and cognition	Climate anxiety, concerns, and optimism both motivate action or inhibit it due to the risk of biased information processing	I. Reframe Communication: (1) reduced consumption as gains (life satisfaction) instead of losses (deprivation of possessions) (2) positive associations with solutions and action-taking (3) present realistic options for taking action (4) feedback tangible proofs of efficacy of individual positive (and cumulative) impacts
Behavioral control and habits	Unsustainable behaviors are often habitual (i.e., repeated, automatic, and context-dependent), and demand time and effort to change	I. Adapt choice architecture (see **Choice Editing**): (5) financial incentives and penalties (see e.g., **Personal Carbon Allowance**) (6) **green nudging** and **ecolabeling** (see (4)) II. Strengthen behavioral control and capabilities: (7) **Education for Sustainable Consumption** (8) role models for sustainable lifestyles
Knowledge and trust	Acceptance of information depends on the information source (e.g., experts vs. peers), dissemination channel, and media coverage	I. Build trust and credibility to overcome misinformation: (9) disseminate factual and practical knowledge (10) implement local sustainability events, actions, and pop-ups (see **Alternative Consumer Cooperatives**, **Grassroots Innovation**, and **Living Labs**) (11) transparent marketing communications and metrics (see **Advertising**, **The Role of Business** and **Greenwashing**)
Social surrounding	**Social norms** can influence behavior effectively if they are widely visible and supported by one's social environment	Display (salient) sustainable social norms: (12) public visibility of voluntary simplicity, downshifting, and other **sufficiency** lifestyles (see (3), (4), and (8)) (13) disseminating (social media) trends like *Underconsumption Core*, "social proof" to overcome **Conspicuous/Positional Consumption** (see (10)) (14) limit visibility of unsustainable norms by restricting ads for climate-damaging products (see **Subvertising**)
Choice architecture	Sustainable product and service alternatives are often inaccessible and/or not affordable	Increase the accessibility and affordability of sustainable choices through: (15) public policies supporting consumers and businesses with financial incentives (see (5)) (16) institutional and corporate policies to improve certification programs and provide transparent, easily identifiable, and consistent labeling
Research methods	Methodological shortcomings and misguided research can result from the increased social desirability of sustainable attitudes and behaviors	Adjust research methodology: (17) use extended measures, longitudinal study designs, and mixed methods with realistic scenarios (18) test attitude-behavior consistency (19) avoid using attitudes to infer behaviors (20) embrace inter- and transdisciplinary approaches for realistic models

Further Reading

Carrington, M.J., Neville, B.A., & Whitwell, G.J. (2010). Why ethical consumers don't walk their talk: Towards a framework for understanding the gap between the ethical purchase intentions and actual buying behaviour of ethically minded consumers. *Journal of Business Ethics*, 97(1), 139–158. https://doi.org/10.1007/s10551-010-0501-6.

ElHaffar, G., Durif, F., & Dubé, L. (2020). Towards closing the attitude-intention-behavior gap in green consumption: A narrative review of the literature and an overview of future research directions. *Journal of Cleaner Production*, 275, 122556. https://doi.org/10.1016/j.jclepro.2020.122556.

Govind, R., Singh, J.J., Garg, N., & D'Silva, S. (2019). Not walking the walk: How dual attitudes influence behavioral outcomes in ethical consumption. *Journal of Business Ethics*, 155(4), 1195–1214. https://doi.org/10.1007/s10551-017-3545-z.

Kollmuss, A., & Agyeman, J. (2002). Mind the gap: Why do people act environmentally and what are the barriers to pro-environmental behavior? *Environmental Education Research*, 8(3), 239–260. https://doi.org/10.1080/13504620220145401.

Lange, F. (2022). Behavioral paradigms for studying pro-environmental behavior: A systematic review. *Behavior Research Methods*, 55(2), 600–622. https://doi.org/10.3758/s13428-022-01825-4.

9 Behavior Change

Oksana Mont and Laura Maria Wallnöfer

Definition

Behavior change within sustainable consumption and lifestyles refers to the transformation of peoples' actions and choices to minimize direct or indirect sustainability impacts and maximize personal and collective well-being.

To reduce these impacts, a combination of efficiency, shift, and sufficiency strategies is advocated (see Box 9.1). Efficiency strategies focus on improved use of per-unit inputs, such as choosing energy-efficient appliances and fuel-efficient vehicles. Shift strategies involve changing from unsustainable behaviors, like eating meat, to more sustainable options, like plant-based proteins (see **Protein Shift**). **Sufficiency** strategies aim to reduce overall consumption levels, such as decreasing or avoiding travel by plane or car.

Behavioral changes to address unsustainable consumption levels and patterns are shaped by people's motivations and the surrounding socio-structural context. Factors such as attitudes, norms, infrastructure, production systems, technology, institutions, and regulations can either hinder or promote sustainable consumption behaviors (see **Social Norms**, **Urban Planning and Spatial Allocation**, **Sustainable Mobility**, **Sustainable Housing**, **Product-Service Systems**, **Information and Communication Technology**, **Choice Editing**, **Co-Benefits of Climate-Policy**, **Political Economy of Consumerism**). Therefore, behavior change interventions employ techniques that target both motivational and contextual elements. These interventions can be designed and implemented by institutional actors (e.g., through regulations or campaigns), corporate actors (e.g., social marketing, see **The Role of Business**), and individual actors (e.g., goal-setting and implementation intentions).

History

Amidst the environmental crises of the 1960s and 1970s, reports such as *Limits to Growth* and campaigns like the first Earth Day brought attention to the implications of post-war mass production and consumption, the rise of consumer society, and its resulting consumer behaviors.

In the 1980s, the Brundtland Report called for new behavioral norms and values change, as products were increasingly linked to aspirational lifestyles (see **Conspicuous/Positional Consumption**). At this time, consumption was increasingly fueled by credit expansion (see **Money, Consumerism**).

In the 1990s, policies and businesses promoted efficiency strategies to improve production processes and product design while raising consumers' environmental awareness and willingness to pay premium prices for eco-efficient products (see **Ecodesign**). Information provision, seen as a key intervention, followed theories linking information, attitudes, values, and behavior in a stepwise change process (e.g., Theory of Planned Behavior). Consequently, tools such as **ecolabeling** and information campaigns gained prominence.

DOI: 10.4324/9781003584056-11

However, the limitations of these interventions soon became apparent with the "**attitude-behavior gap**", contradicting the information-deficit hypothesis. One response was **social practice theory**, which gained prominence in social sciences in the early 2000s. This theory emphasizes the interplay of meanings (e.g., norms), material objects, and competencies (e.g., bodily activities) in shaping people's diets, water use, or commute.

Nudging, popularized in 2008 by Thaler and Sunstein, attempts to make desired behaviors the easy default option through interventions in the physical and social environment, for example, by rearranging the choice architecture and eliciting social norms (e.g., about hotel towel use) (see **Green Nudging**). It is particularly effective when individuals are engaged in habitual activities such as buying daily groceries.

The social context, including peers and social groups, gained prominence as a behavioral determinant, especially through its increased visibility in social media. Consequently, social marketing strategies and prompts were integrated into campaigns and movements like Transition Towns (see **Social Movements**). Many intentional communities began using social media to engage their members and the public in more sustainable living (see **Eco-Communities, Voluntary Simplicity**).

In the 2010s, the focus moved from changing individual behaviors to lifestyles. The Paris Agreement brought the understanding that **1.5°C lifestyles** must ensure maximum well-being with minimal environmental and social impacts (see **Well-being Economy**). The new focus moved beyond efficiency and shift strategies toward **sufficiency** strategies. These aim at (i) staying below planetary boundaries by minimizing environmental impact and (ii) staying above minimum social thresholds of basic human needs satisfaction by maximizing quality of life (Figure 9.1; see **Consumption Corridors, Doughnut Economy, Fair Consumption Space**).

Different Perspectives

Understanding behavior change in sustainable consumption requires considering various perspectives on behavioral determinants, intervention effectiveness, techniques, strategies, and the actors involved. Realizing efficiency, shift/consistency, and sufficiency strategies (Box 9.1) demands the

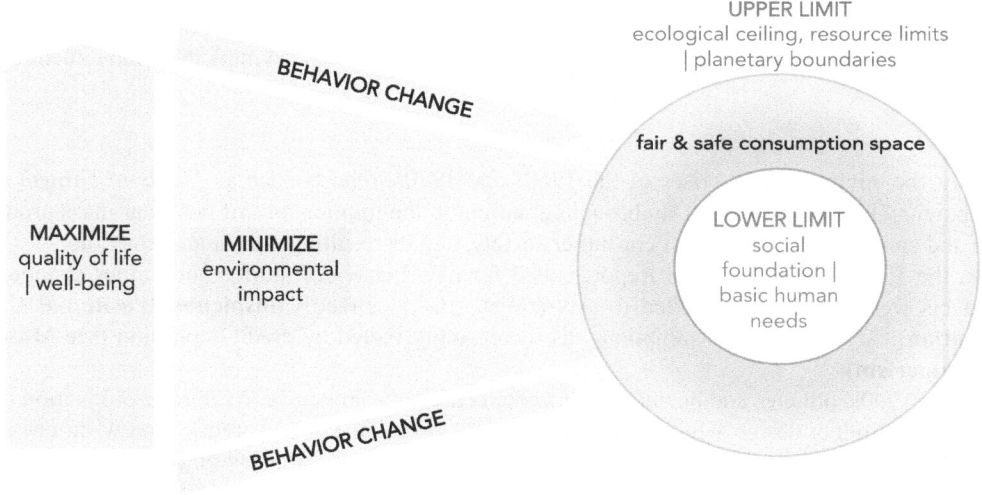

Figure 9.1 Dynamics of behavior change following sufficiency principles and premises

Source: Figure designed by Laura Maria Wallnöfer for this publication

agency of different actors, including individuals, industrial, corporate, and institutional players, to change their behaviors and facilitate change in others. Responsibility for behavior change often shifts among these actors, who hold differing degrees of power over behavioral determinants.

Behavior change and system change are often framed as separate approaches to sustainability, with one focusing on individual actions and the other on technological, built-environment, and regulatory contexts. Therefore, an integrated view of contextual and motivational determinants is essential in the intervention design and strategy implementation. Contextual determinants, such as infrastructure, products, services, economic incentives, and regulations, fall under the influence of institutional or corporate actors (top-down). Motivational determinants, including attitudes, values, **social norms**, and self-efficacy, are often within individuals' influence (bottom-up). Effective interventions help individuals to overcome obstacles, related to money, time, and/or effort, enabling them to adopt desirable behaviors. They focus on behavioral skills, like adhering to implementation plans, rather than relying on less effective determinants like knowledge and beliefs.

Behavior change is limited when motivations – such as the willingness to buy and use efficient products, modal shifts from driving to cycling, and meat consumption reduction – are not supported by structures that are available, accessible, and affordable. Conversely, if institutions and corporations provide these structures but fail to generate sufficient motivation among individuals, behavior change remains inhibited due to the **attitude-behavior gap** or entrenched habits. The bottom-up actions of individuals, households, and communities must thus be accompanied by top-down efficiency, shift, and sufficiency actions in production, infrastructure, and regulations to enable a functioning *structure-agency nexus*.

Applications

Efficiency-, shift-, and sufficiency-oriented behaviors for sustainability require a multi-actor process and evidence-based interventions. Socio-psychological, industrial, and political insights should guide these efforts based on environmental effectiveness, societal acceptance, and scalability to encourage actors to become change agents in their spheres of influence (see Box 9.1).

Box 9.1 Examples of entry points for inducing behavioral change through the target strategies: efficiency, shift, and sufficiency

Strategy	*Efficiency: improving consumption quality*		
Behavior change	Adopting better alternatives of products and services		
Effect	Mitigating negative impacts within the current technological paradigm through green innovation and product and process improvements		
Action fields	Greening Markets: providing efficient products and consumer education, considering gender and income disparities in environmental impact	Communication and **Advertising**: promotion and mainstreaming, considering greenwashing and consumer awareness	**Choice Editing** and Nudging: Rearranging choice architecture to reduce the need for additional information to guide behavior

Interventions	Ecolabels, reframing plant-based diet, gender-connotations, Austria's *Mission 11* energy-saving campaign	Restricting adverts for fossil-fuel-intensive products, EU's **greenwashing** regulation	Meatless Mondays (sustainable option as default), store layout, menu design
Actors	Policymakers: transparent labeling, energy-saving campaigns, and efficiency regulations Corporations: innovate for better input-output ratios and effective label use Consumers: informed choices about resource efficiency and purchase eco-efficient products, supported by certified labels In 2019, 29% of Europeans viewed technological improvements as the best climate mitigation strategy		
Strategy	*Shift: changing consumption patterns*		
Behavior change	Shifting to alternative business models and provision systems		
Effect	Changing consumption patterns by questioning needs and wants, e.g., the shift from ownership to shared use (see **Sharing Economy**)		
Action fields 	Social Practices: habits and social norms (meanings)	Innovative Business Models: circular and **sharing economy** models, enabling individuals to transition from consumers to prosumers	Socio-Technical Systems: infrastructures, and institutions
Interventions	Showers at workplaces for cyclists, gamification, rewards, implementation plans	Sharing apps, financial incentives for circular behaviors (see **Circular Economy and Society**)	Multimodal mobility offers, taxation, sustainable urban planning and procurement practices
Actors	Policymakers: providing infrastructure and enabling regulations, creating an environment conducive to business model innovation. Corporations: lead by mainstreaming sustainable offers, implementing circular and sharing business models, and adding value to waste. Individuals: contribute by participating in sharing networks, providing products to share, and valuing use over ownership.		
Strategy	*Sufficiency: curtailing consumption volumes*		
Behavior change	Absolute reduction in the total levels of resource consumption and associated environmental impact, while fulfilling basic human needs and enhancing well-being		
Effect	Transformative processes at a macroeconomic level toward sustainable living through "beyond the market" solutions		

Action fields	Degrowth and New Economic Order: economic models prioritizing environmental sustainability and social equity over continuous economic growth, emphasizing policy reforms and cultural shifts toward sufficiency	Sufficiency and Sustainable Lifestyles: lifestyles of fulfillment and enhanced well-being without excessive consumption, based on new value systems rooted in equity and sufficiency, enabled by policy and mainstreamed through collective action	Societal Transformation and Sustainability: deepen and accelerate change through an "ecosystem of transformation" comprising deliberate policies, shifts in socio-technical systems, and social practices, supported by autonomous bottom-up actions
Interventions	Alternative measures to GDP, **universal basic services**, shorter working time (see **Work-Life Balance**), carbon caps	Communication Framing of reduced consumption as gain, de-marketing, utilizing windows of opportunity and disruptions to normalize a sufficiency mindset	Participatory beyond-growth conferences, increasing the visibility of changes in behavior, to potentially trigger norm changes and, thus, **social tipping points**
Actors	Policymakers: promoting growth-independent well-being and alternative economic measures (see **Well-being Economy**). Companies: facilitate sufficient production, extended product use, and alternative business models. Individuals: contribute by adopting less resource-intensive lifestyles. Progressive top-down approaches to facilitate radical individual behavior changes were considered the best climate mitigation strategy by 39% of Europeans in 2019, while 14% preferred strong regulation.		

Each strategy uniquely contributes to sustainability transitions, with examples spanning individual behaviors, market and business innovations, and systemic transformations. The interplay of efficiency, shift, and sufficiency strategies highlights the need for concerted efforts across societal, economic, and institutional dimensions.

Further Reading

Jackson, T. (2005). *Motivating sustainable consumption – A review of evidence on consumer behaviour and behavioural change* (Issue December). Available at: http://www.sd-research.org.uk/researchreviews/documents/MotivatingSCfinal.pdf (accessed: 7 January 2025).

Klaniecki, K., Wuropulos, K., & Hager, C. (2019). Behavior change for sustainable development. In L. Filho (Ed.), *Encyclopedia of sustainability in higher education*. https://doi.org/10.1007/978-3-319-63951-2_161-1.

Mont, O., Lehner, M., & Dalhammar, C. (2022). Sustainable consumption through policy intervention – A review of research themes. *Frontiers in Sustainability*, 3. https://doi.org/10.3389/frsus.2022.921477.

Newell, P., Twena, M., & Daley, F. (2021). Scaling behaviour change for a 1.5-degree world: Challenges and opportunities. *Global Sustainability*, 4. https://doi.org/10.1017/sus.2021.23.

Thøgersen, J. (2021). Consumer behavior and climate change: Consumers need considerable assistance. *Current Opinion in Behavioral Sciences*, 42. https://doi.org/10.1016/j.cobeha.2021.02.008.

10 Energy Consumption Behavior

Noel Cass

Definition

Energy consumption arises from almost every human action: the production, distribution, and end-use of goods; the operation of machines and devices; inhabiting buildings with Heating, Ventilation, and Air Conditioning (HVAC) systems; using electrical systems and devices of data storage and retrieval, communication and entertainment; and traveling by different modes of transport. Energy Consumption Behaviour (hereafter ECB) encompasses almost all activities, and historically different approaches have been taken to understanding and addressing it.

ECB is defined as any behavior that involves the consumption of energy. In the context of sustainability, energy consumption typically refers to the use of fuel-based energy, such as electricity or gas, rather than the energy humans expend through physical activities like exercise. This definition is relevant with regard to climate policies that aim to reduce energy use and lower carbon emissions.

History

The term ECB first appeared in the 20th century in studies on energy use across organizations, within sectors or industries, and in buildings. In each area, the insertion of the word "behavior" signaled that studies were about human influences on the consumption of energy, contrasted with purely technical analyses of technological systems and processes, or engineering-focused building research. ECB was not a paradigm of thinking about human action, but a phrase signaling that human behavior introduces "errors" in technical predictions, requiring social science to explain what was happening. The initial social science explanations were dominated by contributions from social psychology. After about 2010, sociological and cultural theories expanded the explanations of human action to include the influences of habit, distinction, **social norms,** and more (see **Moments of Change**). In the past decade, there have been attempts to reconcile all these theories about why and how human activity results in energy consumption, some of which are reviewed below.

Different Perspectives

National- and sectoral-level analyses aggregated energy consumption by, for example, industry, agriculture, or transport, and proposed electrification, energy efficiency, and promoting renewable energy generation as the dominant, non-behavioral policy responses. Over the past 30 years, the research and policy focus has shifted to the "domestic" sector, especially household and individual behaviors, which according to one study are responsible for 74% of UK carbon emissions. This

DOI: 10.4324/9781003584056-12

level of analysis, associated with carbon footprints, has been criticized as absolving industry and the state from their responsibilities and instead "blaming the victim" (see **Household Income Versus Carbon Footprint, Consumer Scapegoatism**).

The behaviorist understanding of energy consumption has dominated this way of thinking, seeing ECB as a series of individual, more-or-less rational choices – buying this fridge or that, driving or taking a bus, turning thermostats up or down – largely driven by knowledge about prices, attitudes to the environment, and "utility seeking". This means that people consciously maximize, for example, comfort, cleanliness, convenience, and speed, while reducing cost. This behavioral understanding of ECB has led to policies aimed at changing Attitudes, Behaviour, and Choices (known as the ABC approach), mostly through price signals or providing more information (see **Behavior Change**).

Social Practice Theory has redefined ECB by de-emphasizing individual rational choice and emphasizing habit and normalized social practices. The majority of everyday energy consumption is not consciously chosen: we largely do what we did yesterday, or what other people do. This implies that reducing energy consumption is most effectively tackled not through *convincing* billions of individuals to do things differently but by changing the contexts of those actions, making less energy-consuming ways of everyday life the default or the easiest option (see **Choice Editing, Green Nudging**). This involves not just changing the devices, vehicles, and infrastructures that "lock-in" higher energy consumption but also addressing social norms: ideally making energy-intensive ways of conducting daily life (e.g., habitual use of air conditioning in homes, offices, and public spaces, use of clothes dryers in households) as socially unacceptable as smoking indoors or drunk driving. Achieving this would require a combination of regulation, infrastructural investments, and supporting alternative ways of living, rather than focusing on persuasion and pricing.

The Needs perspective stresses that ECB is directly related to the satisfaction of basic and universal needs, such as shelter, heat, light, communication, and mobility. However, in highly industrialized societies, this is accomplished in ways that involve the ever-increasing consumption of energy as the satisfiers (technologies) provide more services in socially expected ways, all conditioned by the availability of systems of provision (see **Consumption Corridors, Product-Service Systems**). As the satisfiers (technologies) used to meet these needs offer more services than are strictly required, they foster overconsumption. A departure from overconsumption to sustainable consumption demands a shift from an efficiency-based approach – which focuses on reducing energy use per unit of service for the same absolute level of service – to **sufficiency**-based approaches. Adopting sufficiency in ECB includes the reconsideration of which energy services are required and the socially conditioned nature of related activities and practices. Perhaps even more importantly, we should seek to address the dominant interests that structure our ECB through, for example, systems of provision and norm-shaping **advertising** (see **Political Economy of Consumerism**). These ratchet up with expectations (of comfort, cleanliness, convenience, and speed) as the technologies that provide them proliferate in more energy-consuming ways.

A framework that combines these multidisciplinary understandings is offered by Burger et al. (2015), integrating psychological, economic, consumer behavior, business and political science, and sociological perspectives. Their framework views an individual's choices and routines as being influenced by decision-making heuristics, and individual and social "opportunity space" features, all of which are influenced by policy structures, policies, and politics (see Figure 10.1).

Similarly, Stephenson et al.'s (2010) Energy Cultures framework emphasizes the combined effects of material culture, cognitive norms, and energy practices on ECB (see Figure 10.2).

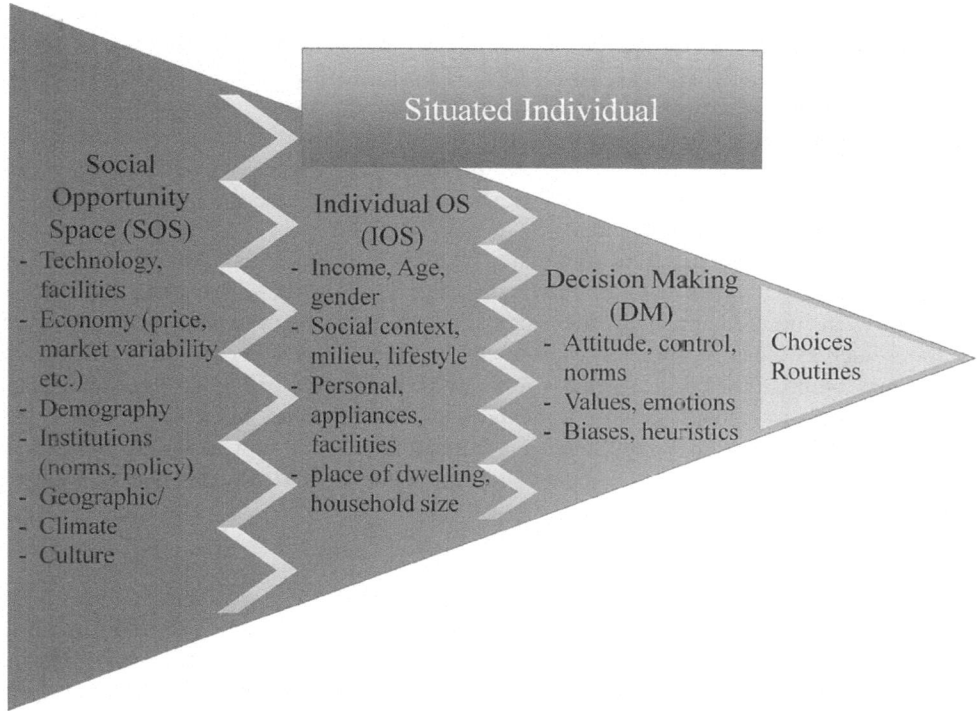

Figure 10.1 Energy consumption behavior framework

Source: reproduced from Burger et al. (2015)

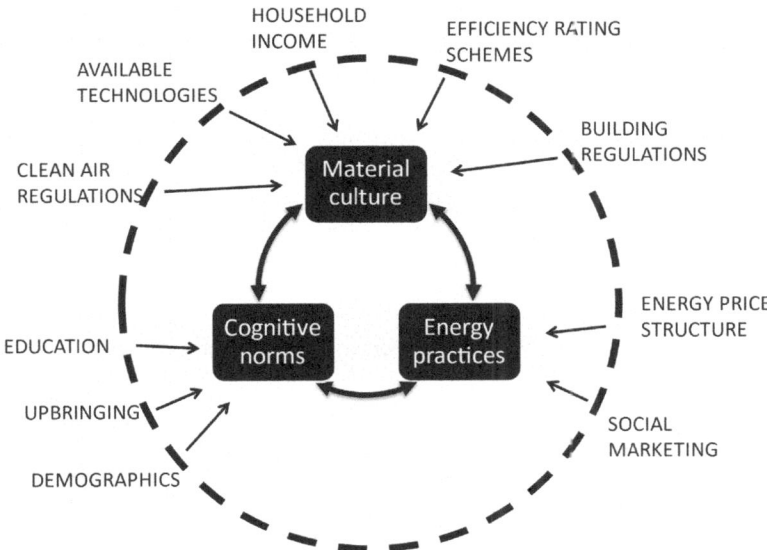

Figure 10.2 Energy cultures framework

Source: Adapted from Stephenson et al. (2010)

Other frameworks emphasize the hierarchy of lifestyle choices (e.g., where to work and live and whether or not to own a car) **as setting** the choice architecture and thereby determining subsequent day-to-day activity and travel choices. The "orders of need satisfiers" model of Brand-Correa et al. (2020), for example, suggests hierarchical yet interlocking influence from systems of provision (the political economy of technologies and infrastructures) and activities (social practices), which use energy or material services provided by specific technologies to satisfy needs. These frameworks and models shift the focus from individual choices to the larger systems, infrastructures, and social norms that lock in certain behaviors.

Considering the above perspectives about choice hierarchies and decision-makers, it is important to recognize different types of ECB in policymaking. As such, one-shot choices about device, appliance, or even infrastructure purchases may structure subsequent choices and thus require interventions such as incentives for more sustainable options. Repeated choices, such as energy use habits, may be pre-structured or heavily influenced by lifestyle or higher-order choices and require interventions to reshape social norms and choice architecture. Lifestyle or higher-order choices about housing locations, work, transport infrastructures, and how to satisfy basic human needs often require interventions by policies while campaigns could, for instance, make use of **moments of change**.

Applications

Using the example of a company and its buildings, technical solutions such as smart HVAC might induce passive consumption. Procurement policies based on information and energy labeling could source energy-efficient technologies, low-impact supplies, and/or raw materials. Behavior change policies would attempt to save energy in everyday work processes (e.g., recycling, switching off lights and devices). A key example of a norm-shifting "policy" is "Cool Biz" in Japan, where the Prime Minister and bosses abandoned business suits for short shirt sleeves in hotter weather. This demonstrated changing norms to save energy for cooling.

Domestically, there is a similar focus on heating/cooling individuals rather than spaces, using fans, blankets, and thermals rather than turning up heating or cooling, and the automation of home technologies, for example, to match intermittent renewable energy availability.

Behavior change programs have encouraged "smarter choices", tending to promote relatively low-effort-low-impact behavior changes such as changing and switching off lightbulbs or lowering thermostat settings. Tackling the more problematic ECB linked to the ubiquitous availability of energy-intensive technologies and the norms of their use remains a key challenge (see **Energy Overshoot**).

Further Reading

Brand-Correa, L.I., Mattioli, G., Lamb, W.F., & Steinberger, J.K. (2020). Understanding (and tackling) need satisfier escalation. *Sustainability: Science, Practice and Policy*, 16, 309–325. https://doi.org/10.1080/15487733.2020.1816026.

Burger, P., Bezençon, V., Bornemann, B., Brosch, T., Carabias-Hütter, V., Farsi, M., Hille, S.L., Moser, C., Ramseier, C., & Samuel, R. (2015). Advances in understanding energy consumption behavior and the governance of its change–outline of an integrated framework. *Frontiers in Energy Research*, 3, 29. https://doi.org/10.3389/fenrg.2015.00029.

Gram-Hanssen, K. (2014). New needs for better understanding of household's energy consumption–behaviour, lifestyle or practices? *Architectural Engineering and Design Management*, 10(1–2), 91–107. https://doi.org/10.1080/17452007.2013.837251.

Shove, E. (2010). Beyond the ABC: Climate change policy and theories of social change. *Environment and Planning A*, 42, 14. https://doi.org/10.1068/a42282.

Stephenson, J., Barton, B., Carrington, G., Gnoth, D., Lawson, R., & Thorsnes, P. (2010). Energy cultures: A framework for understanding energy behaviours. *Energy Policy*, 38, 6120–6129. https://doi.org/10.1016/j.enpol.2010.05.069.

11 Repair

*Sahra Svensson-Hoglund, Jennifer D. Russell,
and Jessika Luth Richter*

Definition

Repair entails the fixing of one or several specific malfunctions or performance issues in order to return a product to proper condition or functioning. Issues requiring repair can be cosmetic (e.g., a scratch or dent), functional (e.g., the product does not turn on), or both (e.g., the product's screen is broken). For every product that is successfully repaired instead of replaced, the need to make a new replacement product is avoided; the substantial waste and environmental impacts associated with material extraction, manufacturing, and transporting the replacement product are avoided. Further, the original product is diverted from becoming waste.

Whereas *general repair* is undertaken to restore functionality, that is, by fixing or replacing, e.g., a broken chain or handle on a bike, *refurbishment* (sometimes referred to as restoration) entails the comprehensive work needed to remove and replace worn and/or broken bike parts and tires, for example, as well as to clean the bicycle, perhaps after prolonged storage. *Maintenance*, on the other hand, involves interventions or fixes that are performed to prevent malfunction or performance issues before they occur, e.g., the cleaning and oiling of the bicycle chain after a ride is completed.

History

For as long as humans have used tools and kept possessions, repair has played an important role in the relationship that we have with material things. In early civilizations, the making and acquiring of possessions required materials and labor that were often limited in supply, and so repair and maintenance played a necessary role in ensuring the care and maintenance of those possessions. The expensive, labor-intensive nature of the products of earlier economies meant that the maintenance and repair of one's things was a valued investment, and an activity commonly undertaken by those who had the skill to do so.

Mass production and the increasing efficiency of industrialization have enabled increased consumption levels and patterns. As production of goods quickly became more efficient, faster, and cheaper, practices and priorities of frugality and thrift instead became obstacles to the world's prosperous, growing economies (see **Consumerism, Political Economy of Consumerism**). In the early 20th century, manufacturers began to explore the intersections of design and product lifetime, including the planned design of product end-of-life (i.e., planned obsolescence), to maintain desired profitability. Over time, these corporate efforts – coupled with marketing narratives focused on "low-prices" and the rise of large-scale retail empires – contributed to a shift in the consumer values driving demand: away from local, high-quality, and long-life products and toward distributed lower-quality, and low-cost, short-life products. The decline of people possessing repair

DOI: 10.4324/9781003584056-13

skills and increasing consumption bias in favor of "new" products led to reduced engagement in repair activities in industrialized economies – despite the continued economic necessity and need for repair. In contrast, in emerging and industrializing economies, repair remains generally widely practiced, since repair skills and an economic case for repair and the consumption of repaired products tend to exist in these contexts.

Different Perspectives

Repair is diverse across products and contexts, with perspectives ranging from *repair as a technical activity* (narrow scope) to *repair as a normalized behavior* within sustainable economic systems (broad scope). Many discussions (particularly at the policy level) focus on the intentional design of products to make them more repairable (see **Ecodesign**) and on the provision of repair necessities (e.g., spare parts, tools, schematics, manuals). Some perspectives position repair within an economic system as a relatively more (or less) affordable alternative to buying a replacement when a malfunction occurs. More broadly, repair advocates (e.g., via Right to Repair initiatives) posit repair as an important part of the evolving economic and **social norms** needed for systems to become more sustainable (see **Circular Economy and Society**).

Manufacturers are conventionally motivated and incentivized to produce and sell new products. Repair – as an activity that may undermine the sale of new products – is often at odds with conventional linear business models, which generate more profit from selling new than from repair services and, at best, may treat repair as merely necessary for fulfilling warranty commitments. Producers increasingly have the responsibility to internalize the risk and waste associated with their products. Such an **Extended Producer Responsibility** can be useful for preventing waste in the first place (see also **The Role of Business**). To counteract the predominant linear business models, calls and action in support of consumers' "Right to Repair" are growing worldwide (see Box 11.1).

Box 11.1   The right to repair

The Right to Repair is a global social movement that has mobilized to assert the legal right of product owners to have fair access to repair, foremost through demanding access to necessities from producers and repairable product design. Products most commonly included within Right to Repair regulations and policies include personal electronics, automobiles, and farm equipment. In many cases, the Right to Repair movement is as much about demanding government action to clarify and uphold consumer rights (e.g., the right of a product owner to make decisions regarding that product) as it is about holding corporations accountable (e.g., the producer's responsibility to practice ecodesign).

Application

When a consumer product requires repair, the product user is responsible for deciding whether repair is pursued (or not). This repair-or-not outcome is often influenced by a wide range of conditions outside the control of the individual (Box 11.2).

With repair commonly defined narrowly as diagnostics and repair, Figure 11.1 takes a more comprehensive view, depicting repair as a diverse multi-step process. It is rarely as linear as depicted: depending on the outcomes at each stage, it might be necessary to return to previous stages, such as when the "Diagnostics" stage points to an issue that the repairer is unable to fix or the appliance does not fit into a person's car at the "Arrangement" stage – both requiring a return to the "Investigation" stage to explore other options. Depending on the conditions encountered at each stage (e.g., an individual's predispositions, the cost of repair, and product design), the repair effort either progresses to the next stage or ceases.

Box 11.2   The elements of repair-or-not

As product designs become more complex (e.g., slim designs and embedded software), the time, labor, and cost required to repair increases. As a result, the economic cost of repair is often surprisingly high, and therefore less attractive to product users when compared with the option for a lower-cost, more easily accessible replacement. Overall, the repair-or-not outcome is determined by a range of conditions encountered in the repair process (Figure 11.1) that can be divided into the following overlapping categories:

- *Technical barriers*, relating to, e.g., product design and the compatibility and availability of spare parts
- *Policy and Legal barriers*, relating to, e.g., the constraints imposed on the manufacturing and distribution of spare parts due to intellectual property protection
- *Economic barriers*, relating to, e.g., the increasing price of a repair relative to the generally decreasing price of a replacement product
- *Infrastructural barriers*, relating to, e.g., transportation and other accessibility challenges of locating a repair site
- *Socio-cultural barriers*, relating to, e.g., consumer habits, preference for new versus repaired, and attitudes toward repair as not worthwhile

---

Some key conditions must be in place for the practice of repair to become mainstreamed: first, product repair must be technically *possible* (e.g., as mandated by the EU's *Ecodesign for Sustainable Products Regulation*); second, the repair must be *preferable* to replacement (e.g., as facilitated via the distribution of monetary repair vouchers in Austria); and third, repair must be *normalized* (e.g., the realization of a cultural shift toward valuing product longevity and rejecting consumerism). This speaks to a synchronized, gradual change in both the supply and demand of both products and repair across the five elements (Figure 11.2).

The repair of durable consumer products can be conducted in different settings: For a high-value and complex product (e.g., a smartphone or large appliance), a professional repairer may be engaged to perform the repair as a commercial, high-quality activity. In contrast, for other products, individuals may engage in so-called Do-It-Yourself (DIY) and Do-It-Together (DIT) formats of repair. Increasingly common, DIY and DIT non-commercial community repair groups, self-identifying as "repair cafes" and "repair networks", for instance, offer individuals community-building and the collaborative guidance, tools, and skills needed to engage in successful repair. Accordingly, repair over replacement, as a central principle for sustainable consumption and **sufficiency**, can lower

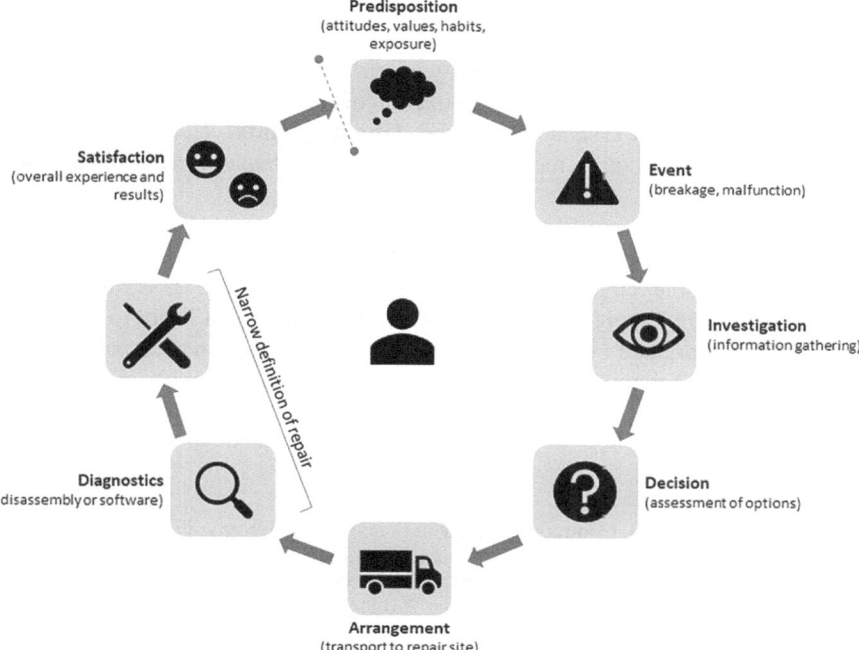

Figure 11.1 The repair process, as experienced from the perspective of the product user

Source: Svensson-Hoglund et al. (2023)

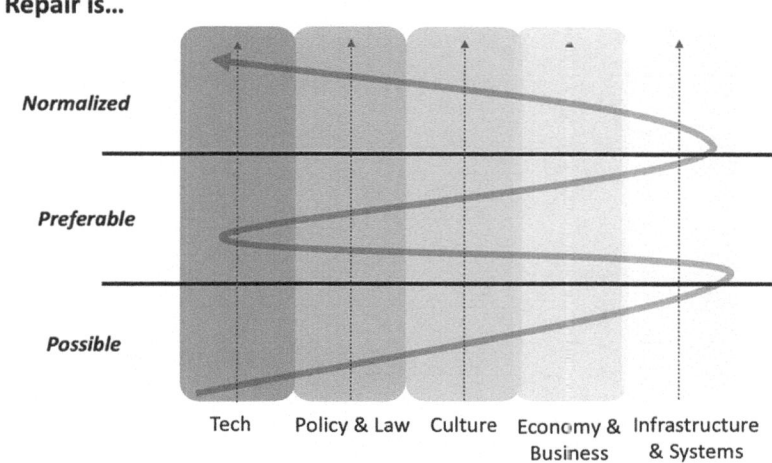

Figure 11.2 The bottom-up process of normalizing repair across the overlapping elements

Source: Svensson-Hoglund et al. (2021)

the cost of living, enhance human and community well-being, and increase material and resource efficiency.

Beyond increased availability and affordability of repair services, many stakeholders who argue for supporting community repair see the normalization of repair and a change toward a more collective "mending mindset" as crucial (see **Social Tipping Points**). Here, there is potential for repair to be a more radical action in moving toward a more sustainable consumption and production system.

The Right to Repair legislative framework primarily addresses the first level of *possibility* to upscale repair activities. However, additional measures are needed to reposition repair as a central activity within a sustainable, circular, and sufficient society (Figure 11.2) – via *preferability* and *normalization* – which are largely missing from the current focus on a Right to Repair. The shifting of consumer behaviors and attitudes requires, at a minimum, increased public awareness of the benefits of repair and assurance of lower-friction opportunities and incentives to engage in the practice (Figure 11.1).

Further Reading

Bradley, K., & Persson, O. (2022). Community repair in the circular economy: Fixing more than stuff. *Local Environment: The International Journal of Justice and Sustainability*, 1–17.

Jaeger-Erben, M., Frick, V., & Hipp, T. (2021). Why do users (not) repair their devices? A study of the predictors of repair practices. *Journal of Cleaner Production*, 286, 125382. https://doi.org/10.1016/j.jclepro.2020.125382.

Parajuly, K., Green, J., Richter, J., Johnson, M., Rückschloss, J., Peeters, J., Kuehr, R., & Fitzpatrick, C. (2024). Product repair in a circular economy: Exploring public repair behavior from a systems perspective. *Journal of Industrial Ecology*, 28(1), 74–86. https://doi.org/10.1111/jiec.13451.

Svensson-Hoglund, S., Richter, J.L., Maitre-Ekern, E., Russell, J.D., Pihlajarinne, T., & Dalhammar, C. (2021). Barriers, enablers and market governance: A review of the policy landscape for repair of consumer electronics in the EU and the U.S. *Journal of Cleaner Production*, 288, 125488. https://doi.org/10.1016/j.jclepro.2020.125488.

Svensson-Hoglund, S., Russell, J.D., & Richter, J.L. (2023). A process approach to product repair from the perspective of the individual. *Circular Economy and Sustainability*, 3, 1327–1359. https://doi.org/10.1007/s43615-022-00226-1.

12 Fast Fashion

Katia Dayan Vladimirova

Definition

Fast fashion refers to patterns of intensified production and consumption of clothes, shoes, and fashion accessories, characterized by shorter fashion cycles, shorter use cycles, and lower prices. These patterns come at the expense of poor working conditions in the supply chain and environmental pollution upstream and downstream.

History

The fast fashion business model emerged in the late 20th century and was based on copying high fashion styles to produce similar garments at a price point affordable to the middle class. In its early days, it was praised for democratizing fashion and making beauty and style accessible to consumers beyond the wealthiest elites.

The model scaled in the 1990s, relocating production away from final consumer markets and allowing brands to drastically reduce prices by externalizing social and environmental costs of production (Niinimäki et al., 2020). To cut expenses, most fashion brands gave up ownership of their factories, as used to be the norm before the 1990s. Instead, they started working with independent suppliers through intermediaries based in the Global South, notably in China, and countries in South-East Asia. The lack of ownership and brands' accountability, along with declined supply chain transparency lead to producers unprotected from macroeconomic shocks. The relocation of supply chains resulted in both distancing (separation of production and consumption decisions) and shading (obscuring production costs) (Princen, 1997).

While there is no reliable data for the global production volumes of fashion, they are estimated to have doubled between 2000 and 2020. Much of this growth can be attributed to fashion brands attempting to predict consumer demand. In the first two decades of the 21st century, fast fashion typically operated through 18-month cycles, in which brands pre-ordered garments based on trend predictions, far ahead of the actual sales. This practice routinely resulted in over-ordering, driving up the volumes of production – and resulting in unsold stock. It is estimated that roughly 20% of apparel produced today is never sold (EEA, 2024). A typical way to deal with unsold merchandise is incineration – a business practice that has recently been forbidden in France.

In response to the intensified supply, two major consumption trends emerged since the early 2000s. On the one hand, consumers started to buy more clothes – roughly doubling consumption volumes between 2000 and 2020. On the other hand, the use time of garments decreased by half over the same period. Rather than treating clothes as an investment, consumers became accustomed to lower prices and constant renewal of collections. As skills of **repair** and clothing

DOI: 10.4324/9781003584056-14

maintenance decreased significantly around the same period, a new culture of disposable fashion emerged among, for instance, the young consumers of the early 21st century, the Millennials (those born between 1981 and 1996).

With the rise of social media commerce, a new era of ultra-fast fashion has begun, disrupting the fast-fashion model and accelerating fashion consumption even further. During the COVID-19 pandemic, a new breed of ultra-fast fashion brands started selling exclusively online, cutting the costs of brick-and-mortar shops and infrastructure, investing heavily in e-commerce, and advertising through influencers on social media. With inexpensive garments ordered online and delivered by plane from (mostly) China, the "sketch-to-consumer" time has been reduced to as little as a few days.

Ultra-fast fashion brands created AI-enabled production in small batches to generate and satisfy micro-trends propagating on social media daily. This reduced the volumes of unsold stock; however, in line with Jevon's paradox, it also resulted in a dramatic increase in the overall volumes produced (see **Rebound Effects**). Notably, Shein, a Chinese ultra-fast fashion giant, has been seen to release over 10,000 new styles and items on their app per day.

Different Perspectives

Plastification of fashion: One of the most important factors that allowed brands to reduce their material costs (and resulted in growing production volumes) is a switch from "natural" (cotton, linen, wool) to oil-based synthetic textiles. According to the organization Textile Exchange, the share of synthetic fibers in the total fiber mix increased from about 30% to over 60% since 1990 – with polyester alone representing 54% of all fibers produced today. The growth of the use of synthetic fibers in fashion drives up demand for oil and chemicals alike.

Moreover, synthetic fashion bought in the Global North generates vast volumes of textile waste in the Global South. Synthetic fibers do not biodegrade and currently, there are no recycling solutions available beyond a pilot phase. EU exports of post-consumer textiles have tripled since 2000. It is estimated that at least 25% of these material flows are textile waste, including non-biodegradable synthetic garments.

Changes in consumer practices and values: The culture of disposable fashion centers around the key premise of devaluation of fashion. Due to the externalization of the environmental and social costs of production, prices of clothing today are not representative of the true cost of production. Consumers buy more clothes than ever, while household expenditure on fashion dropped to a historical low of 4% in 2020. Consumption of fashion is also highly unequal, with the majority of its carbon footprint attributed to the wealthiest consumers in the Global North.

The quality of garments has universally dropped across the fashion industry, while consumers lost knowledge about materials, their properties, and maintenance practices. With free returns offered by retailers, consumers developed new practices such as (i) buying multiple sizes or styles, intending to return most of the ordered items, and/or (ii) buying to wear the clothes once, hiding the tag, and returning the garment later. Online returns today are estimated at 22–37% (EEA, 2024), which has major environmental implications as a large part of these material flows ends up destroyed (see **Product Returns and Right of Withdrawal**).

Social media's double role: Following the rise of social media and ultra-fast fashion, consumer culture is undergoing another transformation akin to the accelerated patterns of consumption in the early 2000s. New **social norms** among Gen Z (born between 1997 and 2012) include not wearing a garment more than once on social media. The unprecedented popularity of "haul" videos (where influencers unpack purchases or gifts from ultra-fast fashion brands) led to most sales for fashion brands shifting to social media commerce. Social media influencers promote a culture of constant novelty, with no attention to the garments' quality or origin.

However, consumption patterns also became fragmented, with a growing proportion of consumers seeking more responsible and sustainable fashion choices. The recent social media trend of "de-influencing", which encourages consumers not to buy things they do not need on social media, is an example of a counterculture and a response to the "influencer fatigue" among social media users.

Applications

Fast fashion's business model will continue to exploit people and nature until there are regulations in place (see **The Role of Business**). Considering the global nature of fashion's value chains, an international agreement would be a necessary step forward to ensure synchronization among diverse stakeholders. It is important to develop not only efficiency measures to decrease the per unit carbon footprint of the system but, most importantly, **sufficiency** measures that would reduce demand for new fashion products and thus reduce the material throughput of the fashion system.

In terms of **behavioral change**, multiple avenues are available to consumers to move toward more sustainable and responsible practices when it comes to fashion: from repairing and using their own garments longer to swapping, renting, and buying second-hand, on to buying new products from local and responsible brands (Vladimirova et al., 2021; see also **Repair**, **Circular Economy and Society**). There is already a growing movement of sustainable fashion enthusiasts who advocate for buying less but better; however, this culture has yet to overtake the mainstream overconsumption patterns. Shifts toward more responsible consumption can be enabled through education and awareness raising (see **Education for Sustainable Consumption**); however, there is also a strong need to support and nurture existing alternatives to fast fashion locally.

Further Reading

Coscieme, L., Akenji, L., Latva-Hakuni, E., Vladimirova, K., Niinimäki, K., Nielsen, K., Henninger, C., Joyner-Martinez, C., Iran, S., & D'Itria, E. (2022). *Unfit, unfair, unfashionable: Resizing fashion for a fair consumption space*. Berlin: Hot or Cool Institute.

European Environment Agency. (2024). The destruction of returned and unsold textiles in Europe's circular economy. *Briefing*. Available at: https://www.eea.europa.eu/publications/the-destruction-of-returned-and (accessed: 8 January 2025).

Niinimäki, K., Peters, G., Dahlbo, H., Perry, P., Rissanen, T., & Gwilt, A. (2020). The environmental price of fast fashion. *Nature Reviews Earth & Environment*, 1(4), 189–200. https://doi.org/10.1038/s43017-020-0039-9.

Princen, T. (1997). The shading and distancing of commerce: When internalization is not enough. *Ecological Economics*, 20(3), 235–253. https://doi.org/10.1016/S0921-8009(96)00085-7.

Vladimirova, K., Iran, S., Barber, J., Blazquez, M., Burcikova, M., Henninger, C.E., Johnson, E., Joyner Martinez, C., Laitala, K., Maldini, I., McNeil, L., Niinimaki, K., Onthank, K., Plonka, M., Sauerwein, M., & Wallaschkowski, S. (2021). Conceptual framework for sustainable fashion consumption within the circular fashion system. *International Research Network on International Fashion Consumption*, June, 2025. Available at: https://www.sustainablefashionconsumption.org/framework.

13 Moments of Change

Lorraine Whitmarsh, Kate Burningham, Vanessa Timmer, Lewis Akenji, and Lisa Mastny

Definition

Many of our everyday behaviors are habits, which are hard to break. But certain periods of profound, rapid disruption in people's lives can act as catalysts for change by disrupting the context of habits. These periods are known as "moments of change" (MoCs). They may be planned or unexpected. MoCs can be divided into two categories: biographical events or "life transitions" – such as relocation, becoming a parent, starting university, and retiring – and exogenous events, such as extreme weather events, infrastructure disruption, economic shock, and political crises. Biographical MoCs operate at the individual or household level (e.g., relocation, parenthood), whereas exogenous MoCs operate at a wider scale (e.g., financial crises, pandemics, droughts). MoCs may remove cues that maintain habits and may change the social, economic, and physical contexts of action, leading the individual to consider alternatives (see **Behavior Change**). This window of opportunity is one in which people are open to making new decisions or finding themselves in a new context that catalyzes or even imposes behavior changes. These windows of opportunity can make behavior change interventions more effective during this period than in more stable times.

History

MoCs have been conceptualized and theorized in different ways across several fields and are also known as "transformative moments", "epiphanies", or "critical moments". Life-course studies have focused on major developmental changes that can reconfigure lifestyles – such as becoming an adult or parent (see **Green Parenting**), or significant changes in work, relationships, group memberships, or finances (see Box 13.1.) – while research in the clinical domain has tended to focus on negative life events. Less work has examined how exogenous events can reshape behavior either directly or via a policy response (e.g., travel disruptions, political unrest, environmental disasters). Similarly, there has long been an interest in how life events shape health behaviors, such as smoking or exercising. One review found that health behaviors were relatively stable over the life course, although some did change. For example, the breakdown of a relationship was associated with a greater likelihood of starting smoking and binge drinking. There is a longer history of behavior change studies noting the effectiveness of targeting moments of transition and habit formation and discontinuity as key times of intervention. However, applying the notion of MoCs to understanding pro-environmental behavior change (e.g., in energy use, travel, buying secondhand) is more recent.

DOI: 10.4324/9781003584056-15

Box 13.1   "Empty nesting" can lead to changing consumption patterns

When children leave home, parents face a new phase of change. Many parents find the transition to an empty nest difficult because it forces them to reconsider their roles and identities. This shift can lead to a feeling of loss of control and purpose for some parents, as they may struggle to find new ways to contribute to their households. Studies have found that consumption patterns change during this transition, as money and time are now spent on family-oriented consumption rather than on productive household tasks such as cooking a meal or doing laundry.

Different Perspectives

While a number of studies suggest that MoCs may provide an opportunity to effectively promote sustainable behaviors, the evidence is mixed as to whether this is always the case (Verplanken & Whitmarsh, 2021). Indeed, there is debate about whether the start and end point of a particular MoC can be identified and, therefore, when (if at all) is an opportune time to intervene. Many MoCs are preceded by a period of planning, during which key decisions are made, and others (e.g., childbirth) may be very stressful, both of which may explain why behavior change interventions targeting an MoC are not more effective (Schäfer et al., 2012). More fundamentally, some have argued that people's lives are always in flux, making it hard to distinguish periods of disruption from times of stability (Burningham & Venn, 2020).

Research highlights that MoCs are extremely diverse in their characteristics, in their effects on different behaviors, and across different groups. A systematic review of MoCs and low-carbon behavior found that no MoC uniformly shifts behaviors to being lower-carbon; effects vary by behavior and situation (see Table 13.1). For example, the effects of physical MoCs (relocation; Figure 13.1) on mobility behavior tend to depend on the physical infrastructure (parking, walkability) of the new home or workplace (see **Sustainable Mobility** and **Urban Planning and Spatial Allocation**). Demographic, social, economic, and physical factors also moderate the effects of MoCs on low-carbon behavior. For example, childbirth influences women's travel patterns and car ownership more than men's, and women's diets are more affected by cohabitation than men's. Taken together, the diversity of MoCs and their behavioral impacts represent a challenge to identifying the most promising groups and times to target for sustainable lifestyle interventions.

Application

Research on MoCs shows that behavior can shift rapidly in response to context change and implies that *when* one intervenes can be as important as *how* one intervenes. By drawing attention to the contextual drivers of behavior, MoC studies reinforce the evidence showing that the most effective interventions are those that target the context of decision-making (i.e., making sustainable actions easier, cheaper, and ideally the default choice), whereas those focused on individual decision-making (e.g., through information provision) are far less effective and may exacerbate inequalities (see **Choice Editing**).

Table 13.1 Different "moments of change" and their effects on low-carbon behaviors

MoC	Mobility	Energy	Diet	Material Consumption	Activism
Biographical MoCs					
Home relocation	+/-	+/-			
Work relocation	+/-				
Migration	+/-				
Transition to adulthood	-		+		
Parenthood	+/-	+	+		
Relationship change	-	-	-		
Retirement	+/-				
Environmental epiphany			+	=	=
Change in employment circumstances	+/-				
Exogenous MoCs					
Financial crisis	+	+	=		-
Natural disaster	+	+			+
Infrastructure disruption	+/-	+			
New infrastructure	+				
Pandemic	+				
Food scare			+		
Terrorism	+				

Note: "+" indicates predominantly positive change; "-" indicates predominantly negative behavior change; "+/-" indicates both a positive and a negative change; "=" indicates no change in behaviors. Blank cells indicate no available evidence.

MoC research highlights the dynamic nature of lifestyle changes and implies that timing matters when trying to intervene to reshape habits. Interventions timed to coincide with an MoC tend to be more effective than when timed to stable contexts and habits likely present a barrier (Verplanken & Whitmarsh, 2021; see **Social Practice Theory**). This has important implications for policymakers and other change-makers wanting to maximize the efficacy of their interventions. However, the evidence base on this is limited primarily to mobility behaviors (see Box 13.2). Thus, more evidence is needed to test MoC interventions targeting other low-carbon behaviors. More evidence is also needed about when exactly the "window of opportunity" to intervene is; while habits take on average three months to form, critical decision-making may actually precede the MoC. When targeting moments in which habits are more malleable, policymakers and change-makers should be mindful that individual responses to these moments (and associated interventions) can vary widely. More research is also needed on how long any observed changes last.

Box 13.2 Supporting the transition to public transport

In one study, researchers delivered an intervention to participants six weeks after they had relocated, providing them with personalized information on public transport use for daily trips as well as a one-day free ticket. As a result, transit use by participants increased from 18% to 47% after relocation. This shows that a small incentive, when combined with personalized information at the right time, can have an impact on people's transport behavior.

Figure 13.1 Relocation. Moving house is one of the most well-studied "moments of change" that can have significant impacts on low-carbon behaviors, such as reducing car use

Source: Image by HiveBoxx, Feb 2022 on Unsplash – https://unsplash.com/photos/woman-in-blue-sleeveless-dress-sitting-on-bed-oM_2BlDTrjM

MoCs are relational experiences. People's lives are "linked" to others, so changes in one person's life can affect those they are close to (see **Social Norms** and **Social Tipping Points**). In addition, people may change their social networks during MoCs in ways that affect their uptake of behaviors. There is no one-size-fits-all approach to encouraging sustainable and low-carbon behaviors during such transitions. Interventions should be targeted to the values (see **Values and Consumption**), needs, and abilities of different groups, including those with high/low environmental concerns, men/women, high-/low-income groups, rural/urban residents, and homeowners/non-owners. The upscaling potential of MoCs is significant given the widespread experience of life transitions and destabilizing external events. MoCs are universal, deeply personal, and yet culturally diverse. Harnessing MoCs to accelerate the uptake of sustainable ways of living holds great promise.

Further Reading

Burningham, K., & Venn, S. (2020). Are lifecourse transitions opportunities for moving to more sustainable consumption? *Journal of Consumer Culture*, 20(1), 102–121 https://doi.org/10.1177/1469540517729010.

Kurz, T., Gardner, B., Verplanken, B., & Abraham, C. (2015). Habitual behaviors or patterns of practice? Explaining and changing repetitive climate-relevant actions. *WIREs Climate Change*, 6, 113–128. https://doi.org/10.1002/wcc.327.

Schäfer, M., Jaeger-Erben, M., & Bamberg, S. (2012). Life events as windows of opportunity for changing towards sustainable consumption patterns? Results from an intervention study. *Journal of Consumer Policy*, 35(1), 65–84. https://doi.org/10.1007/s10603-011-9181-6.

Thompson, S., Michaelson, J., Abdallah, S., Johnson, V., Morris, D., Riley, K., & Simms, A. (2011). 'Moments of Change' as opportunities for influencing behaviour. A report to the department for environment, food and rural affairs. *New Economics Foundation and Defra*, London. Available at: https://orca.cardiff.ac.uk/id/eprint/43453/1/MomentsofChangeEV0506FinalReportNov2011%282%29.pdf (accessed: 8 January 2025).

Verplanken, B., & Whitmarsh, L. (2021). Habit and climate change. *Current Opinion in Behavioral Sciences*, 42, 42–46. https://doi.org/10.1016/j.cobeha.2021.02.020.

14 Quiet Sustainability

Petr Jehlička and Lucie Sovová

Definition

The concept of quiet sustainability is defined as comprising of:

> widespread practices that result in beneficial environmental or social outcomes and that do not relate directly or indirectly to market transactions but [sic] are not represented by their practitioners as relating directly to environmental or sustainability goals. These practices represent exuberant, appealing and socially inclusive, but also unforced, forms of sustainability.
>
> (Smith & Jehlička, 2013: 148)

Despite the undertone of inconspicuousness, the concept of quiet sustainability (QS) represents a departure from the mainstream understanding of sustainable consumption and lifestyles. Specifically, it presents a counterpoint to the so-called ABC model (see **Behavior Change**), according to which attitudes (A) drive behavior (B) which people choose (C) to adopt. While the ABC model remains popular among policymakers and certain scholarly debates on sustainable consumption, it has also been criticized (see **Attitude-Behavior Gap**, **Social Practice Theory**). The model privileges deliberate actions driven by conscious ethical values. As a result, it fails to account for everyday practices driven by motivations that are unrelated to political activism but nonetheless have sustainable outcomes. The notion of QS addresses this gap by highlighting practices characterized more succinctly as "sustainability by outcome rather than intention".

History

Quiet sustainability can be associated with low-resource lifestyles and practices of (self-)**sufficiency**, reuse, and **repair**. While such practices predate the notion of sustainability, it is in the context of sustainability debates that the notion of QS was coined in Smith and Jehlička's (2013) article *Quiet Sustainability: Fertile Lessons from Europe's Productive Gardeners*.

In this chapter, the concept is empirically grounded in the practice of food self-provisioning (FSP), intended here as growing food in home or allotment gardens. In their research in Czechia and Poland, Smith and Jehlička discovered that FSP was a practice popular across all social classes, age groups, and educational and professional backgrounds, in which around half of the population participated. It presented a significant source of fresh fruits and vegetables, covering on average one-third of practitioners' consumption of these foods. This food was typically produced using (near-)organic growing methods, as about half of the practitioners refrained from using synthetic fertilizers and pesticides, and many others tried to minimize their use. FSP presents an extremely short and localized supply chain, where consumers are simultaneously producers (see

DOI: 10.4324/9781003584056-16

Prosumerism, Food Miles). The practice contributes to food literacy and informal environmental education, while research also showed the importance of home-grown food in fostering social connectedness and resilience through informal sharing and exchange.

Originating from the practice of FSP, the notion of QS has gained significant relevance in research on sustainable food practices. Apart from numerous countries in Central and Eastern Europe, a region with widespread FSP, researchers have located QS practices in contexts as diverse as Greece, Colombia, the Faroe Islands, and China. Quiet sustainability is also visible in sharing meals and seeds and non-monetized food distribution from sources like orchards and fields facilitated by digital platforms and social media (see **Sharing Economy, Information and Communication Technology**). Recent debates explore the relationship of QS to more political articulations of **food sovereignty** (Visser et al., 2015, see also Velicu and Ogrezeanu's [2022] critical take on QS in connection to Romania's small peasants "quietly contributing to sustainability" in the shadow of systemic marginalization).

Beyond food, we note an affinity of QS with concepts such as quiet activism, actually existing sustainability, and frugal or inconspicuous innovation (see **Grassroots Innovation**). Recent discussions (e.g. Pungas et al., 2024) have also pointed to other practices that could be framed as QS, such as do-it-yourself (DIY) practices, low-tech climate adaptation strategies on the household level, or sustainable forms of travel and tourism (see **Energy Consumption Behavior, Sustainable Housing, Sustainable Mobility**). Extending QS to profit-oriented activities, the concept has also been applied to the recycling and reuse of car parts at salvage yards. While the goal of these activities is to generate financial profit, they also lead to frugal resource use and unintended environmental benefits. These emerging debates raise questions about the role of (technological) innovation and political mobilization, which are highly relevant for sustainability-oriented action. We thus hope that future research and practice will explore the relevance of QS within and beyond food provision.

Different Perspectives

Quiet sustainability presented a significant turn in the theorization of FSP, a practice previously interpreted as a reaction to economic hardship (Daněk et al., 2022). Conversely, more recent research in Central and Eastern Europe convincingly shows that the quality of home-grown food and the enjoyment of gardening are key motivations for most practitioners. At the same time, environmental motivations rank low in gardener surveys. In this sense, FSP – and other examples of QS – provide a counterpoint to conscious consumerism, which is grounded in environmental awareness and conceptually underpinned by the theory of planned behavior. Similar to the aforementioned ABC model, this theory presumes that individual actions result from conscious choices which are based on rational decisions and which can, in turn, be influenced by providing people with information.

QS's advocacy of the importance of maintaining what already works challenges the presumed connection of sustainable consumption with novelty and change. In the sphere of food, the mainstream version of sustainable consumption – alternative food networks – is implicitly conceptualized as niche innovations for the future, in need of successful "scaling up". In contrast, QS practices are *already* widespread. Nevertheless, instead of being associated with a hopeful future, QS practices are often viewed as traditional and stuck in the past; and for that reason threatened by so-called modernization. This association often results – both in scholarly accounts and in the sphere of public policies – in the negative valuation of QS practices as mere reproduction and maintenance. In contrast, the mainstream variants of sustainable consumption are habitually associated with positively valued qualities such as difference, novelty, creativity, innovation, and transformation. The concept of QS proposes to recast already-existing sustainable consumption practices as valuable and extends the notion of sustainable consumption in a novel direction. The effectiveness of different communication frames about the sustainability contribution of already existing and widely practiced behaviors is thus increasingly tested by communication scientists and environmental psychology scholars.

Applications

While sustainability campaigns often rely on awareness raising and nudging toward responsible behaviors (see **Green Nudging**), QS highlights practices that are *already* contributing to environmental and social goals without needing to be developed, promoted, and propped up by external funding and activist efforts. Instead, these practices are grounded in everyday routines/habits/practices (see **Social Practice Theory**), long-standing traditions, and social relations. This makes them accessible to large swathes of the population, while mainstream notions of sustainable consumption are often criticized for their niche focus on urban elites of the Global North (see **Social Tipping Points**). In contrast to the exclusivity of market-based sustainable consumption, QS presents a radical alternative, as it happens largely outside the market, thus nurturing community economies and contributing to a good life beyond capitalist growth (see **Degrowth**, **Well-being Economy**). Indeed, while sustainable lifestyles might often imply constraints and (self-)limitation, QS is typically associated with exuberance, generosity, and enjoyment (see **Mindfulness**, **Alternative Hedonism**).

Through its broad geographical relevance, QS also presents an important epistemological counterpoint to mainstream notions of sustainable consumption, which, while presented as universally valid, are too often embedded in Western contexts. Indeed, Western discourses frame many of the international debates and policy agendas around sustainability. In this context, acknowledging the relevance of local "quiet" practices and preserving the conditions for their continuation presents not only a sustainability concern but also a matter of epistemic justice.

Further Reading

Daněk, P., Sovová, L., Jehlička, P., Vávra, J., & Lapka, M. (2022). From coping strategy to hopeful everyday practice: Changing interpretations of food self-provisioning. *Sociologia Ruralis*, 62(3), 651–671. https://doi.org/10.1111/soru.12395.

Pungas, L., Kolínský, O., Smith, T.S.J., Cima, O., Fraňková, E., Gagyi, A., Sattler, M., & Sovová, L. (2024). Degrowth from the east – between quietness and contention. Collaborative learnings from the Zagreb Degrowth Conference. *Czech Journal of International Relations*. https://doi.org/10.32422/cjir.838.

Smith, J., & Jehlička, P. (2013). Quiet sustainability: Fertile lessons from Europe's productive gardeners. *Journal of Rural Studies*, 32, 148–157. https://doi.org/10.1016/j.jrurstud.2013.05.002.

Velicu, I., & Ogrezeanu, A. (2022). Quiet no more: The emergence of food justice and sovereignty in Romania. *Journal of Rural Studies*, 89, 122–129. https://doi.org/10.1016/j.rurstud.2021.11.024.

Visser, O., Mamonova, N., Spoor, M., & Nikulin, A. (2015). 'Quiet Food Sovereignty' as food sovereignty without a movement? Insights from post-socialist Russia. *Globalizations*, 12(4), 513–528. https://doi.org/10.1080/14747731.2015.1005968.

15   Voluntary Simplicity

Ana Maria Soares, Raquel Rebouças, and Teresa Heath

Definition

Voluntary simplicity (VS) is a lifestyle characterized by intentionally reducing material consumption, making a conscious effort to live a balanced life, practicing frugality, and cultivating non-materialistic sources of well-being, life satisfaction, and personal growth. The decision to live a simple life may stem from a realization of the unsustainability of overconsumption for both individuals and the planet, feeling disillusioned with materialism and excessive ownership of possessions, and/or wanting to dedicate time and resources to more meaningful and fulfilling activities.

The concept was described by Elgin and Mitchell (1977) as a way of life guided by five core values: material simplicity (frugal consumption; see **Sufficiency**), human scale (a desire for a more human sense of proportion), self-determination (a desire to achieve greater control over one's life), ecological awareness (recognition of the interdependence of people and other species), and personal growth (a desire to explore and develop the inner life).

This lifestyle often involves reducing working hours, which is linked to living a balanced life and a desire to break the "earn and spend" cycle (see **Hedonic Treadmill**, **Work-Life Balance**). Voluntary simplifiers may choose to reduce their hours of paid work to create more free time and find purpose through meaningful activities for themselves and others. There are various trajectories that individuals may take in pursuing a VS lifestyle and different levels of engagement, with some people partially adopting this lifestyle by embracing some practices, and others fully transforming their lives.

Importantly, achieving a simple lifestyle involves more than just cutting back on consumption; it necessitates a deliberate dedication to simplicity. What sets this lifestyle apart is its voluntary nature. Hence, it is important to distinguish between voluntary simplicity and reduced consumption as a response to poverty, limited resources, or due to the requirement of an external authority.

History

Throughout history, many individuals and groups have embraced the idea of living a simpler yet more fulfilling life. The central concepts of intentionally simplifying one's life can be traced back to the teachings of ancient Greek and Chinese philosophers. Henry David Thoreau, who chose to leave his town to live alone in the woods, is a prominent advocate for this philosophy. His book, *Walden,* reflects on the virtues of simple living in nature. In the United States, Thoreau's birthday on July 12 is informally celebrated by voluntary simplifiers as National Simplicity Day.

However, the concept was formally conceptualized by Gregg (1936/2009) as an "inner and outer condition", which entails "singleness of purpose, sincerity, and honesty within, as well as avoidance of exterior clutter, of many possessions irrelevant to the chief purpose of life" (p. 4).

DOI: 10.4324/9781003584056-17

For Gregg, this means "an ordering and guiding of our energy and our desires, a partial restraint in some directions" to achieve "greater abundance" in others.

In recent years, VS has received growing attention in mass media, marketing, and academic discourse. An example is the emergence of social media communities where consumers search for inspiration from like-minded consumers (Table 15.1). Other resources are available on YouTube or websites such as The Simplicity Institute, which aims to foster dialogue on the necessity of shifting from growth-driven, consumer-focused societies, or The Simplicity Collective.

Different Perspectives

Despite being rooted in environmental and social responsibility concerns, VS arguably embraces broader ethical considerations than those traditionally associated with sustainable consumption. Adopting a simpler lifestyle promotes ecological well-being and is closely linked to values such as moderation, thrift, and wisdom, facilitating the practice of justice, generosity, and other virtues (see **Values and Consumption**). Research in psychology and marketing has suggested that traits such as consumer impulsiveness and materialism negatively impact practicing VS, while **mindfulness**, satisfaction with life, and self-efficacy positively impact VS.

Several related constructs may partially overlap with voluntary simplicity as expressions of conscious consumption, including downshifting, mindful consumption, minimalism, frugality, and anti-consumption. Downshifting involves choosing to reduce income and consumption to improve one's quality of life. Mindful consumption entails adopting a conscious mindset and moderating one's behavior to consider the consequences of consumption and ultimately supporting sustainability. Minimalism is a lifestyle and design philosophy focusing on reducing excess and clutter to prioritize what is essential. Frugality is the practice of being economical with resources, particularly money, by prioritizing careful and efficient use of finances. Finally, anti-consumption refers to a resistance to, aversion toward, or even resentment of consumption (see also **Boycott and Buycott**).

While these practices intersect and may be steps toward VS, this lifestyle is distinguished by its commitment to simplifying life for its intrinsic benefits – such as clarity, purpose, and inner

Table 15.1 Social media communities on VS

Social media	Group	# of members	Self-description
Reddit	SimpleLiving group	1.4 million members	"Breaking free of the work/spend/borrow cycle to live more fully, sustainably, and cooperatively"
Facebook	Voluntary Simplicity	2.8 thousand members	"A bunch of people interested in living simpler lives, living slower, getting more quality and happiness out of life, being a bit more green, a bit less consumerist".
Facebook	Minimalism. Simplicity. Frugality	5.7 thousand members	"For those who are determined to declutter their homes, simplify their lives, actions, and thoughts; and if they so desire, choose frugality (not required!). Also, we talk about simplifying decisions, obligations, and more. The final goal is to have a life that focuses on having time for one's true passions by discarding the nonessential possessions and the distractions of this modern world".

peace – rather than merely reducing possessions, cutting expenses, or opposing consumption. It is a holistic lifestyle aimed at enriching life quality by fostering a connection to one's values and a balanced, intentional existence.

Application

VS manifests in deliberate practices of restricting material consumption and is thus intrinsically linked to consumer behavior. Consequently, we can identify VS-related practices across all stages of consumer behavior (Figure 15.1). In the pre-purchase stage, for example, individuals may engage in a more intentional and thoughtful process of identifying needs and evaluating product options, while carefully choosing the items to buy. During the purchase stage, simplifiers may avoid branded items and instead prefer second-hand products, long-lasting items, and sustainable products. In the post-consumption stage, sharing practices or practices of reusing and recycling may be adopted (see **Circular Economy and Society, Sharing Economy, Repair**).

Thus, VS can manifest in a wide range of life activities, consumption experiences, and work preferences. In a netnographic study, eight primary life practices in which VS may be enacted were identified: material and digital decluttering, work, routines, hobbies, eating, clothing, gift-giving, and fitness (see Table 15.2 for some examples).

Given the increasing number of consumers adopting VS and its relevance for sustainability, companies need to consider the managerial implications of this lifestyle when designing their

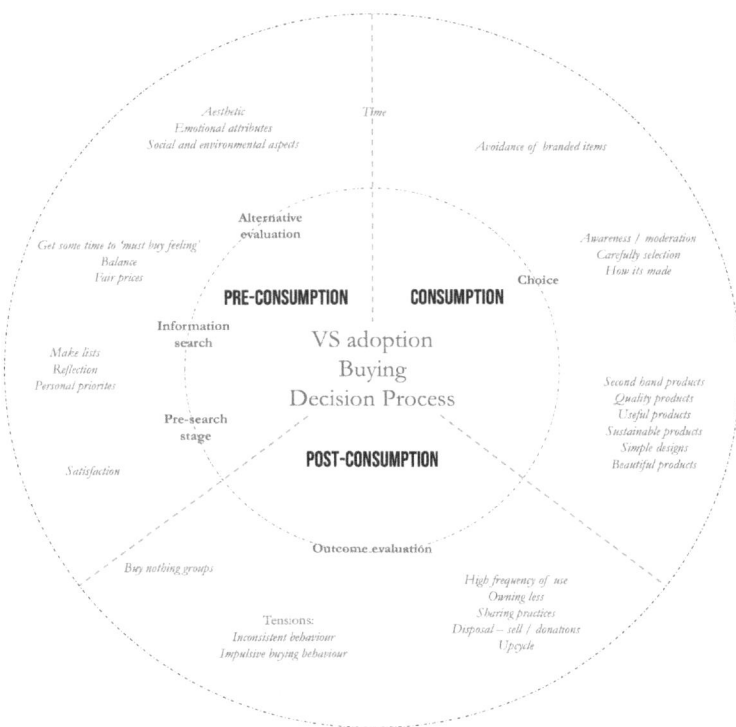

Figure 15.1 VS Buying decision process

Source: Adapted from Rebouças and Soares (2021)

Table 15.2 Examples of practices of VS across life dimensions

Dimensions	Examples of practices
Work	Part-time work, Flexible schedule, Early retirement
Leisure & routines	Outdoor activities, Hobbies, Crafts, Volunteering, Journaling, Meditation/mindfulness practices
Clothing	Capsule/simple wardrobe, Natural materials, Ethical clothing, Rental/second hand/upcycling
Gift-giving	Experience gifts, Consumables, Books, Plants, Local stores
Material and digital decluttering	Books/ebooks/library card, Organization hacks, KonMari method
Eating habits	Simple meals, One meal per day, Eating less meat

business strategies (see **The Role of Business**). Governments and businesses need additional focus on the well-being benefits of reduced materialism – for citizens and national healthcare budgets – and ensure that marketing and **advertising** do not seek to obfuscate the benefits of VS and similar practices.

This calls for measures as basic as limiting advertising and marketing, prioritizing basic needs over luxuries, and restricting practices such as planned obsolescence and the rampant introduction of new product models with no significant improvements in functionality (see **Ecodesign, Extended Producer Responsibility, Greenwashing**). Similarly, characteristics such as durability, reusability, reparability, and multifunctionality need to be reintroduced to help simplify choices.

Further Reading

Aidar, L., & Daniels, P. (2020). A critical review of voluntary simplicity: Definitional inconsistencies, movement identity and direction for future research. *The Social Science Journal*, 1–14. https://doi.org/10.1080/03623319.2020.1791785.

Devenin, V., & Bianchi, C. (2023). Trajectories towards a voluntary simplicity lifestyle and inner growth. *Journal of Consumer Culture*, 23(3), 497–516. https://doi.org/10.1469505221122065.

Elgin, D., & Mitchell, A. (1977). Voluntary simplicity. *Planning Review*, 5(6), 13–15.

Gregg, R.B. (1936 [2009]). *The value of voluntary simplicity*. Wallingford, PA: Pendle Hill.

Rebouças, R., & Soares, A.M. (2021). The consumption behaviour of beginner voluntary simplifiers: An exploratory study. *Journal for Global Business Advancement*, 14(4), 433–452.

16 Mindfulness

Jacob Gordon

Definition

Mindfulness describes a person's capacity to observe inner and outer experiences with non-judgmental attention in the present moment. Although often associated with Buddhist origins, properties of mindfulness can be found in many of the world's spiritual, contemplative, and psychological traditions. Frequently discussed alongside meditation, mindfulness can be understood as a universal and trainable human capacity, while meditation is a method to cultivate mindful states of awareness and enduring personality traits.

The development of mindfulness often instructs placing one's attention on a point of focus, such as the breath or an everyday task, and repeatedly guiding the mind back as it inevitably strays. Such practices may also include techniques of mind/body connection or those meant to generate attitudes of compassion for self and others.

The word mindfulness itself is translated from *sati* in the ancient Pali language, and in Buddhist teaching, it is a pillar of the Eightfold Path, a framework designed to bring freedom from life's **hedonic treadmill** and the suffering it sews. As the practice and investigation of mindfulness has grown around the world, so have questions about the influence that such inner development may have on environmental values, greener lifestyles, and movements toward sustainability. This interest is magnified by mounting concerns over a technology-powered attention economy that manufactures consumer demand, with grave implications for ecological degradation, large-scale mental health issues, and ideological polarization (see **Consumerism**).

History

While Buddhism itself is over 2,500 years old (and other Eastern meditation systems predate it), mindfulness has effectively become globalized in less than a century. The United States and Europe saw their first wave of Asian meditation teachers in the 1960s. In 1979, the first course in Mindfulness-Based Stress Reduction (MBSR) was introduced into scientific and medical settings. Mindfulness is now found in such fields as healthcare, psychotherapy, education, government, business, sports, prisons, and the military (see **Education for Sustainable Consumption**, **The Role of Business**, **Values and Consumption**). Apps and digital media offering instruction have proliferated, with companies like Calm and Headspace attracting millions of subscribers and hefty valuations. This mainstreaming has been met with enthusiasm as well as caution, with some critical that the commercialization of mindfulness detaches it from its mystical roots and moral values.

The popularization of mindfulness and meditation has increasingly intersected with growing environmental alarm, propelling wider philosophical and scientific exploration of the mind's role

DOI: 10.4324/9781003584056-18

in our treatment of the planet. Religious leaders, meditation experts, activists, and researchers offer a variety of perspectives suggesting it should be taken seriously (see **Spiritual Consumption**).

The 14th Dalai Lama has authored books on the climate crisis and engaged in public dialogues with activists. Influential Vietnamese Buddhist teacher Thich Nhat Hanh, especially in his later years, focused extensively on the connection between inner mindfulness and outer sustainability. Jon Kabat-Zinn, the medical professor who developed the original Mindfulness-Based Stress Reduction protocol, frequently suggests that addressing existential matters of war and peace, rights and justice, and ecological stability require inner capacities for compassion, equanimity, and collectivism, all of which can be enhanced through, as he calls it, a mindful "love affair" with present moment reality.

Different Perspectives

Mindfulness itself has already been the focus of an outpouring of scientific study, with positive effects validated for stress, anxiety, depression, addiction, eating disorders, pain management, hypertension, sleep, and other issues. It is often theorized that more mindful people will also lean toward being more Earth-friendly, and now a substantial body of science tests these effects and seeks to find their mediating pathways.

Many results support the hypothesis that mindfulness correlates with pro-environmental actions (though further research would be required to establish causality). A self-reported study of meditators showed that participants tend to have lower levels of **conspicuous/positional consumption** (see **Voluntary Simplicity**) and less susceptibility to normative influences (including **advertising**) due to a healthier self-concept and less conformity to peer expectations. Another found meditation associated with greater belief in climate change. A study of more long-term meditators found them to have higher levels of happiness, connectedness with nature, and more environmental concern, as well as lower levels of land and water use and reduced greenhouse gas emissions.

Meanwhile, approaches such as systems thinking are drawing connections between inner consciousness and global societal phenomena, viewing human damage to the environment as an extension of deeply embedded problematic mindsets that can be better understood, and healed, through an inward view.

The founder of Lund University's Center for Contemplative Sustainable Futures, Christine Wamsler, positions the climate crisis as the manifestation of a collective mind that is suffering from disconnection with itself, others, and nature. Thus, the climate crisis is a relationship crisis, and mindfulness is an important tool for rebuilding connections.

Systems thinker and MIT economist Otto Scharmer founded the Presencing Institute to teach a mindfulness-informed "change management framework" emphasizing "deep listening, co-creation, and an open mind, heart, and will" meant to shift viewpoints from "ego-centric to eco-centric". In 2024, Wamsler, Scharmer, and others presented a report to the Club of Rome pronouncing the critical importance of "the system within" for envisioning and realizing sustainable futures. The paper discusses "the overlooked inner dimension of system change" and offers "the language to advocate for psychological, social, and spiritual factors crucial to sustainable solutions" (Bristow et al., 2024).

Applications

Examples of mindfulness applied to environmental issues are increasingly plentiful. Some courses teach mindfulness-informed insights and lifestyle changes, such as those from the Inner Green Deal organization and the Mind and Life Institute. The University of Wisconsin School of Medicine and Public Health designed a Mindful Climate Action course, instructing behaviors intended to decrease carbon footprints and increase health and happiness (with positive results published in a number of peer-reviewed papers) (see **Behavior Change, Attitude-Behavior Gap**).

Many prominent universities now have mindfulness research centers where topics of human flourishing and environmental health are frequently entwined. A notable addition is the Thich Nhat Hanh Center for Mindfulness in Public Health at Harvard, born in 2023 from an anonymous $25 million gift.

Mindfulness techniques are increasingly being taught in elementary and secondary schools, intersecting with models of social-emotional learning that encourage inner awareness and outer kindness to people and the planet.

Mindful introspection may also benefit political discourse. In the British parliament, since 2013 the All-Party Parliamentary Group on Mindfulness has offered mindfulness courses to lawmakers in both houses of Parliament. Of the hundreds of politicians and staff who have participated, many report enhanced collaboration, including in the face of contradictory opinions, suggesting the potential of these practices to ease gridlock on divisive topics.

And while the United Nations' Sustainable Development Goals are now a widely applied framework, an independent consortium from Sweden has released its own open-source Inner Development Goals meant to help advance the UN's objectives from within, warning that despite humanity's extensive technical knowledge, "we seem to lack the inner capacity to deal with our increasingly complex environment and challenges".

While mindfulness development itself is an ancient technology, new questions are being asked about how it might play a consequential role in surmounting our era's great environmental and social dilemmas. The answers are not yet clear. Some will see mindfulness at best as a sedative for our consumer anxiety and environmental panic, others as a way to temper over-consumption within existing or slowly reforming systems, and yet others as a way to envision and transform into entirely different kinds of societies.

Further Reading

Bristow, J., Bell, R., & Wamsler, C. (2022). *Reconnection: Meeting the climate crisis inside out.* Research and Policy Report. The Mindfulness Initiative and LUCSUS. Available at: www.themindfulnessinitiative.org/reconnection (accessed: 8 January 2025).

Bristow, J., Bell, R., Wamsler, C., Björkman, T., Tickell, P., Kim, J., & Scharmer, O. (2024). The system within: Addressing the inner dimensions of sustainability and systems change. *The Club of Rome.* Available at: https://www.clubofrome.org/publication/earth4all-bristow-bell/ (accessed: 8 January 2025).

Inner Development Goals (2021) Inner Development Goals: Background, method and the IDG framework. Available at: https://drive.google.com/file/d/13fcf9xmYrX9wrsh3PC3aeRDs0rWsWCpA/edithttps://drive.google.com/file/d/1I0ThTPl75h3M6iLzZ7KgYOsiLcrZw3Bw/edit. (accessed: 138 January June 2025).

Kabat-Zinn, J. (2018). *Meditation is not what you think: Mindfulness and why it is so important.* New York: Hachette Books.

Scharmer, C.O. (2018). *The essentials of theory U: Core principles and applications.* Oakland: Berrett-Koehler Publishers.

17 Work-Life Balance

Jared Berry Fitzgerald and Jiayu Huang

Definition

Work-life balance can be described as a state in which individuals feel satisfied with how they allocate their time and energy to work, family, and social commitments. Achieving work-life balance is directly connected to the broader issue of working time, especially for individuals with limited control over their work-time arrangement. Working time reduction is key to enabling a transition to more sustainable economic systems that do not require perpetual increases in consumption and production. Work-life balance is important in shaping consumption and production practices that contribute to reducing environmental degradation as well as improving personal health and life satisfaction (see **Well-being** and **Life Satisfaction Versus Income**).

History

The issue of working time and work-life balance are fundamental to discussions of sustainable development. The term "work-life balance" was coined by Supreme Court Justice Louis Brandeis in the early 1900s with his arguments about scientific management and worker productivity in regard to railroad rate increases. The issue was also very important to John Maynard Keynes, who famously predicted in 1930 that by 2030, people would spend around 15 hours a week at work with a drastic increase in leisure time as their material needs were met.

While Brandeis and Keynes were primarily concerned with issues related to labor productivity, leisure time, and life satisfaction, Juliet Schor's (1992) seminal book, *The Overworked American: The Unexpected Decline of Leisure*, connected the issue directly to broader concerns of sustainable development that included consumerism, inequality and environmental decline. In the book, Schor traces the issue of working time throughout history and notes that working time drastically increased with the rise of capitalism and began to decline in Western countries in the mid-19th century. This homogenous decline continued until the late 1970s with the rise of neoliberal economic policies that favored reduced taxation for corporations and the wealthy, reduced regulations, and emphasized individual responsibility. From then, while working hours continued declining slowly in some countries such as Germany, France, and the Netherlands, they became stable or increased in many others including the United States, Canada, and the United Kingdom. The result is a work-life balance that suppresses leisure time in favor of work. Figure 17.1 visualizes the trends in working time across a selection of developed countries from 1950 to 2023.

Given these trends, the appeal and potential of working-time reduction derives largely from the fact that declines in working time have stalled despite rapid increases in labor productivity (defined as the amount of GDP produced per hour of work). For example, based on data from the OECD, in the United States, it is now possible to reproduce the 1970 standard of living in less than half the

DOI: 10.4324/9781003584056-19

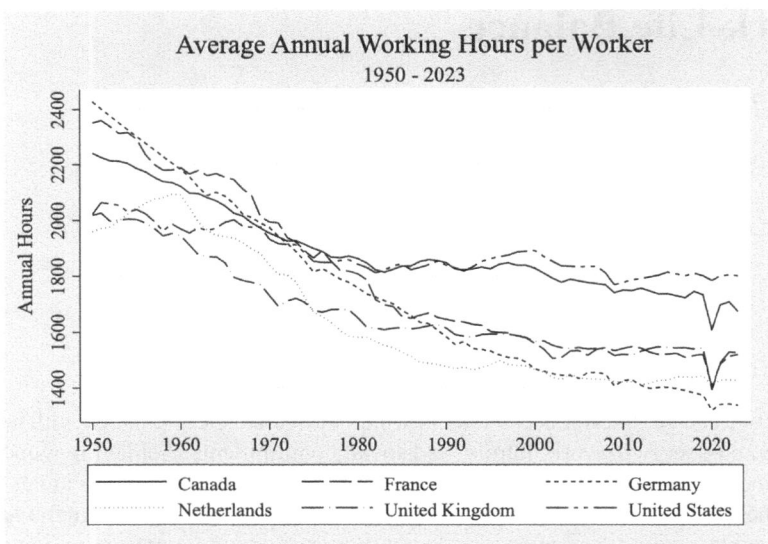

Figure 17.1 Average annual working hours per worker, 1950 to 2023

Source: Conference Board's, Total Economy Database (2024)

time it took that year, due to improvements in labor productivity. While there was a 119% increase in labor productivity in the United States, there was only about a 7% decrease in average hours worked. This suggests that there is much room to reduce hours and improve work-life balance.

Different Perspectives

Proponents of a working-time reduction suggest that it has the potential to be a triple-dividend sustainability policy that enhances *social*, *economic*, and *environmental* well-being (see **Well-being Economy**). Starting with social well-being, reducing working time can improve health outcomes and life satisfaction. There are many reasons for this. The first is that the structure of work, especially the problem of over-work, often produces greater levels of stress and work-related anxieties, which are then translated to greater incidence of health problems. These include a lack of sleep, heart disease, or cancer, as well as a higher likelihood of unhealthy coping behaviors, such as tobacco use, illicit drug use, or alcohol consumption. Time poverty, or the lack of free time, is an important issue here. The adverse effects of a lack of free time include a lack of exercise, sleep, and quality time with family and loved ones, as well as increased odds of unhealthy eating, particularly fast food.

The first economic benefit of reduced working hours is how it can mitigate unemployment. Instead of needing to create new jobs to absorb displaced workers, available work could be spread out among workers. This is particularly important in connection to **degrowth**, to avoid large disruptions to employment. There are also implications for unemployment in a growth economy, particularly with the rapid rise of artificial intelligence and automation.

Finally, working time is also understood to affect environmental outcomes through two pathways, known as scale and compositional effects. The scale effect is how longer working hours contribute to overall economic growth. Longer working hours lead to more production, consumption, and overall GDP growth. The compositional effect is how working time structures the composition of household consumption. Similar to the time poverty approach when considering the relationship between working time and health, time-stressed households are more likely to opt for consumption choices that might save time but are more ecologically damaging. One clear example of this is in transportation, where the more time-intensive options (biking, walking, or public transportation) are much better environmentally compared to the time-efficient option of driving a car (see **Sustainable Mobility**).

It should be noted that there are several unresolved issues in this area of research. The first is the issue of **rebound effects**. It is possible that systematically reducing working hours will lead to greater consumption. For example, if the free time that comes from reduced working hours leads to more vacations or travel, it could result in greater environmental impacts. Another unresolved issue is the feasibility of achieving a working time reduction. Not only are many businesses fundamentally opposed to reducing working time for fear that it could lead to declining profits, but there is the issue of inequality as well. While reducing working hours, and thus income is possible for high-income workers, it is not as feasible for those at the bottom of the income ladder who struggle to make ends meet. Similarly, it is unclear how working-time reduction may differentially benefit men and women. Aside from the gender pay gap, women are still expected to perform more household labor than men so time off from the formal labor market may not have the same benefits for women as for men (see **Gender**). Thus, a working-time reduction must be accompanied by other structural changes that address broader issues of inequality as well.

Application

Conditions and structures of work are evolving quickly, with various initiatives introduced to improve work-life balance, showing their potential sustainability benefits. A prominent example is the four-day week trial, where participating organizations maintain pay at 100% while giving employees a meaningful reduction in work time (4 Day Week Global, 2024). Results from two landmark trials conducted in Iceland, one with 2500 public sector workers (2014–2019) and another with 440 workers (2017–2021), showed that workers took on fewer hours and enjoyed improved work-life balance, greater well-being, while maintaining productivity. Similarly, a 2022 UK trial, comprising 61 companies and about 2900 workers, reported enhanced work-life balance and well-being, and healthy growth of company revenue over the six-month trial period.

In addition to these positive influences, evidence gathered from the United States and Canada involving 41 companies between 2022 and 2023 demonstrated the impact of reduced work time on sustainable consumption. Fewer employees commuted to work by car during the trial, and 42% of participants engaged more in environmentally friendly activities, such as buying ecofriendly products, recycling, and walking and cycling. No "travel rebound" effect was identified among these participants (e.g., by traveling more in their extra free time). The trials have been running on six continents, and their successful stories support the feasibility and sustainability implications of reduced work hours.

It is also important to note that in addition to work time reduction, there is a trend toward flextime and remote work, which was accelerated by the COVID-19 pandemic and has been increasingly adopted by more workplaces for its perceived benefits to work-life balance. These work-life balance initiatives are expected to affect individuals' interactions with material objects, such as encouraging sustainable household practices and their involvement in personal fulfillment and social relationships, including developing personal hobbies and broader public engagement.

Further Reading

4 Day Week Global. (2024). *Research – 4 day week global*. Available at: https://www.4dayweek.com/research (accessed: 2 June 2024).

Fitzgerald, Jared B., Givens, Jennifer E., & Briscoe, Michael D. (2024). Working time and the environmental intensity of well-being: A cross-national analysis of high-income OECD Countries, 1970–2019. *Sociology of Development*, 1–28. https://doi.org/10.1525/sod.2023.0048.

Schor, Juliet B. (1992). *The overworked American: The unexpected decline of leisure*. Basic Books.

The Conference Board. (2024). "Total Economy Database Data." Available at: https://www.conference-board.org/data/economydatabase/index.cfm?id=27762 (accessed: 2 June 2024).

Victor, Peter A. (2019). *Managing without growth: Slower by design, not disaster*. 2nd ed. Edward Elgar Publishing.

18 1.5-Degree Lifestyles

Viivi Toivio, Luca Coscieme, and Lewis Akenji

Definition

The concept of a 1.5-degree lifestyle refers to living at a carbon footprint level that is consistent with the global carbon budget. The goal of limiting global warming to less than 1.5°C relative to pre-industrial levels comes from the aspirational target set by governments in the Paris Agreement on climate change in 2015 to avoid the most severe impacts of climate change.

The concept is designed to understand and influence how policies, business practices, and provisioning systems shape patterns of consumption toward more equitable ways of meeting the needs of people within ecological limits. The lifestyle perspective takes into account the most carbon-intensive consumption domains of transport, nutrition, housing, consumer goods, services, and leisure activities. Considering the average per capita carbon footprint of high-income countries, achieving 1.5-degree lifestyles generally implies substantial reductions in carbon-intensive activities (see Box 18.1), in particular, flying, car use, housing, and meat consumption (see **Sustainable Mobility, Protein Shift**).

Focusing on lifestyles implies also considering non-economic, non-market and non-consumptive aspects of our lives, such as cultural norms, values, social circles, infrastructure, public policy, and

---

Box 18.1   Current lifestyle carbon footprints and targets

The 1.5-degree lifestyle approach highlights a major gap between current carbon footprints and the targets needed to limit global warming to 1.5°C (Figure 18.1). Currently, average lifestyle carbon footprints far exceed the global average target of 2.5 tons per capita by 2030, with affluent nations contributing disproportionately. The disparity between high- and low-income countries is particularly stark. In high-income countries, per capita carbon footprints range from 8 to 14 tCO2e per year, driven by greater energy use, private transportation, and goods consumption. In contrast, lower-income countries have an average footprint of approximately 2 tCO2e per person per year. Despite their minimal contributions to global emissions, these low-income nations suffer most from climate impacts. Bridging this gap requires urgent action in high-income countries to adopt changes that significantly reduce emissions while simultaneously supporting low-income nations in developing low-carbon infrastructures and improving living standards with only limited increases in their carbon footprints (see **Climate Justice**).

DOI: 10.4324/9781003584056-20

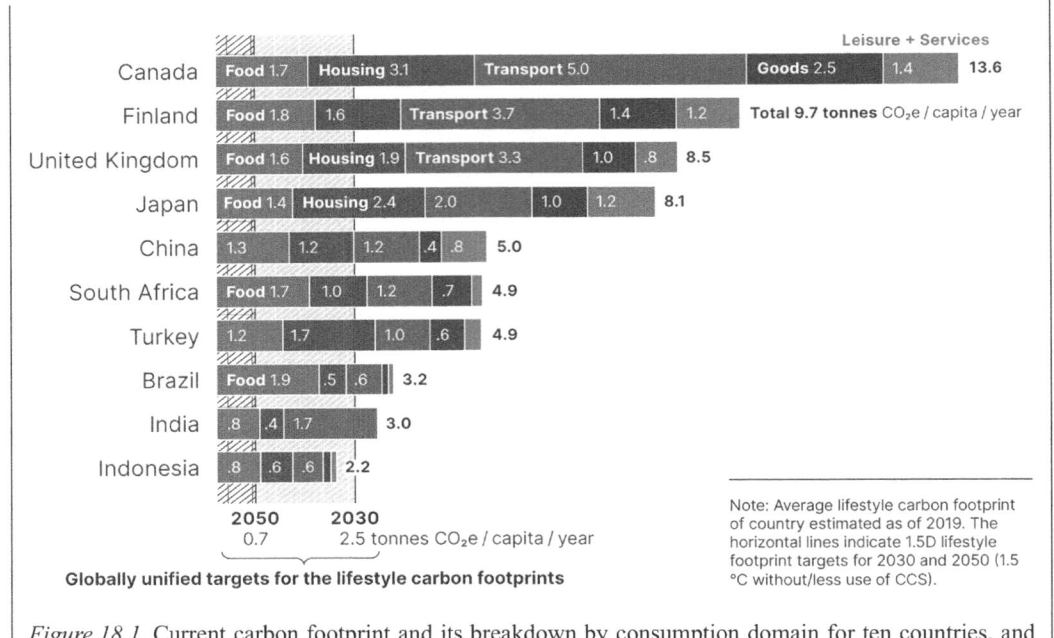

Figure 18.1 Current carbon footprint and its breakdown by consumption domain for ten countries, and globally unified targets for 2030 and 2050

Source: **Akenji et al. (2021)**

others (see **Social Norms, Values, and Consumption, Urban Planning and Spatial Allocation, Co-Benefits of Climate Policy**).

History

Early international discussions about sustainable lifestyles trace back to the 1992 Earth Summit in Rio de Janeiro, where Agenda 21 highlighted the importance of sustainable consumption and production. The specific focus on 1.5-degree lifestyles emerged in the mid-2010s, following the Paris Agreement in 2015 to mitigate climate change. Recognizing that technological innovations alone would not suffice to meet this target, researchers and policymakers began exploring the role of individual and collective lifestyle changes.

The 1.5-degree lifestyles concept was first introduced by Lewis Akenji in 2018 at a meeting convened by the Institute for Global Environmental Strategies and KR Foundation, exploring the link between behavior and lifestyle changes and climate change mitigation. The report he conceived and co-authored, *1.5-Degree Lifestyles: Targets and Options for Reducing Lifestyle Carbon Footprints* (IGES et al., 2019), analyzed the Paris Agreement's implications from a lifestyle perspective, unlike most studies focusing on production- and technology-based solutions. It set the first global per-capita lifestyle carbon footprint targets for 2030 to 2050, linked to the 1.5-degree target. It also introduced the "lifestyle carbon footprint" indicator, a consumption-based measure of greenhouse gas emissions. This consumption-based emissions accounting method was used to set targets, assess the current situation, and identify the necessary high-impact lifestyle changes (see **Consumption-Based Accounting**).

The follow-up 2021 report from Hot or Cool Institute, *Towards a Fair Consumption Space*, expanded the principles and strategies needed to achieve sustainable lifestyles within the framework of global equity and justice. It emphasized integrating sustainable lifestyle practices into broader policy frameworks aimed at reducing resource consumption. Central to this approach is ensuring that necessary lifestyle changes contribute to environmental sustainability and well-being (including equity and justice) for all while staying within the planetary boundaries. The space between the upper and lower boundaries of such lives is referred to as the **fair consumption space**.

Different Perspectives

The 1.5-degree lifestyles idea stems from the scientific evidence that achieving the 1.5-degree target of the Paris Agreement requires both technological innovations and significant changes in consumption patterns. The approach focuses not just on individual consumption behavior, but also on broader structural changes affecting consumption, emphasizing a holistic responsibility for change across individuals, industry, institutions, and other players. It integrates different areas of lifestyle to understand what trade-offs and synergies exist across different consumption choices. For example, living in smaller, more energy-efficient dwellings can reduce the amount of resources needed for construction, heating and cooling, and the number of goods needed to fill the space (see **Stocks Versus Flows**).

While most carbon accounting footprints use financial expenditure data by citizens, the 1.5-degree lifestyles approach counts the actual physical consumption by households. This tedious methodology has several advantages. It reveals social patterns, such as inequality in consumption, gaps between needs and access to necessities (poverty), and unequal distribution of resources between different social groups. It is also decoupled from the market price of goods. The approach thus brings a social dimension into the analysis of climate change and emphasizes that addressing climate change is also strongly linked to addressing inequality. It also shows that factors such as **gender**, age, geography, and income affect the lifestyle carbon footprint (see e.g., **Household Income Versus Carbon Footprint, Carbon Inequality**). Effective climate policies need to take these differences into account and address carbon inequalities to ensure equitable emission reductions.

Several attempts have been made to complement the 1.5-degree lifestyle analysis – focused on climate change mitigation – with analysis of impacts on biodiversity impacts. This contributes to a more comprehensive lifestyle perspective. These efforts conclude that both must be integrated to fully address the environmental footprint, as actions beneficial for reducing carbon emissions may not always protect biodiversity.

Application

The 1.5-degree lifestyles reports are among the most detailed and influential science-policy publications on the nexus of climate change mitigation, consumption, and inequality. They have been used for public communications by leading international and local media, influenced research design by leading scientific bodies and public institutions, and have been applied by policymakers at the highest levels including the United Nations, European Union, and national levels (see Box 18.2).

Box 18.2 Sample impacts of 1.5-degree lifestyles analysis on research, communications, and policy

a. *Public communications*: Recognizing the combination of its scientific approach and practical application, the analyses have been used to communicate the need for sustainable consumption and lifestyles to the mainstream. Analyses have been covered by leading global and local media, including the BBC, The Times, Reuters, CBC News, the South China Post, the Financial Times, Forbes, and Bloomberg News.

b. *Public policy*: The 1.5-degree lifestyles approach has been used at the highest levels of government. For example, it was used to provide expert testimony to the Environment and Climate Change Committee of the UK House of Lords and reflected in its government action recommendations in the official report *In Our Hands: Behaviour Change for Climate and Environmental* Goals. The European Parliament has referred to analysis using the 1.5-degree lifestyles approach and focused on the fashion industry to call for reductions in volumes of textile production in the "European Parliament resolution of 1 June 2023 on an EU Strategy for Sustainable and Circular Textiles" (see **Fast Fashion**).

c. *Scientific research*: The first report formed the basis for the low-carbon lifestyles chapter of the UN Environment Programme *Emissions Gap Report 2020*. The second was used for UNEP's first policy brief on the topic *Enabling Sustainable Lifestyles in a Climate Emergency*. Findings from the report were featured in the first (2021) edition of Future Earth's annual report *10 New Insights in Climate Science*. Analysis of 1.5-degree lifestyles led to the report by the *Cambridge Sustainability Commission on Scaling Behaviour Change*.

The approach allows for a broader framing of climate mitigation interventions by adding the human dimension. It brings to the surface issues such as global inequalities in consumption and meeting basic needs. Hotspot analysis can point to high-impact consumption that does not improve well-being. It directs attention to where investments in renewable technology could result in burden-shifting: whereby improvements in one place result in negative impacts elsewhere (see **Energy Overshoot**). Recognizing these as foundations of social tensions and addressing them from an equitable and well-being perspective brings broader acceptance and legitimacy to the tough actions needed to address climate change.

Using the results of a 1.5-degree lifestyles analysis to design policy interventions or investments in society could also help avoid **consumer scapegoatism**. By focusing on the provisioning systems that allow or restrict people to access goods and services, it shifts policy interventions (i) to widen beyond individual consumers (who may not have the agency to challenge the system), (ii) to put more emphasis on the design of production and distribution systems (see **Product-Service Systems, Ecodesign, Repair**), and (iii) to broaden the palate of opportunities beyond the marketplace, toward alternatives such as collective provisioning, sharing and **grassroots innovation** (see also **Foundational Economy, Sharing Economy, Quiet Sustainability**).

1.5-degree lifestyles are useful in the design of **sufficiency** measures. Gap analysis between the remaining carbon budget and the 1.5-degree target shows the remaining carbon budget for food, transport, housing, consumer goods, and services. This is useful in determining what to prioritize to meet well-being needs within this budget. This information can be used for **choice editing**. From an equity perspective, this carbon budget and reduction targets can be combined with an understanding of fundamental human needs to estimate what is a fair consumption space for all (see also **Personal Carbon Allowance, Consumption Corridors, Doughnut Economy**). The information can also be used to direct where **universal basic services** can be provided.

The 1.5-degree lifestyles approach responds to questions about how the results of integrated assessment models used in the Intergovernmental Panel on Climate Change (IPCC) assessment reports reflect the global political economy, international trade flows, and social justice (see **Political Economy of Consumerism, Climate Justice**). The issue of unfairness has long dogged the IPCC negotiation process owing to the territorial-based emissions accounting methods used to allocate national responsibility for climate change. The 1.5-degree lifestyles approach allocates responsibility for emissions to those who enjoy the material benefits of production. In this regard, the results of analysis from consumption-based emissions accounting could be used as an alternative or complement to the territorial-based accounting approach used by the IPCC.

Further Reading

Akenji, L., Bengtsson, M., Toivio, V., Lettenmeier, M., Fawcett, T., Parag, Y., Saheb, Y., Coote, A., Spangenberg, J.H., Capstick, S., Gore, T., Coscieme, L., Wackernagel, M., & Kenner, D. (2021). *1.5-degree lifestyles: Towards a fair consumption space for all*. Berlin: Hot or Cool Institute.

IGES, Aalto University, and D-mat ltd. (2019). *1.5-degree lifestyles: Targets and options for reducing lifestyle carbon footprints. Technical report*. Hayama, Japan: Institute for Global Environmental Strategies.

IPCC. (2022). *Climate change 2022: Mitigation of climate change. Working Group III contribution to the sixth assessment report of the Intergovernmental Panel on Climate Change*. Intergovernmental Panel on Climate Change.

UNEP. (2020). *Emission gap report 2020*. Nairobi: United Nations Environment Programme. https://doi.org/10.18356/9789280738124.

UNEP. (2022). *Enabling sustainable lifestyles in a climate emergency*. Paris: United Nations Environment Programme.

Cluster II

Concepts, Frameworks, and Applied Theories

19 Freedom of Choice

Stephan Lorenz

Definition

Freedom of choice in an affluent society refers to the ability to individually consume material possessions and services and to shape lifestyle patterns on the basis of one's preferences, either with or without any regard for societal and environmental implications. In contrast to this simple idea of consumer sovereignty, however, important social expectations and functions of social inclusion are linked to consumption decisions. These include the expectation of the reliable availability of consumer goods, the permanent renewal of supply, and the possibility of social comparisons. The latter encourages expressions of individuality, group affiliation, and distinction from others in terms of lifestyle and status. The result is a consumer society dynamic with consequences for social and environmental sustainability.

History

The gradual expansion of consumer opportunities took on a new dynamic with the transition from feudal agrarian to modern industrial-capitalist societies, as early critics like Henry David Thoreau noted. For the bourgeois classes, freedom meant above all freeing themselves from traditional and feudal constraints, producing and consuming independently of them, and increasingly being able to exert political influence. The emergence of the Social Question – of rich and poor social classes and their conflicts – in the 19th century shows that this freedom did not apply to broad wage-dependent classes while the expansion of consumption became attractive to them as a desirable form of participation. However, they were also increasingly dependent on it. The less people produced their own necessities, as was the case in agrarian societies, the more modern money-based consumption became the normal and unavoidable means of provision.

With widespread mass consumption, affluent societies emerged in Western countries from the middle of the 20th century. Assumptions by contemporaneous observers like John K. Galbraith or David Riesman, that an end to poverty and permanent consumerism was in sight, proved to be wrong. Paradoxically, consumerism instead continued to gather pace. It was no longer just the new consumer goods that were desirable, but also the ability to choose became a value in itself (see **Choice Editing**). The promises of freedom of choice permeate almost all areas of life – democratic politics, partner choices, family forms, careers, as well as leisure activities or religious affiliations (see **Consumer-Citizen**). Many activities in these areas are in turn largely realized through consumption – through various equipment and technical devices, health and education courses, dating services, media use, etc. (see **Convival Technology**). Consumption is therefore the central means of shaping everyday life. What does this mean in terms of sustainability?

DOI: 10.4324/9781003584056-22

Different Perspectives

Firstly, freedom of choice can be deployed as the basis for making decisions in favor of sustainable alternatives. The guiding questions of critical consumers are typically: Do I really need this? Can I take responsibility for the social and environmental consequences of my consumption? There are numerous ways to make such decisions. Just three decades ago, it was difficult to buy organic food in Western industrialized countries, but today you can find organic and **fair trade** products in every supermarket. You can also make sure that packaging is kept to a minimum, that energy consumption is low, and that technical devices are easy to **repair**. Decisions in favor of green electricity, the use of bicycles, and ecoclothing are further examples, as is simply buying less, for example, clothing, or not buying some things at all (see **Voluntary Simplicity**). This also forms the basis for a gradual transition to collective engagement, such as participation in **boycott or buycott** activities.

These options undoubtedly exist, but they soon reach limits. This is obvious if the budget does not allow for expensive alternatives or if there are no alternatives offered. Irrespective of this, individual consumption decisions are typically only made at the end of supply chains and can therefore have little direct influence on numerous preceding decisions in production, processing, and trade. In addition, supply chains are usually cloudy and complex, making it difficult to make informed decisions. The situation is similar where there are too many options, such as with telephone or insurance tariffs. The freedom to choose in all areas of life leads to decision-making requirements everywhere, which can also be experienced as overwhelming demands (see **Choice Paralysis**).

Secondly, the dynamics of consumption are driven forward for two reasons. On the one hand, the ever-increasing variety of options means that the ideas and expectations of constantly renewed consumption are also constantly expanding. Supply and technical innovations are engaged in a mutually stimulating race with desire-driven demand. On the other hand, social comparison provides orientation. Consumer choices serve to keep up-to-date, thereby expressing group affiliation on the one hand and setting oneself apart on the other, be it individually, from other communities, or in terms of social status (see **Conspicuous/Positional Consumption**). Establishing such consumer communities typically goes hand-in-hand with driving material wealth standards, for example, regarding new clothes, bigger cars, or whole neighborhoods with cost-increasing houses/rents, better schools, health, and other infrastructures. This consumption dynamic creates social pressure and, in ecological terms, fuels the consumption of resources and energy and the generation of waste. Sustainability requires limits to freedom of choice here.

Thirdly, the importance of the value of freedom of choice in affluent societies makes another aspect relating to social sustainability understandable. Being forced to refrain from choice is experienced individually as massive social exclusion. It excludes people from being able to make simple everyday decisions like everyone else (see **Fair Consumption Space**). The food pantries established in recent decades are prototypical places of such exclusion (see Box 19.1). Being able to shape everyday life through consumption is essential for social inclusion in the consumer society. Accordingly, social sustainability requires the expansion of choice for the socially disadvantaged.

Box 19.1   Food banks as hotspots of social exclusion

Food banks collect food for charitable distribution. This assistance is used by those whose consumption status is severely restricted, meaning that they are unable to meet their living requirements and numerous everyday lifestyle options by shopping in stores like everyone else. But for food bank users, exclusion does not just mean having fewer choices than others. They even have to gratefully take what others no longer want, what is left over from the excessive choices of others in an affluent society.

Social inclusion means more than survival. The extent to which the experience of exclusion is determined by consumer status and freedom of choice can also be seen in the simulation of "normal" shopping at food banks. Their users are often referred to as customers and the outlets as stores. In some cases, the outlets are organized in the form of stores, thereby creating quasi-options for choice. This recognizes how important this kind of normality is for social inclusion. It remains a simulation, however, for those who are not normal customers and who only make use of the assistance for exactly this reason.

Applications

Concluding a sustainable future is not easy because the analysis of freedom of choice suggests (i) its necessity in part, (ii) its limitation in part, and (iii) its expansion in part. Within the limited framework of individual action, which can be strengthened collectively through civil society initiatives and associations (see **Social Movements, Alternative Consumer Cooperatives,** and **Grassroots Innovation**), a responsible use of choice is required. For more far-reaching answers, the affluent society's foundations of consumer choice must be questioned – in terms of scientific analyses, societal practice, and politicization. Certainly, the socially disadvantaged should have more opportunities to make their own decisions about their lifestyles. However, if the standards of social participation are geared toward unsustainable consumption dynamics, the core of sustainable development must be to put the brakes on destructive drivers of consumption (see **Doughnut Economics**).

The central value of freedom of choice would have to be reinterpreted to detach it from consumption, and so would the close link between consumption and the social functions of comparison, that is, belonging and distinction. How this can be achieved comprehensively is still unclear. Valuing ease of repair and long-term reliability instead of novel choice in consumption comes to mind. The aspect of social comparison, however, is even more challenging. Available ideas and activities include trying out alternative lifestyles in social niches, some elements of which may then spread beyond this (see **Eco-Communities**). Also, there is a spectrum of political intervention options for regulating unsustainable offers. This is, however, often perceived publicly as a restriction of the fundamental value of freedom of choice. Another promising perspective is to develop supply infrastructures that are publicly accessible to all (see **Foundational Economy**), thereby withdrawing them from consumer dynamics (see Box 19.2). After all, attempts to contain or undermine the value of freedom of choice and its social functions are shaking the foundations of today's consumer societies, which makes massive resistance always to be expected.

Box 19.2   Water infrastructures: collective consumption for reduced social and environmental pressure

Infrastructures are physical and social networks that provide material goods and services for all. In any case, the use of infrastructure as collective consumption forms the preconditions for individual consumption. If, for example, the water supply is available to everyone as a matter of course and at a low cost, no status differences arise. Since the social function lies in general and equal access, this removes the basis for social comparison and status struggles from consumption. The resulting slowdown in consumption dynamics prevents social and ecological pressure.

The expansion of water and sanitation services since the middle of the 19th century was one of the first municipal infrastructures in Europe to improve the general quality of life. With deregulation policies since the 1980s, increased profit interests, often achieved through savings in personnel and maintenance and system renewal, have been promoted. This calls into question the general and equal provision of services and, therefore, the mentioned advantages regarding pressure prevention. In some European countries, however, a tendency toward remunicipalization can be observed.

Further Reading

Bauman, Z. (2000). *Liquid modernity*. Cambridge: Polity Press.
Bourdieu, P. (1984). *Distinction. A social critique of the judgement of taste*. London et al.: Routledge.
Foundational Economy Collective. (2018). *Foundational economy. The infrastructure of everyday life*. Manchester: Manchester University Press.
Lambie-Mumford, H., & Silvasti, T. (Eds.). (2020). *The rise of food charity in Europe*. Bristol: Policy Press.
Thoreau, H.D. (1854). *Walden; or, life in the woods*. Boston: Ticknor and Fields.

20 Social Practice Theory

Mary Greene

Definition

Social practice theory (SPT) examines how shared routines and activities shape our daily lives and societies, focusing on actions within social contexts. Sitting at the intersection of key theoretical debates in the social sciences about what drives social action and behavior, it seeks a middle ground in the long-standing "structure-agency" debate. This debate refers to the tendency of approaches to emphasize either individual agency (the capacity of individuals to act independently) or social structures (such as political economy) as the primary drivers of behavior. In contrast, SPT focuses on how agency and structure interact within social practices and their everyday performances.

While definitions of social practices vary, they are generally understood as consisting of interconnected social, material, and bodily elements, see Box 20.1. These include materials (objects, tools, infrastructures), competencies (skills, knowledge), meanings (understandings, cultural norms, and values), and rules (regulations, procedures).

Box 20.1 Examples of social practices

Social practices are dynamic interactions between individual agency and the social and material structures or contexts that form societies. For example, food shopping involves personal skills (agency) and the infrastructure of stores and supply chains (structure), with each continuously influencing the other. Individuals are considered active "carriers" of practices. Many practices, such as car driving or supermarket shopping exist long before the individual starts to perform it. However, individuals transform existing practices through their participation. For instance, the practice of car driving exists in society, and individuals learn and adopt it. However, as more diverse groups take up driving, the practice itself evolves.

Practices themselves are interconnected and form bundles or complexes that make up the rhythm of everyday and social life. They hold key implications for sustainable consumption and lifestyles. For instance, travel, work, food shopping, cooking, and eating are interrelated practices that together shape how we use resources (see **Choice Editing** and **Stocks Versus Flows**). Changes in one practice can lead to shifts in related ones. For example, adopting zero-waste grocery shopping can alter shopping routines, food storage at home, and waste disposal methods.

Social practices are shared within groups and communities and can change over time due to new technologies, cultural shifts, or policy changes. For instance, home heating has often evolved

DOI: 10.4324/9781003584056-23

from using hearth fires to central heating systems, leading to higher indoor temperature standards and increased resource use. This practice continues to change with renewable energy sources, better insulation, and cultural movements for energy conservation (see **Sustainable Housing** and **Energy Overshoot**). Social practices also vary across cultures, reflecting different material contexts, norms, and values (see **Social Norms** and **Values and Consumption**). For example, commuting in some cultures involves mainly driving cars, while in others, bicycles or public transportation are more common, highlighting diverse infrastructures, histories, and cultural norms surrounding transportation (see **Sustainable Mobility**).

History

The roots of SPT can be traced back to 19th- and 20th-century sociology and philosophy and the works of scholars such as Karl Marx and Martin Heidegger. Marx emphasized that human actions are inherently social. He argued that to grasp their meaning and intelligibility we must consider the social and historical contexts in which they occur. Heidegger explored the nature of human existence and how our being-in-the-world is shaped by everyday practices. Such insights laid the groundwork for understanding human actions as deeply embedded in social and material contexts.

In the mid-20th century, scholars like Pierre Bourdieu and Anthony Giddens expanded on these ideas. Bourdieu introduced "habitus", the ingrained habits and skills developed from life experiences, highlighting that actions are deeply embedded in social and cultural contexts. Giddens' theory of structuration emphasized that social practices shape and are shaped by the social contexts and structures in which they occur.

In the late 20th century and early 21st century, SPT became more defined and prominent with the work of scholars like Elizabeth Shove, Andreas Reckwitz, and Theodore Schatzki. Shove applied SPT to sustainable consumption, showing how practices like showering and heating are influenced by materials, competencies, and meanings. Reckwitz distinguished between practices as abstract entities and their actual performance. Schatzki explored the ontology of social practices and their role in social life, emphasizing their interconnectedness and embeddedness in broader contexts. Recent work by Shove and others has further developed the idea of interconnected practices, demonstrating how SPT can be used to study and change larger social phenomena such as economic systems and cultural trends (see **Generational Consumption Differences (in China)**).

With the application of SPT to contemporary challenges such as sustainable consumption and public health, researchers are exploring how practices can be changed to promote more sustainable lifestyles and how policies can support these changes. This has made SPT a potentially valuable tool for understanding and promoting sustainable practices in various areas of life.

Different Perspectives

SPT is not a single framework but a collection of diverse approaches. Scholars differ, for instance, in how much they emphasize material versus non-material elements in shaping social practices. Some focus on more tangible components like technologies and infrastructures, arguing that these are crucial in structuring practices (see **Urban Planning and Spatial Allocation** and **Information and Communication Technology**). Others emphasize cultural and bodily factors like dispositions, meanings, skills, and **social norms**, suggesting that these are equally important in shaping actions.

A key debate within SPT is how practices evolve. Some scholars, like Bourdieu, primarily focus on the stability of practices and how established routines resist change. Others, like Shove, focus on how practices adapt due to policy interventions, technological innovations, and cultural shifts.

Contrary to the misconception that SPT ignores change, many scholars are interested in how practices evolve, which is essential for designing policies that support sustainable change.

Critics often argue that SPT neglects individual agency by focusing too much on social structures. However, SPT acknowledges that individuals are active participants who can drive innovation and transformation within practices, despite the emphasis on broader social and material contexts.

Another critique is that SPT oversimplifies the complexity of social practices by breaking them into distinct elements. Proponents, however, argue this analytical approach helps to better understand and intervene in practices by identifying key components and their interactions.

SPT has also been critiqued for not sufficiently addressing power, inequality, and justice within practices. Critics say it often overlooks how practices can exclude certain groups. Advocates counter that these issues can and should be integrated into SPT analyses, as examining who participates in practices and why can provide insights into social inclusion and exclusion.

Finally, some argue that SPT focuses too much on everyday life and isn't suited for larger systems or political economy dynamics (see **Political Economy of Consumerism**). However, as Schatzki and Shove have recently sought to demonstrate, SPT offers valuable perspectives on understanding large systems by examining how everyday practices aggregate and accumulate over time to influence broader social, economic, and political dynamics. There's also growing recognition of the need to apply SPT beyond Western contexts to understand diverse practices across cultures and societies.

Applications

Applying SPT in sustainability governance offers new strategies for promoting sustainable lifestyles by understanding behavior holistically and identifying effective intervention points. SPT moves beyond individual choices or technology-centric solutions to consider interacting social, material, and cultural factors, explaining why certain practices are resistant to change and why individual-focused approaches frequently fail (see **Attitude-Behavior Gap**). Analyzing these elements helps identify leverage points for impactful policies (see **Social Tipping Points**). These leverage points have the potential to support systems change from the bottom up.

Shifting the focus from individual behaviors to social practices has significant policy implications and researchers are developing governance approaches based on SPT. Interventions should aim to transform practices themselves rather than just changing purchasing decisions. Box 20.2 provides information on two such approaches – the practice intervention framework and the "Change Point" toolkit.

Box 20.2   Two approaches to change practices – the practice intervention framework and the "Change Point" toolkit

The practice intervention framework developed by Spurling and colleagues provides a practical lens for societal and policy applications of SPT. It shifts focus from individual **behavior change** to the underlying elements that sustain practices, identifying three key avenues for promoting sustainable lifestyles: recrafting elements, substituting practices, and changing how practices interlock (see Spurling and Blue's edited volume in the Further Reading for details).

Recrafting elements involves modifying the materials, competencies, and meanings within a practice to make it more sustainable. For example, in household energy use, this might involve introducing energy-efficient appliances, promoting conservation norms, and developing new skills (see **Energy Consumption Behavior**). Recrafting practices involve not just providing new technologies but ensuring that these are integrated into everyday routines and supported by cultural shifts and new knowledge.

Substituting practices replaces less sustainable practices with more sustainable alternatives. Encouraging cycling or public transport, for instance, requires not just promoting alternatives but creating infrastructure and cultural acceptance, making the sustainable version the most convenient and attractive option.

Changing how practices interlock recognizes that practices are interconnected. For example, commuting is linked to work, food shopping, and time management. Sustainable change might involve rethinking work policies and urban planning to promote sustainable commuting and reduce car travel (see **Urban Planning and Spatial Allocation**).

The "Change Points" toolkit (Hoolohan & Browne, 2020) is another notable example of how SPT can be operationalized to design interventions. This toolkit offers a structured method for collaboratively planning interventions with policy, industry, and societal stakeholders by examining the social, material, and institutional elements that shape everyday practices. It guides users through a series of steps to identify and target key points within practices where interventions can be most effective.

Integrating SPT into policies allows stakeholders to design flexible, long-term strategies targeting underlying systems and structures for sustainable change. Concrete steps can include collaborative practice-based research to map existing practices, developing multi-element practice-focused interventions, engaging stakeholders in the design and implementation of these, and continuously monitoring and adapting strategies to achieve desired outcomes (see **Living Labs**). In this way, leveraging SPT has the potential to lead to more effective, context-sensitive, and enduring solutions to unsustainable lifestyles. This makes SPT a useful, systematic, tool for anyone looking to promote sustainable living, including policymakers, activists, and educators.

Further Reading

Hoolohan, C., & Browne, A.L. (2020). Designing design thinking for practice-based intervention: Co-producing the change points toolkit to unlock (un)sustainable practices. *Sustainability,* 12(1), 1–20. https://doi.org/10.1016/j.destud.2019.12.002.

Shove, E., Pantzar, M., & Watson, M. (2012). *The dynamics of social practice: Everyday life and how it changes.* Sage Publications. https://doi.org/10.5324/njsts.v1i1.2125.

Spaargaren, G., Weenink, D., & Lamers, M. (Eds.). (2016). *Practice theory and research: Exploring the dynamics of social life.* London: Routledge.

Spurling, N., & Blue, S. (Eds.). (2016). *Social practices, interventions and sustainability: Beyond behaviour change.* London: Routledge.

Warde, A. (2005). Consumption and theories of practice. *Journal of Consumer Culture,* 5(2), 131–153. https://doi.org/10.1177/1469540505053090.

21 Rebound Effects

Eva Alfredsson and Mikael Malmaeus

Definition

The term "rebound effect" refers to an unintended increase in the demand for energy and/or resources as a result of steps being taken to reduce that demand, for example, through efficiency improvements. The rebound effect can be explained by the following process: Technological progress makes the production process more efficient. Less energy and sometimes also resources are needed to produce the same amount of products and services. However, because of increased efficiency the costs of production decrease, which also can reduce the cost of the final product. A price decrease normally leads to increased consumption of the now-cheaper product or other products. As the demand increases, so does energy consumption and resource use. Thus, some of the energy savings from the efficiency improvement are lost. A rebound effect of (say) 10% means that 10% of the energy efficiency improvement initiated by the technological improvement is offset by increased consumption. Other examples of rebound effects include how building new roads to improve traffic stimulates more car use, or how voluntary reductions in some activity (like traveling) result in increased spending on some other (environmentally harmful) activity. The rebound effect can even backfire, which means that it is larger than the initial efficiency gain, resulting in a net negative efficiency gain.

History

The idea of rebound effects goes back to the suggestion, by economist and philosopher William Stanley Jevons (1865), that improved efficiency of coal-fired steam engines would not result in less but more use of coal and therefore contribute to a more rapid depletion of England's coal reserves. This became known as the Jevons paradox. The reason for this is that higher efficiency means a lower effective cost of coal, which stimulates the diffusion of coal-using technology throughout the economy.

In recent times the rebound effect has been the subject of considerable research. This has deepened our understanding of the dynamics of the rebound effect. Today it is widely acknowledged that there is a direct effect (first-order effect), but also indirect (second-order effects) and economy-wide effects (see Figure 21.1).

The direct rebound effect is caused by increased efficiency leading to lower prices of a product which in turn leads to an increase in demand for the same product. This effect may offset some or all of the savings made from increasing efficiency. A typical example is more fuel-efficient cars which lead to lower fuel costs, but often also lead to the car being driven more often and further (see **Moral Licensing**).

DOI: 10.4324/9781003584056-24

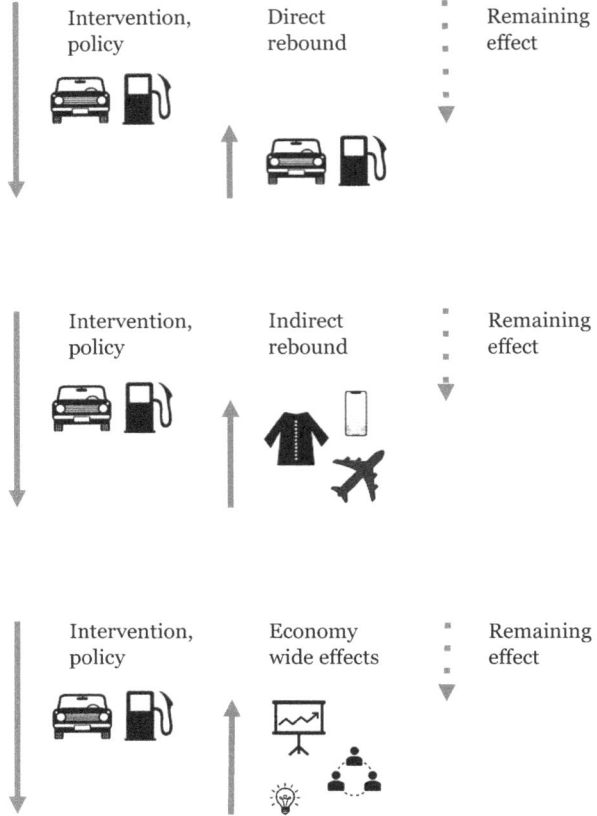

Figure 21.1 Direct, indirect, and economy-wide rebound effects

Source: Malmaeus, M., Nyblom, Å., Mellin, A., Hasselström, L., & Åkerman, J. (2021). *Rekyleffekter och utformning av styrmedel* (in Swedish). IVL Report B2410

The indirect effect is similarly caused by increased efficiency leading to lower prices, but instead of increasing the demand for the same now-cheaper product, the accrued savings are spent on additional consumption of other products or services. As with the direct effect, the indirect rebound effect may offset some or all of the initial savings.

A rebound effect from technology-based efficiency increases is especially observed when energy is a significant factor in production and when it replaces other factors of production, such as labor. The size of the rebound effect can be partial, and only offset part of the gain; or it can sometimes be total, which means that all the efficiency improvements are offset (Figure 21.2).

The rebound effect has been observed and studied using various metrics. Energy is the most frequently studied, but effects regarding GHG emissions, resource use, and other metrics are also analyzed.

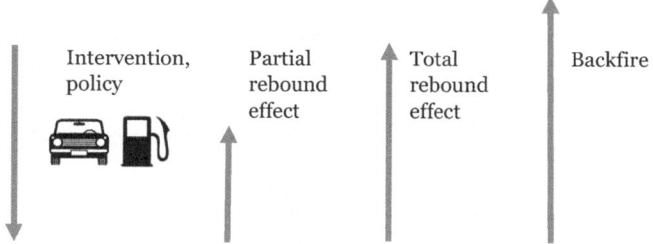

Figure 21.2 Three sizes of rebound effects

Source: Malmaeus, M., Nyblom, Å., Mellin, A., Hasselström, L., & Åkerman, J. (2021). *Rekyleffekter och utformning av styrmedel* (in Swedish). IVL Report B2410

Different Perspectives

While the rebound effect from an environmental and resource perspective is a negative side effect, it is a positive effect from a neoliberal economic perspective. The efficiency improvements free up resources that can be used in other parts of the economy and the rebound thus contributes to economic growth.

Some authors even find it likely that a synergistic relationship exists between economic growth and energy consumption, with "each causing the other as part of a positive feedback mechanism" (Sorrell et al., 2009). Researchers have identified the existence of a circular feedback process within which increasing time lags come into play: a quick response (direct rebound), a slow mechanism (indirect rebound), and a long-term restructuring process that affects the overall economic structure (general equilibrium effects). Ayres and Warr (2002) describe resource consumption as both a driver of growth and a consequence of growth and represent the growth mechanism as a positive feedback cycle between consumer demand, industrial investment, declining unit costs, and lower prices for consumers.

Application

Because the rebound effect can be substantial, and even backfire, it is important to include it in the analysis and evaluations of policy (see **Co-Benefits of Climate Policy**). Without this, the net effects will not be estimated correctly. An example of a policy instrument that takes the rebound effect into account is a cap-and-trade policy. The trade seeks to reduce emissions efficiently while the cap fixes the level of emissions, thus preventing rebound effects within individual businesses or other parties covered by the policy. The cap-and-trade policy does not however prevent system-wide rebound effects. The only way to handle these is through reduced overall consumption and production levels (see **Sufficiency**).

Further Reading

Allan, G., Gilmartin, M., McGregor, P.G., Swales, J.K., & Turner, K. (2009). Modelling the economy-wide rebound effect. In H. Herring & S. Sorrell (Eds.), *Energy efficiency and sustainable consumption*. London: Palgrave Macmillan. https://doi.org/10.1057/9780230583108_4.

Ayres, R., & Warr, B. (2009). Energy efficiency and economic growth: The 'Rebound Effect' as driver. In H. Herring & S. Sorrell (Eds.), *Energy efficiency and sustainable consumption*. London: Palgrave Macmillan. https://doi.org/10.1057/9780230583108_6.

Jevons, S.W. (1865). *The coal question; an inquiry concerning the progress of the nation, and the probable exhaustion of our coalmines*. London: Macmillan and Co.

Sorrell, S., Dimitropoulos, J., & Sommerville, M. (2009). Empirical estimates of the direct rebound effect: A review. *Energy Policy*, 37(4), 1356–1371. https://doi.org/10.1016/j.enpol.2008.11.026.

van den Bergh, J.C.J.M. (2011). Energy conservation more effective with rebound policy. *Environmental and Resource Economics*, 48(1), 4358. https://doi.org/10.1007/s10640-010-9396-z.

22 Moral Licensing

Aitor Marcos

Definition

Moral licensing, also known as self-licensing, occurs when previous virtuous or moral behaviors make individuals feel entitled to act in a way they would not permit themselves otherwise. This sense of entitlement to engage in behaviors normally considered unethical is a psychological mechanism that arises when individuals' prior virtuous actions enhance their self-image, making them feel good about themselves. As a result, individuals allow themselves to act in a less virtuous way without feeling guilty or compromising their image as a moral person. When pro-environmental behavior is seen as a virtuous act, it can lead to moral licensing in subsequent decisions.

Moral licensing involves a sequence of two behaviors: a good deed followed by a bad or less good action in the opposite direction. This balancing behavior makes moral licensing a particular case of "negative behavioral spillover". The umbrella term "behavioral spillover" is used when one action influences another, either positively – further promoting the first action – or negatively – permitting oneself to deviate from the first action (see Table 22.1).

When it comes to sustainable consumption, after making a seemingly environmentally conscious purchase (e.g., buying an electric car), individuals may feel justified in making less sustainable choices later, such as indulging in excessive consumption (e.g., having more than one car) or engaging in wasteful practices (e.g., driving the car more often). For example, as illustrated in the moral licensing scenario in Table 22.1, a person who decides to take the train instead of flying may feel proud of their sustainable choice. This pride can enhance her moral self-conception and create a sense of entitlement, potentially leading them to justify subsequent unsustainable behaviors, such as using more heating than they typically would if they had not made the prior sustainable choice.

History

The concept of moral licensing originated in social psychology and was devised by Monin and Miller (2001) as part of their research on prejudice against people of color. Their experiments demonstrated that people are more willing to express prejudiced attitudes when their past behavior has established them (in their own eyes) as moral, unbiased individuals. Monin and Miller's findings challenged the prevailing assumption that moral consistency is ubiquitous. Previously, the social psychology literature held that people show consistent behavior – leading to positive behavioral spillovers – to avoid cognitive dissonance (i.e., the mental discomfort experienced when present actions do not align with past actions).

Despite originating in social psychology, the concept of moral licensing has spread to other fields, including consumer choice and marketing. These fields had already considered that initial pro-environmental actions might lead to neglecting further efforts, borrowing terms

DOI: 10.4324/9781003584056-25

Table 22.1 Different types of spillovers in moral behavior

		Behavior 2 (Heating)	
		(+) Mindful use of heating	(-) Excessive use of heating
Behavior 1 (Transport)	(+) Train	**(+) (+) Moral consistency** *I've taken a train instead of flying, let's keep up the good work and not turn the heating up too much*	**(+) (-) Moral licensing** *I've avoided taking a flight today, I deserve to turn the heating up a bit more*
	(-) Flight	**(-) (+) Moral cleansing** *I've already polluted a lot today through flying, best not to turn the heating up too much when arriving home*	**(-) (-) Moral consistency** *I've already polluted a lot today, so what does it matter if I turn the heating up a bit more*

from economics like the **rebound effect** (i.e., increased consumption following efficiency improvements) or providing other psychological explanations like the single action bias (i.e., thinking a single corrective action sufficiently reduces risks, preventing future action). Moral licensing complemented these perspectives as a particular case of negative behavioral spillover.

Research has shown that moral licensing is not always a within-domain phenomenon: an initial pro-environmental behavior can affect a future action in an unrelated area, and vice versa. For example, in a virtual shopping experiment, participants who bought ecofriendly products were later greedier in an unrelated money-sharing task and more likely to steal money compared to those who bought regular products (Mazar & Zhong, 2010). Moreover, recent research showed that merely reflecting on past climate-friendly actions can justify unsustainable choices in the present. For example, in one experiment, participants who had abstained from flying in the past two years felt less guilty about their meat consumption when reminded of their prior sustainable behavior before – rather than after – being asked about meat consumption.

Different Perspectives

Meta-analytic evidence shows that the moral licensing effect is small to medium in size, slightly below the average effect size in social psychology. Whether this effect impacts a majority of people who have performed previous good deeds or remains a marginal phenomenon largely depends on framing and behavioral ambiguity.

Moral licensing occurs because people tend to pursue multiple, sometimes conflicting goals (e.g., trying to eat more organic food, while also trying to save money on groceries). The framing of past actions can either justify pursuing conflicting goals or reinforce consistent behavior toward a goal. For example, the initial behavior of buying organic food due to a personal commitment to eating sustainably is less likely to lead to moral licensing than buying organic food just because there was a convenient discount at the supermarket. Influencing the level of abstraction is what makes framing effective: thinking concretely about past moral behavior (e.g., "I take the time to compost my food waste") focuses attention on the act itself, making moral licensing more likely, while thinking abstractly about the same action (e.g., focusing on the pro-environmental values that inspired the action) leads to more consistent behavior.

Apart from framing, licensing effects also depend on the extent to which behaviors appear superficial or guided by normative values (see **Social Norms**). Superficial initial behaviors, like bringing a reusable bag to the supermarket, have low diagnosticity – meaning they reveal little about an individual's environmental commitment. As a result, superficial initial behaviors might not establish an individual's morality, preventing them from feeling licensed to transgress in future behaviors. Conversely, behaviors that seem to be motivated by personal values facilitate moral licensing by reducing suspicion of immorality. Investigating this behavioral ambiguity has led researchers to wonder whether initial pro-environmental behavior serves to reinterpret future negative actions ("moral credentials") or simply to counterbalance them ("moral credits") (see Box 22.1).

Box 22.1 Two models of moral licensing from the observer's perspective

Moral credits

This model posits that individuals accumulate moral credits (as in a moral bank account) from past actions to justify subsequent negative behavior; maintaining an overall positive moral balance. The moral "license" is thus a form of capital or credit acquired in a previous behavior used in subsequent behaviors. Under a moral credits model, the licensed behavior is unmistakably bad and the individual's behavior will be interpreted as inconsistent. For example:

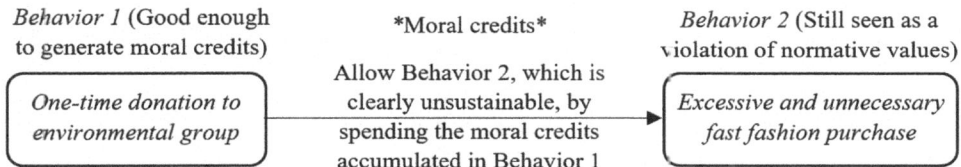

Behavior 1 (Good enough to generate moral credits)

One-time donation to environmental group

Moral credits
Allow Behavior 2, which is clearly unsustainable, by spending the moral credits accumulated in Behavior 1

Behavior 2 (Still seen as a violation of normative values)

Excessive and unnecessary fast fashion purchase

Moral credentials

This model suggests that the initial moral act influences how subsequent behavior is interpreted. The moral credentials model is a plausible explanation in situations where the licensed behavior is superficial and the initial behavior is unmistakably good. This casts doubt on whether the second behavior is immoral at all, considering the person's track record. Under the credentials model, the morality of the licensed behavior is unclear, and without more information, individuals' behavior might still be interpreted as consistent. For example:

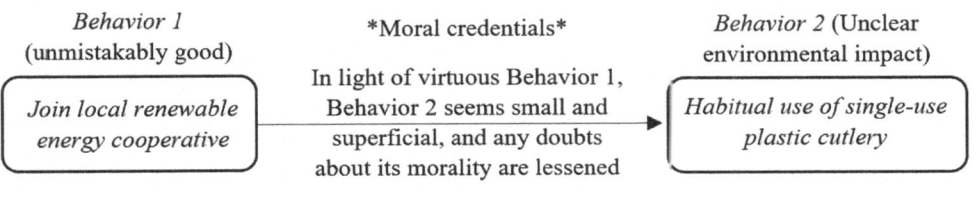

Behavior 1 (unmistakably good)

Join local renewable energy cooperative

Moral credentials
In light of virtuous Behavior 1, Behavior 2 seems small and superficial, and any doubts about its morality are lessened

Behavior 2 (Unclear environmental impact)

Habitual use of single-use plastic cutlery

Application

The behavioral inconsistency caused by moral licensing widens the gap between pro-environmental attitudes and actual behaviors (see **Attitude-Behavior Gap**). Shaming is a common tactic to encourage behavior change by inducing guilt for not conforming to a social norm. However, this can lead people to choose behaviors that minimize guilt with the least effort (e.g., purchasing carbon offsets) rather than undertaking more effective actions (e.g., avoiding flying). Moral licensing can exacerbate green consumerism by making individuals feel justified in skipping sufficiency behaviors after performing easier, efficient behaviors that reduce their guilt and make them feel morally allowed to maintain their consumption levels.

From a policymaking perspective, while guilt can be a powerful motivator, trying to leverage consumers' self-image tends to backfire (see Box 22.2). Consumers need a choice architecture that encourages behavioral consistency without exploiting guilt (see **Choice Editing**). Once people start making sustainable choices out of personal responsibility, moral licensing can be avoided by emphasizing the ongoing nature of their environmental commitment rather than viewing individual actions as isolated accomplishments.

Box 22.2   Three pro-environmental interventions that resulted in moral licensing

Weekly feedback on water consumption → More electricity consumption

In a field experiment in 154 apartments, Tiefenbeck et al. (2013) found that taking part in a water conservation campaign reduced residents' water usage, but also increased their electricity consumption. The researchers concluded that the water conservation campaign may have made people feel they had already done their part for the environment, leading them to be less careful about electricity usage.

Salient carbon taxes → Increased demand for taxed products

In a series of survey experiments, participants knowingly paying a carbon tax felt less guilty about their prospective purchases, and more licensed to consume carbon-intensive products. In this study, Hartmann et al. (2023) showed that salient carbon taxes indicated as part of the product price were less effective in curbing demand than hidden ones – even though the price increase was the same in both cases.

Promote ecolabeled products → Stick to resource-intensive activities

Data from EU-27 countries shows that **ecolabeling** is linked to higher (rather than lower) resource consumption. Survey data from the United Kingdom also indicates that a willingness to pay for ecolabeled products is linked with increased resource use. Based on these two interrelated studies, Barkemeyer et al. (2023) concluded that consumers might use the purchase of ecolabeled items to morally justify continuing other resource-intensive activities, rather than as a first step toward genuine behavior change. For wealthy consumers, buying ecolabeled products is a more convenient and simpler pro-environmental action than making significant lifestyle changes. Consequently, overly promoting ecolabeled products can be risky, as wealthy consumers may wrongly perceive them as guilt-reducing items, ultimately increasing their overall environmental impact.

Overall, pro-environmental interventions are less likely to lead to moral licensing when individuals think abstractly, focusing on their ongoing commitment rather than on the result of a specific action. Interventions that strengthen environmental self-identity are particularly effective, as individuals who strongly identify with a cause are less likely to exhibit moral licensing. Therefore, to avoid licensing, interventions should encourage individuals to reflect on how their actions align with their values, assess the extent of their pre-existing identification with those values, and find ways to enhance this connection (see e.g., **Education for Sustainable Consumption**).

Further Reading

Barkemeyer, R., Young, C.W., Chintakayala, P.K., & Owen, A. (2023). Eco-labels, conspicuous conservation, and moral licensing: An indirect behavioural rebound effect. *Ecological Economics*, 204, 107649. https://doi.org/10.1016/j.ecolecon.2022.107649.

Hartmann, P., Marcos, A., & Barrutia, J.M. (2023). Carbon tax salience counteracts price effects through moral licensing. *Global Environmental Change*, 78, 102635. https://doi.org/10.1016/j.gloenvcha.2023.102635.

Mazar, N., & Zhong, C.B. (2010). Do green products make us better people? *Psychological Science*, 21(4), 494–498. https://doi.org/10.1177/0956797610363538.

Monin, B., & Miller, D. T. (2001). Moral credentials and the expression of prejudice. *Journal of Personality and Social Psychology*, 81(1), 33.

Tiefenbeck, V., Staake, T., Roth, K., & Sachs, O. (2013). For better or for worse? Empirical evidence of moral licensing in a behavioral energy conservation campaign. *Energy Policy*, 57, 160–171. https://doi.org/10.1016/j.enpol.2013.01.021.

23 Risk Perception

Thomas Webler

Definition

Risk perceptions are personal opinions about risks. A risk is the chance that a specific threat will cause a specific harm. Threats may be natural (hurricanes), human-made (pesticides), or both (flooding amplified by anthropogenic climate change). Anything that people care about can be "at risk" – or threatened with harm – including one's life; mental or physical health; the lives, health, or livelihoods of others; one's reputation or property; biodiversity; lifestyle; or values such as freedom, justice, or democracy.

Individuals' risk perceptions can conflict with experts' assessments and lead to conflicts. For instance, the Intergovernmental Panel on Climate Change (IPCC) reports include expert assessments that the risks from climate change are very high. However, some individuals may agree that the risks are high but still give low priority to the issue. Because risk perceptions are subjective, it is not possible to know if people are representing their opinions truthfully or not. For instance, oil company executives who downplay the risks of climate change may privately believe the risks to be high but cannot publicly say it. Or their perception of risk from climate change may be genuinely biased because they are simultaneously focusing on risks to the oil and gas industry. People's risk perceptions may also differ from experts' because people value things that were not included in risk assessments. For example, while experts assess the risks of industrial agricultural products as low, people may rate them high because of a concern for social justice for farmworkers or impacts on pollinators. In other instances, risk perceptions differ because people have local knowledge about risks that experts were not aware of (see **Food Sovereignty**). For instance, experts warn against drinking unpasteurized milk, but local people may know and trust local farmers to produce safe raw milk.

History

In the 1970s Paul Slovic and colleagues began to measure people's perceptions of risk using surveys. They had lay people and risk experts rank the risks of 30 different threats and found significant differences. For instance, people thought the risk of nuclear power was very high, while experts calculated it to be relatively low. At the same time, people thought the risk of driving a car was low while experts rated it high. These results pointed to the most important finding of risk perception research: to non-experts, risk is more than a calculation. Lay people also care about things that technical risk assessors do not or cannot measure. Table 23.1 summarizes factors that shape risk perceptions.

Risk perception is a product of logical reasoning and emotions. Fear, anxiety, and dread can magnify the perceived severity of risks. Conversely, optimism bias can lead us to underestimate certain risks, fostering a false sense of security.

DOI: 10.4324/9781003584056-26

Table 23.1 Factors that shape perceptions of risk

My perception of risk is likely lower when...	*My perception of risk is likely higher when...*
I am in control (driving a car)	I am not in control (climate change)
The risk activity is voluntary (mountain climbing)	The risk activity is involuntary (air pollution)
The risk activity is familiar (bicycling)	The risk activity is unfamiliar (cryptocurrency trading)
The harm is delayed (cancer from smoking)	The harm is immediate (gunshot)
The harm is limited in scale and size (house fire)	The harm is catastrophic (nuclear power plant meltdown)
The harm is familiar (broken bone)	The harm is dreaded (cancer)

When confronted with a problem and a solution, people will likely compare their perceived risks of the problem to those of the solution. For instance, when policymakers propose to mitigate the driving forces of climate change by decarbonizing the global energy system or promoting high-density housing, some fear the solutions more than the problem.

Different Perspectives

Many have pointed out that it is difficult for people to wrap their heads around systemic risks such as climate change or the risks associated with changing to a zero-growth economy. When uncertainty is high, some people imagine the worst, others the best. Confirmation bias also comes into play. Even small changes such as buying an electric vehicle present uncertainty (consider the hype over "range anxiety") that leads people to perceive the risks of switching to EVs as very high.

Why do some people fear policies to address climate change or mitigate consumption? The science of risk perception has something to say about this. Opinions about risk are formed in the minds of individuals; hence they are shaped by experiences, hopes, fears, knowledge, and worldviews. To begin with, some people focus only on themselves and their closest family members. They do not believe they have a responsibility to help others. Others see themselves as part of a large community of extended family, friends, and neighbors. This simple difference leads to strong conflict over fundamental beliefs such as the role of government to limit people's choices or behavior (see **Choice Editing** and **Freedom of Choice**).

Psychologists have also found patterns (aka *heuristics*) in people's perception processes that help explain why risk perceptions differ. For instance, the *availability heuristic* is a common trait. It simply means that recent events have more impact on your perception of a risk than do older events. For example, if it was extremely hot last week, you will be more attentive to a message about global warming than if the weather had been unusually cool.

We also know that people tend to be more concerned about risks than benefits. Given a choice of installing heat pumps in their home or continuing to use fossil fuels for heating, people tend to focus on the fear that heat pumps may not work sufficiently well on a few extremely cold days, even if on the vast majority of days they work perfectly well and save energy, reduce costs, and reduce greenhouse gas emissions.

Application

Differing risk perceptions complicate policymaking about the critical issues of our time. Convincing people to support policies that reduce greenhouse gas emissions through lifestyle changes is a challenge in part because many people resist change (see **Attitude-Behavior Gap** and **Behavior Change**). It is further complicated by industries that amplify fears. For instance, attempts by some states to encourage homeowners to replace gas stoves with electric induction stoves were met with a public relations campaign by the gas industry.

A major debate is how risks should be regulated (see **Co-Benefits of Climate Policy**). Should regulatory officials make decisions based on what people perceive the risks to be, or should they base decisions on technical risk assessments? In a democratic society, one could make the case for using public resources to address the concerns of voters, even if experts rate those concerns as low. But others point out that spending tax revenue to reduce small risks is irresponsible.

The concept of risk perception encompasses a multifaceted interplay of cognitive, emotional, cultural, and societal factors. Risk is not merely about the objective assessment of probabilities and potential outcomes but also about how these assessments are subjectively interpreted, weighted, and acted upon.

Further Reading

Kahneman, D. (2011). *Thinking, fast and slow*. New York: Farrar, Straus, and Giroux.
Kasperson, R.E., Renn, O., Slovic, P., Brown, H.S., Emel, J., Goble, R., Kasperson, J., & Ratick, S. (1988). The social amplification of risk: A conceptual framework. *Risk Analysis*, 8(2), 177–187. https://doi.org/10.1111/j.1539–6924.1988.tb01168.x.
Lee, T.M., Markowitz, E.M., Howe, P.D., Ko, C.-Y., & Leiserowitz, A.A. (2015). Predictors of public climate change awareness and risk perception around the world. *Nature Climate Change*, 5, S. 1014–1020. https://doi.org/10.1038/nclimate2728.
Slovic, P. (1987). Perception of risk. *Science*, 236(4799), 280–285. https://doi.org/10.1126/science.3563507.
Sunstein, C. (2009). *The cost-benefit state: The future of regulatory protection*. American Bar Association.

24 Living Labs

Julia Backhaus, Edina Vadovics, and Marc Dijk

Definition

The European Network of Living Labs (https://enoll.org/) currently defines Living Labs as "open innovation ecosystems in real-life environments based on a systematic user co-creation approach that integrates research and innovation activities in communities and/or multi-stakeholder environments, placing citizens and/or end-users at the centre of the innovation process". Living Labs facilitate interactions between stakeholders to drive innovation and address real-world challenges. As such, they are ideally geared toward innovations that are both locally embedded and potentially scalable. Within them, typically all actors of the quadruple helix – research organizations, policy-makers and public actors, business and industry, as well as civil society – come together to work on complex societal challenges by collaboratively developing and testing possible solutions (see Box 24.1). Living Labs generally share the following four characteristics: (1) a *transdisciplinary approach* to research and knowledge creation; (2) an *iterative, experimental design* committed to *learning* and *reflexivity*; (3) a long-term orientation toward *societal transformation* and an accompanying interest in *transferability* or *scalability*; and (4) a focus on a *real-life setting*.

These characteristics can be considered the lowest common denominator across Living Lab research and practice. Notably, they are not explicitly directed at sustainability-oriented change. In view of this omission and in recognition of research on real-world laboratories, especially in the German-speaking research community, we propose to integrate the now widely recognized quintuple helix model in Living Lab research and practice, to ensure an explicit consideration of the environment and society-nature interactions, as shown in Figure 24.1. Thereby, Living Labs can become conducive to the experimental advancement of sustainable consumption and lifestyles.

History

The Living Labs concept originated in the 1990s in the field of technological innovations, specifically in computer-human interaction. William Mitchell at MIT conceptualized them as physical spaces to observe how people interact with technological prototypes in natural or specially equipped settings. Compared to previous (technology) demonstration projects, Living Labs – which soon after also appeared in Europe – primarily focused on user-centered innovation to accelerate application, implementation, and marketization. They also occasionally served as platforms for science communication to showcase high-tech, interconnected, immersive environments.

Living Labs evolved to include experimental approaches to address complex societal challenges concerning housing, mobility, energy, food, or social inequalities. Some focus on specific domains, such as Slovenia's *Green Point Living Lab* (https://itc-cluster.com/green-point/), which works on sustainable food system innovations aligned with the EU's Green New Deal. It consists of Demo

DOI: 10.4324/9781003584056-27

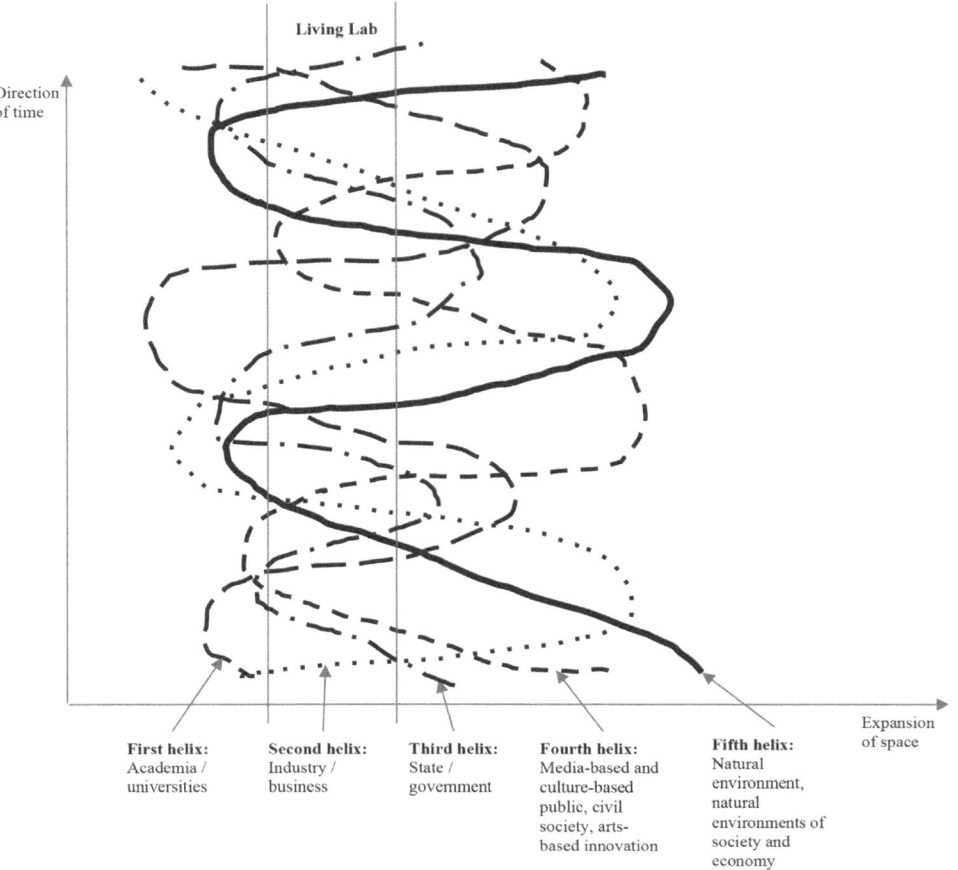

Figure 24.1 Visualization of quintuple helix interactions in and beyond a Living Lab

Source: Authors' own conceptualization based on Carayannis, E.G., & Campbell, D.F. (2014). Developed democracies versus emerging autocracies: arts, democracy, and innovation in Quadruple Helix innovation systems. *Journal of Innovation and Entrepreneurship*, 3(1), 12

Farms for testing real-life agricultural innovations and the *Food Supply Living Lab*, which focuses on advanced food processing, delivery, and consumption within a regional ecosystem (see **Food Miles**, **Community Supported Agriculture**, **Food Sovereignty**). Other Living Labs follow a cross-sectoral approach, like Ireland's *Dingle Creativity and Innovation Hub* (https://dinglehub. com/), aiming to create a sustainable and vibrant community on the Dingle Peninsula – inhabited by 12,000 people and visited by one million annually. This Living Lab aims to foster year-round, well-paid jobs through a diverse ecosystem of stakeholders combining local resources with new opportunities, promoting entrepreneurial growth, and engaging with the community, research, industry, and policymakers.

These are only some of the Living Labs currently part of ENoLL. Since its foundation in 2006, ENoLL has recognized over 480 Living Labs worldwide as its members and currently has 160 active members.

Different Perspectives

Living Labs relate to and can be found across a broad area of research and action initiatives – from testbeds for technical innovations to grassroots initiatives for social change toward more sustainable ways of living (see **Grassroots Innovation, Eco-Communities**).

They operate between user-centered design and citizen-led transformation, offering spaces to explore problems and test solutions. However, they risk "solutionism", where predefined pathways dominate. Three main tensions Living Labs face include: (i) controlled experimentation versus open co-creation; (ii) learning from failure versus showcasing success; and (iii) local embedding versus scalability. With limited, project-based funding, Labs often struggle to preserve participatory processes for deeper local embedding and critical reflection. Building in monitoring, evaluation, and research activities can result in a successful critical reflection that facilitates higher-order learning where the participants collectively reframe the problem and reconsider their own perspectives and assumptions.

The balance between openness and predetermined pursuits depends on who makes decisions, which boundaries are socially constructed and upheld, and which stakeholders, perspectives, and solutions are involved or represented in a Living Lab. This raises issues of innovation management and research policy as well as of participation and democracy. In relation to sustainable consumption and lifestyles, Living Labs may seek to foster sustainable energy prosumption in individual houses, neighborhoods, or entire cities by involving and connecting key actors (see **Prosumerism**). This can involve top-down governed activities to implement and test technologies like hydrogen infrastructure or bottom-up participatory processes, leaving infrastructural choices to local owners and residents. Even if Living Labs usually aspire to be participatory and democratic, various forms of justice (e.g., procedural, recognition, distributional) should be monitored, and not assumed to be there automatically (see **Climate Justice**).

Living Labs are able to facilitate higher order and technical learning, including the re-definition of problems and changing interpretive frames; shifts in consumption patterns; governance innovations; and helping communities tackle sustainability challenges. Significant efforts go into evaluating Living Labs, with increasing emphasis on co-evaluation, involving stakeholders throughout the process. ENoLL has been instrumental in formalizing these evaluations, providing frameworks to measure effectiveness, improve methodologies, and enhance the scalability of innovations across regions.

Research on Real-world Labs (RwLs) highlights diverse impacts, categorized into societal and individual changes, governance transformations, and physical modifications. However, an important knowledge gap remains about understanding (i) whether and how a deeply transformative process occurs and (ii) the barriers and enablers for experiments and labs to drive system change.

Living Labs often center on technological innovation. More participatory and pluralist approaches could enhance the social robustness of change, address democratic deficits, and improve sustainability-related outcomes by considering justice, plurality, and equality. Urban Transformation Labs and Thinking Labs exemplify this shift: Thinking Labs foster open deliberation to inform policy and decision-making, while Urban Transformation Labs provide platforms for citizens, businesses, municipalities, and researchers to collaboratively experiment with cross-sectoral transformations.

Application

The concept of Living Labs continues to inspire numerous research programs, projects, and civic initiatives. There have been projects funded by the European Commission (e.g., *ENERGISE Living Labs*; Box 24.1) that followed a practice-based approach to everyday experimentation, focusing

---

Box 24.1   Practical examples and applications of the concept Living Lab

ENERGISE Living Labs

The *ENERGISE Living Labs* (https://energise-project.eu/livinglabs) aimed to promote sustainable home energy use by targeting **social practices** related to heating and washing laundry in eight European countries (see also **Energy Consumption Behavior**). Each country organized two labs using the same tools and challenges, based on different approaches to participant involvement. That enabled the ENERGISE project team to determine that community involvement outperformed individualistic approaches to stimulating change and that experimentation with everyday practices can lead to lasting, impactful changes.

SmartQuart

SmartQuart (https://smartquart.energy/english-abstract/) is the first Living Lab for the energy transition, funded by the German Federal Ministry for Economic Affairs and Climate Action, to test innovative technologies under real conditions. The project brings together key stakeholders (public actors, energy providers, users, and researchers) to demonstrate and evaluate energy-optimized neighborhoods for decentralized energy and heating solutions. It aims to provide evidence for the technical and economic feasibility of integrating energy, heating, and mobility sectors within and between neighborhoods. Located in three German cities, the project includes a variety of neighborhood types, from dense urban areas to rural settings, representing typical German communities.

Smart Urban Mobility Meta Lab (SUMMALab)

To facilitate learning between regions and accelerate the scaling of promising solutions, the project is conducted in Amsterdam, The Hague, Delft, and Rotterdam. The focus areas of *SUMMALab* (https://summalab.nl/) include last-mile solutions like autonomous shuttles, door-to-door options like car-sharing, and changes in urban infrastructures such as parking areas and access restrictions. To support and scale these experiments, *SUMMALab* develops tools for business models, technical feasibility, upscaling, and mobility and economic impacts, addressing the need for structured development and community involvement.

---

on understanding and reconfiguring the routines and habits that shape consumption. However, these approaches remain sparse and exploratory to date, while most Living Labs focus on changing infrastructures, systems of provision, and regulatory frameworks. The "EU Mission: Soil Deal for Europe, 100 Living Labs and Lighthouse projects by 2030" aims to establish collaborative, real-life experimental spaces where multiple partners, including researchers, farmers, foresters, land managers, and NGOs, co-create innovations for healthy soils at territorial, landscape, or regional scale. Urban sites, smart city initiatives, and societal deliberation are further currently prominent application areas of the concept.

Urban living labs (ULLs) are multi-actor platforms for co-creation, real-life experiments, and joint learning in cities. They focus on innovating urban practices including governance and planning to address sustainability challenges. ULLs involve diverse local stakeholders and create context-sensitive knowledge, though the scalability of their innovations varies. Examples in the city of Amsterdam include an energy lab, a lab for mixed urban spaces, and two labs for healthy urban environments. The "meta-lab" approach aims to enhance the impact of ULLs by connecting and aligning learning processes across multiple labs.

Smart City Labs are collaborative spaces that integrate advanced technologies to improve (the use of) infrastructure, mobility, energy, and public engagement in cities. Through real-life experimentation and data-driven insights, they seek to foster sustainable urban development, optimize resource management, and enhance the quality of life for residents. An example is the platform *smart.aachen* (https://smart.aachen.digital/public/en), which connects and supports smart city projects in Aachen, Germany, toward reaching climate neutrality by 2030. These projects include sensors monitoring the water quality in a local pond, a mobility dashboard guiding motorized citizens past traffic to the nearest available parking spot, and the life-saving use of unmanned aerial vehicles in rescue operations.

Thinking Labs are another variation. They are safe spaces for multi- or single-stakeholder groups to exchange ideas about and co-create solutions to specific problems in a facilitated setting. Examples include the European CIMULACT (http://www.cimulact.eu/) and EU 1.5° Lifestyles (https://onepointfivelifestyles.eu/) projects, both of which employed Thinking Labs extensively in multiple countries, also allowing for cross-comparison.

As these and other applications show, Living Labs can facilitate less consumerist lifestyles and community-driven resilience in various ways. By engaging stakeholders in co-creating and testing sustainable systems of provision and practices for localized food systems, minimalism, collaborative consumption, sharing economies, resource conservation, and circular economies, Living Labs can encourage and enable changes that reduce consumption and foster sustainability, especially if an explicit focus on environmental impacts and concerns is maintained throughout the process.

Further Reading

Compagnucci, L., Spigarelli, F., Coelho, J., & Duarte, C. (2021). Living labs and user engagement for innovation and sustainability. *Journal of Cleaner Production*, 289, 125721. https://doi.org/10.1016/j.jclepro.2020.125721.

Engels, Franziska, Wentland, A., & Pfotenhauer, Sebastian M. (2019). Testing future societies? Developing a framework for test beds and living labs as instruments of innovation governance. *Research Policy*, 48 (9), 103826. https://doi.org/10.1016/j.respol.2019.103826.

Schäpke, N., Wagner, F., Beecroft, R., Rhodius, R., Laborgne, P., Wanner, M., & Parodi, O. (2024). Impacts of real-world labs in sustainability transformations: Forms of impacts, creation strategies, challenges, and methodological advances. *GAIA – Ecological Perspectives for Science and Society. Special Issue*, 1–2024. https://doi.org/10.14512/gaia.33.S1.2.

Scholl, C., de Kraker, J., & Dijk, M. (2022). Enhancing the contribution of urban living labs to sustainability transformations: Towards a meta-lab approach. *Urban Transformations*, 4(1), 7. https://doi.org/10.1186/s42854-022-00038-4.

Vadovics, E., Richter, J.L., Tornow, M., Ozcelik, N., Coscieme, L., Lettenmeier, M., Csiki, E., Domröse, L., Cap, S., Puente, L.L., Belousa, I., & Scherer, L. (2024). Preferences, enablers, and barriers for 1.5°C lifestyle options: Findings from Citizen Thinking Labs in five European Union countries. *Sustainability: Science, Practice and Policy*, 20(1). https://doi.org/10.1080/15487733.2024.2375806.

25 Convivial Technology

Roxana Bobulescu and Nilo Coradini de Freitas

Definition

Convivial technology is technology that is easily accessible, manageable, adaptable, and sustainable and that encourages autonomy. It allows collaboration, requires no heavy infrastructure, and uses very few natural resources. It is a form of self-production and self-use. Users of convivial technologies can master, change, and **repair** them without recourse to experts. Institutionalized forms of convivial technologies include repair cafés, urban gardening, community supported agriculture, local exchange systems, and **eco-communities**.

The author who first proposed the notion, Ivan Illich (1973: 20), spoke of convivial "tools", rather than technology. By "tool", the author meant instruments that transmit users' intentions (like a hammer, for instance), but also included institutions like education, justice, and healthcare. The purpose of qualifying tools as "convivial" was to establish a methodology that would allow evaluation of the characteristics of the instruments regarding their tendency to maintain that dynamic and whether they would tend to turn means to ends. The issue at hand is that certain tools reinforce a tendency to become ends in themselves (like smartphones), fostering ever-growing usage, and making people dependent upon the tools, as opposed to being interdependent toward other people. In this case, society's tooling could be described as *industrial* or *manipulative*, as opposed to *convivial*.

History

Illich first presented the idea of conviviality within a generalizing theory of a process he had been criticizing in the early 1970s. Namely, this was a process of transmission of technology and reconfiguration of geopolitical domination that took place after World War II, under the name of *development*. Through this process, the notions of "developed" and "under-developed" economies or societies were first advanced. Such a process foresaw the subsequent environmental debates that took place around the United Nations and prompted other similar responses, such as the appropriate technology movement (see **Grassroots Innovation**).

Technology became omnipresent, and went through exponential growth, shaping our lifestyles, and demanding many resources. How can we reconcile it with the need to reduce our ecological footprint and minimize resource consumption? In the world of "technoculture", can we make technology more consumer-friendly, by reappropriating it in forms more suited to our planetary limits and human needs?

In the 1970s, Illich helped to set up a center in Mexico that would become an intellectual hotspot against the said conception of development in the Latin American region, called *Centro Intercultural de Documentación* (CIDOC). The debates developed into favoring technical characteristics that empowered users, similar to the right-to-repair movement, and favoring certain

DOI: 10.4324/9781003584056-28

personal postures, such as **voluntary simplicity**. The **Degrowth** movement would then mobilize the concept and related debates in the 21st century.

Different Perspectives

Some authors evaluate the conviviality of certain technologies differently, particularly regarding digital technologies (see **Information and Communication Technology**). For instance, Marco Deriu (see Further Reading) pointed out that some people considered computers to be convivial tools, and showed similarly enthusiastic attitudes toward the internet. These stances emphasize freedom from technocrats in the sense that, with online forums and other means of exchange, people could now learn and develop their initiatives free from the control of teachers and other regulatory authorities, and participate in self-established communities of interest and learning.

However, other scholars see the issue differently from Deriu, including Illich himself in his late works. This second position would sooner see digital technologies as manipulative on the basis that in their programming, they limit the range of possible usages, and reduce the users to operators at best, and to subsystems at worst. In his late work, Illich emphasized the embodied character of interpersonal relations he was trying to defend with the idea of conviviality, in which people face each other, as opposed to simply communicating through cybernetic means.

Application

For the tooling of society to maintain a convivial character, three criteria are to be respected, as summarized by Samerski (2018). Firstly, *how* to use tools must be intuitive enough that preparatory certification provided by specialists is not necessary. Secondly, it must be up to the user *if and when* they would like to use them – if technocratic elites or societal structures render use obligatory, the tool ceases to be convivial. Finally, the tool must *serve the purposes of the user* and not the other way around.

We may consider conviviality to be a relevant criterion for assessing the sustainability of lifestyles enacted in different societies. Specifically, the Matrix of Convivial Technology (MCT) developed by Andrea Vetter can be used to assess the conviviality of a technology. In the MCT, convivial technologies are socio-technical solutions defined around five core dimensions: relatedness, accessibility, adaptability, bio-interaction, and appropriateness. "Relatedness" refers to what technology brings about *between* people. "Accessibility" refers to who can produce or use it, where and how. "Adaptability" relates to how independent and linkable it is. "Bio-interaction" refers to the role of the materials used in ecological cycles. Finally, "appropriateness" refers to the relation between input and output considering the context. These five dimensions are correlated to four stages of a life cycle: materials, production, use, and infrastructure.

A common example is to contrast bicycles with cars as a means of transportation (Table 25.1). The car needs not only an industrialized fuel – be it electricity or gasoline – but also a wide set of institutional arrangements, including the issuing and enforcing of driver's licenses, speed checks on roads and highways, engineers, mechanics, and so on. The bicycle does need the industry to produce, but on top of running on human energy, it can be handled successfully by amateurs in their everyday use, repair, and customization for different needs.

However, bicycles can have their conviviality "taken away" from them, if we propel them by electricity, or depending on the design of the policy for public transportation. If bicycles are proposed within a system that is connected to monitoring devices, be they fixed around the city or carried around by users in their pockets, as in the form of smartphones, the technology "bicycle" fails to meet some criteria of conviviality. Firstly, it becomes less accessible, as it now needs batteries,

Table 25.1 Convivial technologies: bike versus car

Convivial technology criteria (+/-)	Bike: convivial	Car: non-convivial
Relatedness	(+) Autonomy, amateur use, democratic use, grassroots innovations	(-) Creates scarcity, limits autonomy by contractual regulations, supports consumerism
Accessibility	(+) Low-cost, easy to use, easy to understand	(-) Drivers licenses, speed checking, highway tolls
Adaptability	(+) Easy to repair and modify	(-) Repairable by experts (engineers, mechanics)
Bio-interaction	(+) Supports health, improves air quality, non-violent	(-) Polluting, increases toxic waste, safety risks
Appropriateness	(+) No fuel, sustainable mobility, less resource-intensive, allows joyful time	(-) Industrial fuel, infrastructures (highways, etc.), resource-intensive, unpleasantly time-intensive

chips, and satellites for the system to work properly, and is probably behind a subscription-based paywall. Secondly, the possibilities for tinkering with the bicycle are now behind contractual regulations: one can no longer fix or customize the bicycle without breaching contracts. Finally, many features that collect behavioral data now serve the purposes of third parties that can profit from such data, as opposed to the purposes of the user.

In the age of exponential development of artificial intelligence through machine learning, could we imagine a "convivial AI"? Using the MCT to assess its conviviality, Marion Meyers identifies some key limiting factors to the conviviality of AI: (a) the high complexity of the devices and the technical opacity; (b) the environmental impacts throughout the supply chain and the use of Rare Earth Elements in electronics; (c) the size of the infrastructure needed (big data centers, huge energy consumption).

Finally, the 2024 European directive on the "right to repair" is an important step toward convivial technologies, as manufacturers cannot resort to contractual clauses or techniques restricting independent repairers' access to second-hand or 3D-printed parts (see also **Ecodesign** and **Extended Producer Responsibility**). The directive also provides access to repair courses, technical information, and participatory repair spaces.

Further Reading

Deriu, M. (2014). Conviviality. In *Degrowth: A vocabulary for a new era*, pp. 79–82. Routledge. Available at: https://www.taylorfrancis.com/chapters/edit/10.4324/9780203796146–20/conviviality-marco-deriu (accessed: 13 December 2022).

Illich, I. (1973). *Tools for conviviality*. New York: Harper & Row.

Meyers, M. (2023). *A degrowth perspective on artificial intelligence – Analysing the appropriateness of machine learning in a degrowth context*. Master thesis, ETH Zurich. Available at: https://www.research-collection.ethz.ch/handle/20.500.11850/622669 (accessed: 3 January 2025).

Samerski, S. (2018). Tools for degrowth? Ivan Illich's critique of technology revisited. *Journal of Cleaner Production*, 197, 1637–1646. https://doi.org/10.1016/j.jclepro.2016.10.039.

Vetter, A. (2018). The matrix of convivial technology – Assessing technologies for degrowth. *Journal of Cleaner Production*, 197, Technology and Degrowth, 1778–1786. https://doi.org/10.1016/j.jclepro.2017.02.195.

26 Beauty

John de Graaf

Definition

According to the Oxford English Dictionary, beauty is "a combination of qualities, such as shape, color, or form, that pleases the aesthetic senses, especially the sight". In the context of this book, we might explore the relationships between beauty and consumption. Here we are referring to things like the beauty of nature and of human design in architecture – that is, the aesthetic quality of our surroundings.

A focus on creating beautiful environments for human beings might be the goal that can turn our heads away from overconsumption and might create more space for play and relaxation. In the US context, both progressives and conservatives love their flower gardens, and both flock to our national parks and other scenic areas.

First of all, beauty can make us happier. Many studies show that people who report living in beautiful communities enjoy greater life satisfaction than those who feel the places they live are unattractive. A Gallup study, "Soul of the Community", found that the beauty of their surroundings was among the top three factors that bind people to their communities. Despite this, questions about beauty are not included in even the most comprehensive international indices of well-being (see **Well-being Economy**).

Secondly, beauty often (but not always) promotes sustainability. In communities with more trees, parks, green spaces, and intimate varied architecture, people walk more and drive less. They slow down and show more appreciation for their surroundings. They have less inclination for mindless consumerism (see **Urban Planning and Spatial Allocation**). Park visits also can contribute to health. In 1857, Frederick Law Olmsted argued for parks "at points so frequent and convenient they would exert an elevating influence on all the people . . . cultivating taste and lessening that excessive materialism of purpose in which we are so cursedly absorbed". However, big houses with large lawns and perhaps artificial grass may be beautiful, but they are usually very unsustainable.

Adding beauty to a neighborhood can also reduce crime. A study in Philadelphia found that cleaning up blighted areas and replacing vacant lots and buildings resulted in a 30% drop in homicide and other crimes and a 50% reduction in mental illness.

History

We cannot say when the concept of beauty emerged in human history. Early wall paintings may reveal a sacral dimension and an appreciation of beauty. What we do know is that early Greeks valued beauty in art, architecture, and philosophy (see Box 26.1). Beauty is present in most civilizations and cultures, although "beauty for beauty's sake" may be more recent, dating at least from the early Renaissance. Beauty also shows up in science and mathematics.

DOI: 10.4324/9781003584056-29

"Beauty will save the world", claimed Dostoevsky and Solzhenitsyn. "Everybody needs beauty as well as bread", wrote John Muir. Might it be that a new focus on protecting the beauty of nature and designing our communities "with an eye to the effect made up on the human spirit by being continually surrounded with a maximum of beauty", as Thomas Jefferson suggested, could help turn the richer nations of the world to a more sustainable future?

Box 26.1   Beauty in architecture, art, and nature

The Cathedral of Notre Dame testifies that our love of beautiful architecture dates back to at least the 12th century. Versailles has drawn visitors since 1661. Even more remarkable, our love for art predates the Louvre by 30,000 years in the Chauvet Cave and by 17,000 at Lascaux. That's art. What about nature's beauty? In 1336, Petrarch became the first to climb mountains for the view when he ascended Mont Ventoux and vividly described the dazzling glories of the Alps from its summit.

Different Perspectives

Needs: The psychologist Abraham Maslow considered the need for beauty to be among the highest needs in his well-known hierarchy. He viewed beauty as an urgent need for many psychologically healthy adults and children. Maslow thought there were people whose need for beauty was so great they grew mentally ill without its presence in their lives.

Basic instincts: One problem is that many do not take beauty seriously, often thinking of it only in terms of the enhancements of cosmetics, another form of **consumerism**. Others argue that beauty is simply subjective, "in the eyes of the beholder". Many studies have shown this to be false. The love of beauty is one of our most universal instincts and a result of eons of evolution. Life-supporting landscapes appear beautiful to us. A study of landscape and architectural photographs by thousands of UK residents has shown remarkable agreement as to which are more beautiful.

Poverty and beauty: Some claim that beauty is a distraction from more pressing issues such as social justice, the environment, war, and peace. But, as Harvard philosopher Elaine Scarry has shown, beauty makes us more generous, more tolerant, more inclined toward justice, kinder to one another, and more supportive. There is, in fact, something elitist in the opposite claim – it suggests that the poor do not need or care about beauty. But someone should have told that to the thousands of millworkers in the 1912 textile strike in Lawrence, Massachusetts, who carried banners reading "WE WANT BREAD AND ROSES, TOO!". As marginalized as they were, these women understood that their lives were even poorer without beauty – and the time to appreciate it.

Application

In the United States, beauty's true Renaissance came in the 1960s, a tumultuous decade, rocked by struggles and shaped by the will and skill of President Lyndon Johnson. Johnson wished to unify the United States – polarized then as now – around stewardship of its beauty. "Beauty", Johnson said, "must not be just a holiday treat, but a part of our daily life". The value of beauty "does not show up in the Gross National Product", Johnson explained, "but it is one of the most important components of our true national income, not to be left out because statisticians cannot calculate its worth".

More than half a century later, it's time to take beauty seriously again. If a Green New Deal might make us more sustainable, a *Beauty New Deal* could bring us together in pursuit of a proven contributor to happiness. Such a New Deal might include:

- Greater public support for artists, writers, poets, and performers.
- Expanding parks, wilderness areas, and open spaces, while strengthening protections from commercial encroachment.
- Establishing an International Civilian Conservation Corps for beautification and environmental restoration projects, funded by a global capital transaction tax.
- Hosting a United Nations Summit on Natural and Architectural Beauty.
- Encouraging urban beautification, including planting millions of flowering and shade trees.
- Supporting "Renaissance Zones" using grants and tax incentives for beauty-led economic development in poor communities.
- Launching an international multi-university, cross-disciplinary research project on the value of beauty, followed by a communications campaign to circulate its findings.

No one can guarantee that a politics of beauty will save the world. But it might well be an antidote to consumerism, so it's worth a try.

Further Reading

de Graaf, J. (2024). *Towards a politics of beauty*. Front Porch Republic. Available at: https://www.frontporchrepublic.com/2024/05/toward-a-politics-of-beauty/ (accessed: 14 June 2025).
Gallup. (2010). Knight soul of the community 2010. *John S. and James L. Knight Foundation*. Available at: https://knightfoundation.org/sotc/ (accessed: 8 February 2025).
Local Futures (Economics of Happiness). (2016). *Sandra Lubarsky: The importance of beauty* [Video]. YouTube. Available at: https://www.youtube.com/watch?v=GCsYeFgMGG8 (accessed: 8 January 2025).
Reynolds, F. (2017). *The fight for beauty: Our path to a better future*. London. Oneworld Publications.
Scarry, E. (1999). *On beauty and being just*. Princeton: Princeton University Press.

27 Stocks Versus Flows

Dominik Wiedenhofer

Definition

The concept of societal material stocks refers to the long-lived biophysical basis of society, from individual buildings, cars, machinery, or computers to entire settlements and infrastructure systems. Because those stocks lock in production and consumption patterns over years to decades, their role in transformations toward sustainable consumption patterns in everyday practices and the high-level provision of services to ensure well-being is crucial (Figure 27.1; see **Social Practice Theory**, **Well-being Economy**).

For example, safe and affordable housing first requires constructing a building and its water and electricity supply networks which turn flows of construction materials into long-lived material stocks. Then, energy flows are needed to heat and cool living spaces, based on the thermal performance of the building and the inhabitant's demand for thermal comfort. Buildings also require maintenance, repairs, and component replacement. The building's location, with its access to mobility infrastructure and places people want to reach, determines the mobility practices of its inhabitants and thus locks in future energy flows for transport (see **Sustainable Mobility, Urban Planning and Spatial Allocation**). Those mobility practices then shape the required material stocks and flows in infrastructure systems and vehicles, including upstream energy and material flows in industry and construction. Each cycle of use, maintenance, and **repair** causes waste and emissions. At the end of life, a building might be demolished, causing further waste flows which might be recycled, landfilled, or otherwise disposed of.

From a systems perspective, the concept of societal material stocks is usually pragmatically defined as covering those products used and maintained for more than a year (Figure 27.1). Multiple aspects are measured, for example, physical and functional units of service provisioning such as kg of mass, m² of living space, purpose, or economic value and ownership. In the literature, societal material stocks are also called in-use stocks, manufactured and fixed capital, anthropomass, technomass, technosphere, built environment, infrastructure, or artifacts. The concept of stocks is also used in the natural sciences, for example, to quantify carbon stored in forests, or to measure resource deposits. Stocks of natural resources are also valued in economic terms by economists, who then refer to them as natural capital. Those "natural" stocks exist without humans investing work into their creation and maintenance (Figure 27.1).

Flows of material resources cover metals and ores, biomass, non-metallic minerals, fossil fuels, and water, which are extracted from the Earth System and its natural stocks of resource deposits (Figure 27.1). Material flows are processed into various products by industry using stocks of machinery, infrastructure, and energy. Products may be "consumed" within a short time, for example, for energy provision in the form of motor fuels or electricity, as drinking water, packaging material, or fertilizer. Each of these products eventually turn into waste and emissions. Alternatively, products

DOI: 10.4324/9781003584056-30

that are used for a longer time accumulate as societal material stocks, creating unavoidable lock-ins and path dependencies (see above). Multiple aspects of resource flows can be measured, such as mass, value, environmental impacts, purpose, and ownership. However, in contrast to stocks, flows are always measured per unit of time, commonly per year.

History

Since the beginning of the 20th-century economists have dealt with societal stocks, focusing on their monetary market-based transaction value. They address questions of investment, value depreciation, technological capacity, and substitution. They usually view the economy as separate from the environment and hardly acknowledge limits to the growth of the economy as valued in monetary terms (see **Ecological Economics** and **Degrowth**).

Alternatively, the crucial role of societal material stocks for sustainable production and consumption was already identified in the 1960s. For example, Kenneth Boulding's seminal essay on the "Economics of the coming Spaceship Earth" argues that

the ultimate measure of the success of the economy is . . . the nature, extent, quality, and complexity of the total capital [*material*] stock . . . what we are primarily concerned with is stock maintenance, and any technological change which results in the maintenance of a given total stock with a lessened throughput [*of material and energy flows ultimately turning into waste and emissions*] . . . is clearly a gain.

This alternative systems-based perspective on society-nature interactions views society and its economy as highly interdependent parts of the Earth System, subject to natural laws and confined to a materially closed, but energetically open planet Earth (Figure 27.1; see **Steady-State Economy**). Several research approaches emerged from this perspective, aiming to identify opportunities for providing the services required for well-being with a reduced energy and material demand.

System dynamics, for instance, incorporates material stocks and flows to describe a system's resilience and potential transition pathways, as exemplified in the 1972 *Limits to Growth* work by Meadows et al. Life cycle assessments (LCA) analyze and compare the environmental impacts of products and service provisions, often by allocating or depreciating stocks over time. Environmentally-Extended Input-Output Analysis (EE-IOA) models how supply chains link resource flows through production to consumption around the world, but mostly struggle to incorporate material stock dynamics. Material and energy flow analysis (MEFA) analyzes societal material stocks and resource flows of socio-economic systems, from specific production and consumption systems to the urban, national, and global economy. Recently, macroeconomic and integrated assessment models have also been extended toward covering biophysical stock-flow dynamics (Wiedenhofer et al., 2024). These efforts are found in the related fields of **Ecological Economics**, Input-Output Analysis, Social Metabolism, Life Cycle Assessment, Industrial Ecology, Complex Systems Research, and Sustainability Science.

Different Perspectives

The mainstream economic perspective on societal material stocks and flows has several critical limitations for sustainable production and consumption research and practice. Firstly, market-based economic transaction values might fluctuate strongly, without any physical change to the actual material stock. For example, the collapse of the US housing market in 2009–2010 resulted in catastrophic losses of monetary value, destabilizing the global economy; however, the physical

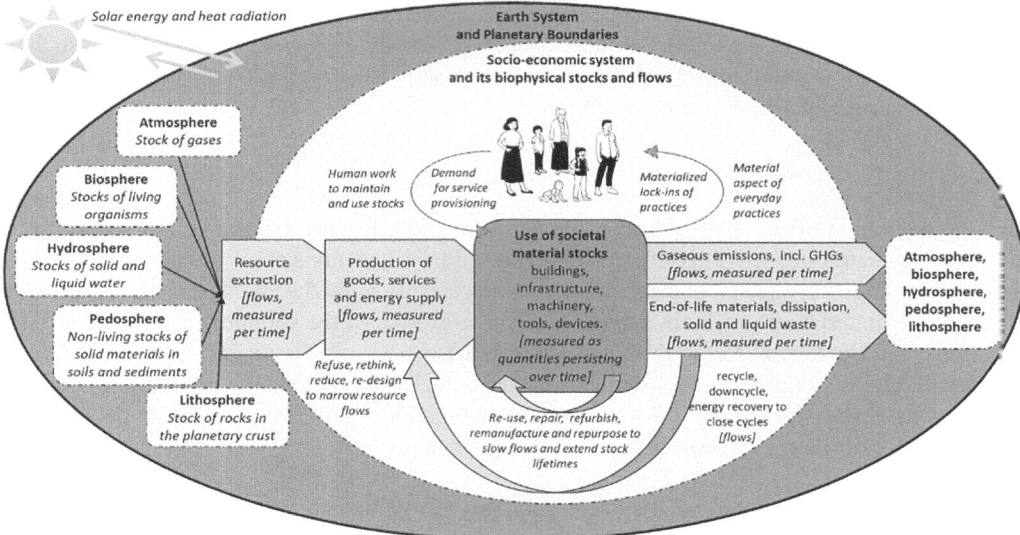

Figure 27.1 Conceptualizing society-nature interactions from a systems perspective explicitly addressing stock-flow relations. Socio-economic systems extract flows of resources from "natural" stocks of the Earth System through socially organized work. Those resource flows are transformed into goods and services by various industries, which can be conceptualized to serve service provisioning, and/or as the material basis of everyday practices. **Circular economy** strategies then aim to narrow, slow, and close material cycles, as shown in the lower part of the figure, which should help mitigate environmental impacts. The system boundaries between society and nature are analytical choices grounded in epistemologies of complex systems theories, aligned with the system of national accounts

Source: Produced by the author

buildings and their potential for service provisioning for well-being did not change (see **Foundational Economy**). Similarly, the value of a building also often increases, without any actual physical changes to it. Secondly, use values are context- and actor-specific, which do not directly translate into the monetary transaction values of service provisioning for human well-being. For example, safe and affordable housing for low-income people is "worth" much less in monetary terms, than a villa for the super-wealthy. Thirdly, monetary depreciation over time, as used in bookkeeping, implies decreasing the value of a stock, while physical functionality and realized lifetimes differ substantially. For example, the moment someone buys a new car, it immediately loses substantial monetary value due to now being second-hand, while the actual car is still the same.

The alternative systems-based perspective on society-nature interactions focuses on a biophysical and stock-flow consistent understanding of sustainable production and consumption, inequality, service provisioning for well-being, as well as resource efficiency and the potential criticality of specific resources. This perspective emphasizes (i) *strong sustainability* and the intrinsic value of nature in different contexts, and (ii) the need for deep structural transformations to achieve internationally agreed-upon climate and biodiversity protection targets. It advances a biophysical perspective on society-nature interactions, distinct but complementary to monetary valuation. Analogous to any biological metabolism, it views societies as having to successfully organize energy and material flows to maintain, expand, and use its societal material stocks. Because most

Table 27.1 Policy-relevant insights and future research avenues addressing societal material stocks and flows

Topic	Insight
Global stock dynamics drive resource use patterns	Societal material stocks have increased 26-fold over the last 115 years and continue to grow across the world. About half of global resource extraction is used to build and maintain material stocks. Another quarter are fossil energy carriers utilized for energy provisioning, and the remaining quarter is used for food provisioning. A very small share goes to dissipative uses such as lubricants, fertilizer, or chemicals used in agriculture.
Energy-materials-GHG nexus	About one-third of the global primary energy supply and subsequent GHG emissions are generated by the production and maintenance of material stocks. The remaining two-thirds are used to utilize material stocks of buildings, infrastructure, machinery, and other devices.
Resource-efficient everyday life and social practices	Service provisioning and everyday life practices rest on specific stock-flow combinations, which can be more or less ecoefficiently designed and utilized (see **Choice Editing**). Because stock-flow relations are controlled by societal actors, and organized via property rights and economic relations, questions of power, justice, inequality, and fairness need to be addressed. This is relevant to understanding the malleability and transformation options of provisioning systems and practices to achieve sustainable production and consumption patterns.
Material stocks as leverage points for more sustainable resource use	Mitigating resource use and emissions requires halting the further expansion of stocks, especially in higher-income countries with already substantial accumulated material stocks. As long as material stocks are growing, more primary virgin resources are required, because recycling can only utilize end-of-life waste from stocks built years or decades earlier. Those end-of-life stocks are necessarily much smaller in a growing system. Stabilizing stocks requires radically re-designing and densifying existing settlement structures, and stopping any new construction on previously not built-up land. This would halt soil sealing and associated environmental problems, such as loss of fertile land and resilience against natural disasters.
Creating positive lock-ins	Creating positive lock-ins into highly ecoefficient service provisioning systems is a key challenge for a sustainable future. For lower-income contexts requiring better and more material stocks to achieve minimum decent living standards, they must avoid resource-intensive, car-dependent, and sprawling settlement patterns. Higher-income contexts need to rapidly transform their already-existing stocks.
Toward a Sustainable Circular Economy	A sustainable circular economy needs to narrow, slow, and close socio-economic material cycles (see **Circular Economy and Society**). This should mitigate energy use and GHG emissions, as well as other associated environmental pressures and impacts. While a perfectly circular economy is thermodynamically impossible, it can still serve as a useful benchmark to strive for. Narrowing material cycles include **sufficiency** and **ecodesign**, building standards, and infrastructure planning. Slowing cycles require lifetime extensions of existing stocks, for example via re-use, **repair**, and refurbishment. Closing cycles require improved recycling systems, as well as restorative land use practices improving soil health and biodiversity.
Stocks are necessary, but will always cause flows	Even stabilized material stocks require continuous inputs of primary materials and energy for their use, maintenance, and service provisioning. Providing for minimum universal social standards around the world also requires a certain amount of material stocks to still be built. However, there are always unavoidable losses and waste by-products during production and end-of-life waste collection, which cannot be fully recovered nor completely recycled, including quality reductions due to impurities, alloying, and complex chemical properties of products.
Criticality of materials and metalloids	Metals and metalloids are crucial for various modern technologies required for renewable energy systems and digitalization. Nowadays, more than two-thirds of all known elements are used, with little recovery nor recycling occurring. Many metals of high concern are used for highly specialized applications with no effective substitutes.

societal stocks are long-lived, they create lock-ins and path dependencies, as in the example about buildings and mobility above (see **Urban Planning and Spatial Allocation**). The composition, magnitude, and patterns of the biophysical society-nature interactions (i.e., the social metabolism) therefore determine society's environmental pressures and impacts.

Applications

Table 27.1 presents insights into topics developed from making stocks and flows explicit in sustainable consumption and production research.

Further Reading

Charpentier Poncelet, A., Helbig, C., Loubet, P., Beylot, A., Muller, S., Villeneuve, J., Laratte, B., Thorenz, A., Tuma, A., & Sonnemann, G. (2022). Losses and lifetimes of metals in the economy. *Nature Sustainability*, 5(8), 717–726. https://doi.org/10.1038/s41893-022-00895-8.

Haberl, H., Wiedenhofer, D., Pauliuk, S., Krausmann, F., Müller, D.B., & Fischer-Kowalski, M. (2019). Contributions of sociometabolic research to sustainability science. *Nature Sustainability*, 2, 173–184. https://doi.org/10.1038/s41893-019-0225-2.

Pauliuk, S. (2018). Critical appraisal of the circular economy standard BS 8001:2017 and a dashboard of quantitative system indicators for its implementation in organizations. *Resources, Conservation and Recycling*, 129, 81–92. https://doi.org/10.1016/j.resconrec.2017.10.019.

Wiedenhofer, D., Smetschka, B., Akenji, L., Jalas, M., & Haberl, H. (2018). Household time use, carbon footprints, and urban form: A review of the potential contributions of everyday living to the 1.5 °C climate target. *Current Opinion in Environmental Sustainability, Environmental Change Assessment*, 30, 7–17. https://doi.org/10.1016/j.cosust.2018.02.007.

Wiedenhofer, D., Streeck, J., Wiese, F., Verdolini, E., Mastrucci, A., Ju, Y., Boza-Kiss, B., Min, J., Norman, J., Wieland, H., Bento, N., Godoy León, M.F., Magalar, L., Mayer, A., Gingrich, S., Hayashi, A., Jupesta, J., Ünlü, G., Niamir, L., Cao, T., Sugiyama, M., & Wilson, C. (2024). Industry transformations for high service provisioning with lower energy and material demand: A review of models and scenarios. *Annual Review of Environment and Resources*, 49. https://doi.org/10.1146/annurev-environ-110822-044428.

28 Food Miles

Kristof Rubens and Tessa Avermaete

Definition

The term "food miles" or "food kilometers" refers to the distance between the place where food is produced and the place where it is consumed. Food miles are measured in tonne-kilometers, representing the transport of one tonne of goods by a given transport mode (e.g., road, rail, air, sea, inland waterways, pipeline) over a distance of one kilometer. It is a proxy indicator for the environmental impact of transporting food from farm to fork. Greenhouse gas (GHG) emissions result from the dependence on fossil fuels for much of this transportation.

History

The concept of "food miles" came to the fore in 1993 when a report was published in the United Kingdom called *The Food Miles Report – The dangers of long-distance food transport* (Paxton, 1994). Food miles emerged to help consumers make more environmentally conscious food purchase decisions to overcome the information deficit hypothesis (see **Behavior Change**). It is linked with the promotion of local food by some governments, environmental groups, and agricultural and other sector organizations. The low food miles of local food are often touted as an environmental advantage (see Box 28.1). Food miles are also used in education due to its didactical properties (see **Education for Sustainable Consumption**). It is an easy-to-use and seemingly straightforward concept and, as such, has gained much media attention.

Box 28.1   The campaign "Local food . . . is miles better"

In 2006, the British agricultural business magazine *Farmers Weekly* started a campaign titled "Local food is miles better". The campaign put forward seven reasons for reducing food miles. The first reason was that "food miles harm the environment". In addition, arguments regarding freshness, security, seasonality, quality standards, transport costs, and impact on Global South countries were also mentioned.

They called on people to sign a petition urging supermarkets to stock, promote, and label locally produced food to reduce food miles and support local producers. They also ran a competition for schools, challenging 7- to 11-year-olds to design low food miles lunch boxes using seasonal and local food. David Cameron, then leader of the Conservative Party, and the secretary of state at the Department for Environment Food and Rural Affairs David Miliband (Labour), supported the campaign.

DOI: 10.4324/9781003584056-31

The concern represented by food miles is linked to the feeling of disconnectedness with where our food comes from (see **Food Sovereignty**). Distrust of the global food system and preferential feelings toward local products play a role. These feelings were fueled during the COVID-19 pandemic. Through the heuristic of "avoiding food miles", the consumer gets a sense of gaining control over their food provisioning. Feeling more connected is also part of related initiatives and activities like the Slow Food movement and locavorism (see also **Community Supported Agriculture**).

Different Perspectives

The concept of food miles is presented as a measure of the environmental impact of transporting the food we consume. In addition, GHG emissions are the commonly used metric due to their dependence on fossil fuels (see **Consumption-Based Accounting**). However, many other factors contribute to food's environmental impact and GHG emissions (see **Protein Shift**). Before reaching our plates, our food is produced (rearing and growing), stored, processed, packaged, transported, prepared, and served.

Although the concept of food miles was initially embraced as an instrument to inform consumers about the environmental impact of food, scientists have criticized the concept's very narrow approach to such a complex issue. Today, LCA (Life Cycle Assessment) is accepted as a more robust method for estimating the environmental impact of food.

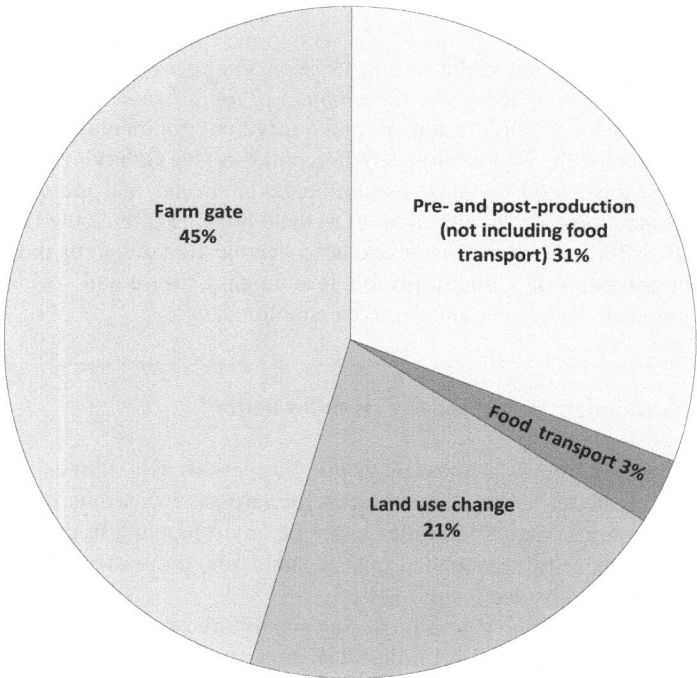

Figure 28.1 Transport is only a small part of the overall picture of agricultural emissions

Source: Based on Sutton et al. (2024)

Food transport poses an ecological burden but this is not as great as sometimes assumed. Research shows that the share of food transportation in the total GHG emissions from food is relatively low on average (3.1%) (see Figure 28.1). There is, however, a marked difference in the share of food transport between high-income (7.1%), middle-income (3.2%), and low-income countries (0.4%).

The exact GHG emissions of food transportation depend largely on the mode of transport. Air transportation has a much higher environmental impact than other modes of transport (Figure 28.2). Although food transported by air supports the use of food miles as an environmental criterion, it accounts for only 0.16% of total food miles (Figure 28.3).

In many cases, importing from regions that are better suited to cultivating particular products (more efficiently for reasons of seasonable production), as opposed to local production in greenhouses (which could require large amounts of fossil fuels), would offer greater environmental benefits. Although "fewer food miles" present an environmental benefit, without proper context, it can result in an incorrect prioritization of consumer actions (see Box 28.2). This has led commentators to suggest "first focus on what you eat (e.g., more plant-based, less food waste), not where it comes from or how it is produced" (Richie, 2020).

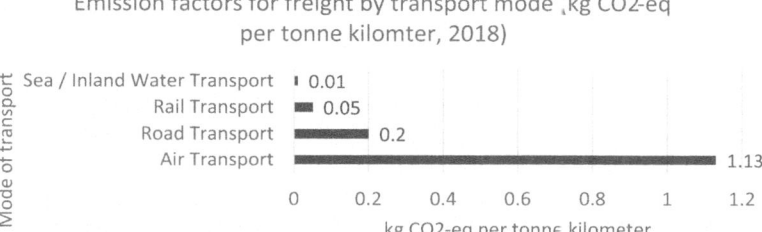

Figure 28.2 Air transport is the most emissions-intensive transport mode for food

Source: Richie (2020) based on Poore, J., & Nemecek, T. (2018). Reducing food's environmental impacts through producers and consumers. *Science*, 360(6392), 987. https://science.sciencemag.org/content/360/6392/987

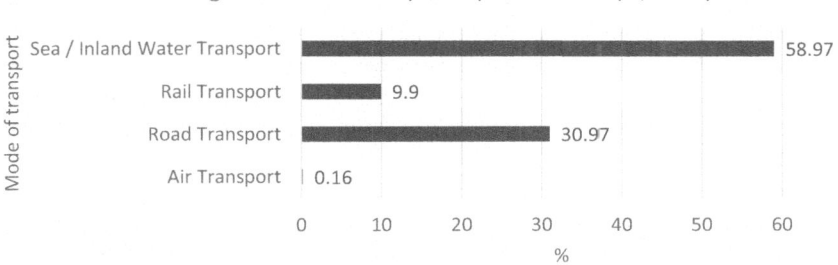

Figure 28.3 Air transport represents a relatively small portion of global food miles

Source: Richie (2020) based on Poore, J., & Nemecek, T. (2018). Reducing food's environmental impacts through producers and consumers. *Science*, 360(6392), 987. https://science.sciencemag.org/content/360/6392/987

Box 28.2   Consumers' perspectives on "food miles" and "buying local"

Although other mitigation options have greater potential to reduce environmental impact, "reducing food miles" or "buying more locally sourced food" are at the top of the list for consumers. When EU consumers were asked in 2019 what actions they have taken in the past six months to protect the environment, almost half (42%) of the respondents "bought local products", the third most popular option. When asked which is the most important characteristic of sustainable food, almost a quarter of the respondents answered "local and short supply chains". Eating locally produced food is also associated with a healthier diet, less packaging, sustainable agriculture, less food waste, and support of the local economy. These associations can be true and explained by the "home country bias" but are not necessarily so in practice.

Food miles are relevant in light of other sustainability aspects. Reducing food miles often involves buying more locally sourced products and supporting local producers and their livelihoods. This may require paying a premium price. The benefits will likely be larger if coupled with shorter supply chains, which could mean more economic power for a smaller number of participants in the food chain. From a consumer point of view, this can also result in a stronger sense of connection and appreciation for these local producers.

Inevitably, some food products must be imported because they cannot be cultivated in the region of consumption. In some parts of the world, production possibilities are limited. Researchers also highlight the importance of food trade for global food security. International trade is considered a key component of climate change adaptation. An interconnected food system, which connects local, regional, and global food systems, may be preferable in terms of food security and resilience.

Applications

Numerous (online) self-help guides promote food miles as a selection criterion to lower the environmental impact of food purchases. Periodically, media outlets emphasize this message in lifestyle articles. In reports or online resources, you can find examples of product food miles being compared.

When communicating a product's environmental performance, transport is included in calculating an ecoscore or level of GHG emissions. However, there are few examples of supermarkets directly informing consumers about food miles. While some attempts at environmental labeling highlight food miles, these efforts have resulted in prototype concepts rather than widespread adoption (see **Ecolabeling**).

The label "local" is often used in general communication and retail but is not always substantiated. Moreover, "local" can have shifting meanings depending on how far away a product may be produced and the consumers' perception. Checking the country of origin is put forward as a tool to estimate food miles.

Some examples indicate the mode of transport in food retailing. Products transported by air can carry the label "by air", sometimes in combination with a "CO_2" or "carbon" notation or red color. However, research has shown that this can backfire, as it might be perceived as "fresh", "high quality", or "high end" and thus be more appealing to consumers.

Better examples can be found in, among others, the Netherlands, Sweden, Switzerland, the United Kingdom, and Ireland where in the early 2020s several supermarkets and food service companies announced they would stop importing food products by air. This can be considered an action to reduce food miles while also delivering great environmental benefits.

Although food miles as an environmental sustainability concept is contested, it is still being used in communication by industry, NGOs, and popular media when discussing sustainable food consumption. Given the limited cognitive space available for the consumer to consider factors when making food decisions (see **Choice Paralysis**), messages on food miles can add to confusion and distortion around environmentally motivated priorities when purchasing food.

Further Reading

Janssens, C., Havlík, P., Krisztin, T., Baker, J., Frank, S., Hasegawa, T., Leclère, D., Ohrel, S., Ragnauth, S., Schmid, E., Valin, H., Van Lipzig, N., & Maertens, M. (2021). International trade is a key component of climate change adaptation. *Nature Climate Change*, 11, 915–916. https://doi.org/10.1038/s41558-021-01201-8.

Paxton, A. (1994). *The food miles report – the dangers of long distance food transport.* London: Sustainable Agriculture, Food and Environment (SAFE) Alliance.

Richie, H. (2020). You want to reduce the carbon footprint of your food? Focus on what you eat, not whether your food is local. *Our World in Data.* Available at: https://ourworldindata.org/food-choice-vs-eating-local (accessed: 14 May 2024).

Stein, A.J., Santini, F. (2021). The sustainability of "local" food: A review for policy-makers. *Review of Agricultural, Food and Environmental Studies*, 103, 77–89. https://doi.org/10.1007/s41130-021-00148-w.

Sutton, W.R., Lotsch, A., Prasann, A. (2024). Recipe for a liveable planet – Achieving net zero emissions in the agrifood system. In *Agriculture and food series*. Washington, DC: World Bank Group. Available at: http://hdl.handle.net/10986/41468 (accessed: 1 January 2025).

29 Sufficiency

Jihoon Min and Caroline Zimm

Definition

The concept of sufficiency in the context of consumption and lifestyles is based on having and doing "enough" for a good life, for what matters. Here, "enough" is both the opposite of "excessive" or "wasteful" – creating a sense of a ceiling – and a synonym for "adequate" or "appropriate", creating a floor. As consumption choices are connected to diverse effects related to natural (i.e., energy, material, water, land) and human resources, they collectively exceed planetary boundaries, endangering livelihoods and the earth's sustainability. Thus, sufficiency specifically asks for a reduction in wasteful consumption through lifestyle and behavioral changes, driven by both individual actions and institutional changes (see **Behavior Change**, **Choice Editing**, and **Product-Service Systems**). At the same time, many people worldwide still lack adequate access to services and goods for their livelihoods.

Along these lines, Chapter 9 of the Sixth Assessment Report of the Intergovernmental Panel on Climate Change (IPCC) summarizes sufficiency as *"avoiding the demand for materials, energy, land, water, and other natural resources while delivering a decent living standard for all within the planetary boundaries"*. It thus aligns with the concepts of **doughnut economics** or safe and just **consumption corridors**. This "avoiding" should not imply sacrifice but describes a new vision of a good life. While sufficiency aims at a reduction of energy and use of materials, it does not necessarily equate to austerity and is at its core still connected to a better quality of life and well-being (see **Well-being Economy**).

Sufficiency is different from the other two guiding strategies for sustainability: Efficiency and Consistency. Consistency aims at adopting processes and technologies in line with natural processes (e.g., "cradle-to-cradle" design) (see **Circular Economy and Society**). Efficiency aims at increasing output per input. Therefore, while these two focus on the characteristics of technologies and engineering systems, sufficiency is a normative concept asking us to reduce the impacts of our economic activities through lifestyle and behavior changes and secure "sufficient" human well-being for all.

History

Key elements of sufficiency – satisfaction through moderation, harmony with surroundings – have been the base principles of diverse communities across the globe for a long time, as reflected in concepts like **Ubuntu, Buen Vivir**, or ān fèn zhī zú (安分知足). Thailand has put forward the "sufficiency economy" as a driving principle for the country's development since the 1997 Asian financial crisis, with sustainability based on moderation at its core.

In the West, since the 1960s, many scholars like Kenneth Boulding, Ernst F. Schumacher, Herman Daly, and Wolfgang Sachs have suggested broader ideas of sufficiency as an ethical

DOI: 10.4324/9781003584056-32

responsibility for the planet. By noting that the incremental and relative (input to output) nature of efficiency is bound to be "incompatible with nature" as consumption increases, Sachs proposed sufficiency ("Suffizienzrevolution") as an alternative principle to efficiency (see **Steady-State Economy**). Thomas Princen called for a normative sufficiency approach to integrate sustainability concerns into our economic activities, aiming to remedy the environmental impacts that threaten the regenerative capacity of ecological systems and, consequently, human well-being.

Recently, the topic has gained momentum in academia, with more quantitative insights on planetary boundaries and minimum well-being thresholds. Following these insights, sufficiency has entered policy discussions, so far mainly in the European Union.

Different Perspectives

Sufficiency is about recognizing the essential needs of people and acknowledging how they differ based on geographies or socioeconomic contexts (see **Fair Consumption Space**). It encompasses the sufficientarian view in ethics that everyone should be above certain levels of services to lead a dignified life, while setting aside the complex question of how to define the "certain minimum level".

So far, the sufficiency literature has concentrated more on the context of high-income populations, with their affluence and structural/cultural wastefulness, and thereby on absolute material consumption reduction. Low-income countries might therefore see sufficiency as prescriptive and neglectful of the development needs and aspirations of poorer societies (a form of "bullying", according to Monyei et al. (2019)).

Sufficiency at an individual level is often understood as the need to "restrain" consumption levels, either voluntarily or following regulation. It is frequently linked to concepts about lifestyle choices such as minimalism, **voluntary simplicity**, or **alternative hedonism**. According to Riefler et al. (2024), individuals associate voluntary consumption reductions with both losses and gains on a personal, social, and universal level, which are relevant to consider in the design and promotion of sufficiency strategies. While individuals are gradually adopting practices of sustainable consumption, the transformative impact of this might be limited if efforts on macro and meso (institutional) levels do not complement them. The focus of sufficiency at these levels is on facilitating societal norm changes and fostering collective and structural actions at the local, national, and international levels required to support beneficial sufficiency policies and provisioning systems (e.g., land-use planning, repurposing of buildings, redesigning transport systems; see **Urban Planning and Spatial Allocation**, **Social Tipping Points**, **Social Norms**, **Political Economy of Consumerism**, **Foundational Economics**).

The circular economy also aligns with sufficiency, by reducing the resource throughput of a society using strategies to minimize the use of new resources and reintegrate materials in production and consumption systems, including, for example, extending product lifetimes and fostering reuse and repair (see **Extended Producer Responsibility, Product Returns and Right of Withdrawal, Repair**). Similarly, sufficiency is also reflected in the "avoid" and "shift" components of *avoid-shift-improve* strategies that aim at reducing environmental impacts.

This concept is closely related to (and can be seen as an element in implementing) **degrowth**, which has been coined in high-income countries as the antithesis of the macroeconomic growth paradigm. Degrowth advocates for a shift away from GDP growth as a primary policy objective, aiming for a transformation of the entire economic system toward steady-state or contracting economies. This transformation consequently has implications beyond the production and consumption system. Sufficiency discussions must thus be extended to topics such as implications for labor and financial markets and business models (see **The Role of Business**). The contradictions between sufficiency and capitalism and growth present a key challenge in efforts toward sustainable consumption and lifestyles.

Application

Sufficiency, often through more structural changes, ultimately provides an opportunity to avoid or reduce resource-intensive behavioral, institutional, and infrastructural lock-ins, resulting from, for example, technological and financial path dependencies. Short-term policy measures have often focused on quick non-structural interventions implemented in the wake of energy crises (e.g., the energy sobriety plan in France or energy conservation plans in European and Asian countries, focusing on turning off lights, putting limits on thermostat settings, lowering car usage). However, policies involving more systemic changes are mostly absent.

Recognizing this, a *Sufficiency Manifesto* was recently signed by many European organizations, calling for more fundamental shifts toward sufficiency in the EU's policy agenda. Sufficiency policy options widely discussed include:

1) reducing individual car use or air travel through improved public transport and spatial planning, the latter especially in cities (see **Urban Planning and Spatial Allocation**, **Choice Editing**);
2) introducing circular economy measures (lowering resource intensity and reducing waste) during the lifecycle of goods and services (e.g., right to repair, industry compatibility standards);
3) optimizing existing building uses;
4) financial interventions such as taxes, subsidies, or tariff systems to discourage excessive activities in resource-intensive consumption/production (e.g., progressive taxes); see **Sustainable Finance, Money**.
5) regulations prohibiting or rationing excessive activities (e.g., rationing of flying, ban on short-haul flights, ban on single-use products, advertising bans) (see **Personal Carbon Allowance**).

One of the trickiest and most sensitive debates in discussions of sufficiency is how to operationalize and ultimately achieve what it intends: How will consumption be distributed within or across countries and time, accounting for technological change, human needs, and place- and time-specific conditions? Who can decide what is enough across massive heterogeneity in societies? Once such decisions have been made, who can enforce them? These questions hint toward an acceptability challenge once discussions become concrete, related to diverse questions of justice (see **Climate Justice**) and concern about restricting desired living standards and sacrificing well-being. For sufficiency to become an integral part of overall systemic change beyond transient projects and initiatives, these challenges have to be overcome.

Further Reading

Jungell-Michelsson, J., & Heikkurinen, P. (2022). Sufficiency: A systematic literature review. *Ecological Economics*, 195, 107380. https://doi.org/10.1016/j.ecolecon.2022.107380.

Mathai, M.V., Sachs, W., & Lorek, S. (2023). Editorial: From an ethic of sufficiency to its policy and practice in late capitalism. *Frontiers in Sustainability*, 4. https://doi.org/10.3389/frsus.2023.1324319.

Monyei, C.G., Jenkins, K.E.H., Monyei, C.G., Aholu, O.C., Akpeji, K.O., Oladeji, O., & Viriri, S. (2019). Response to Todd, De Groot, Mose, McCauley and Heffron's critique of "Examining energy sufficiency and energy mobility in the global south through the energy justice framework". *Energy Policy*, 133, 110917. https://doi.org/10.1016/j.enpol.2019.110917.

Princen, T. (2003). Principles for sustainability: From cooperation and efficiency to sufficiency. *Global Environmental Politics*, 3(1), 33–50. https://doi.org/10.1162/152638003763336374.

Riefler, P., Baar, C., Büttner, O.B., & Flachs, S. (2024). What to gain, what to lose? A taxonomy of individual-level gains and losses associated with consumption reduction. *Ecological Economics*, 224, 108301. https://doi.org/10.1016/j.ecolecon.2024.108301.

30 Consumption Corridors

Antonietta Di Giulio, Rico Defila, Doris Fuchs,
Sylvia Lorek, and Marlyne Sahakian

Definition

The concept of Consumption Corridors (CC) provides a framework for sustainable consumption governance. It suggests achieving sustainability by developing and implementing corridors of consumption, defined by consumption minima and maxima. The lower boundaries are meant to allow every individual to satisfy their needs, and thus to live a life they value by determining what every individual must have access to. The upper boundaries are meant to prevent the consumption by individuals or groups from inhibiting or affecting the well-being of other individuals, living now or in the future. To that end, they determine thresholds that if (quantitatively or qualitatively) trespassed adversely impact the quality of life of other individuals by putting others' minima at risk. The space between these boundaries is what is referred to as a consumption corridor (Figure 30.1). This space leaves room for individual life plans and choices, for individual freedom (see **Freedom of Choice**). The concept of CC posits quality of life and justice as criteria to define minima and maxima of consumption. Both the lower and the upper limits refer to satisfiers (see Box 30.1).

The concept of CC offers a new way of organizing societies and their production and consumption systems, to ensure a good life for all people at present and in the future. It is not a single policy measure but a framework for developing policies. While suggesting far-reaching changes and acknowledging the necessity of protecting the natural environment, it has a strong focus on well-being and justice and neither adopts a narrative of renunciation nor imposes specific lifestyles. It offers a narrative for sustainable consumption governance that explicitly focuses on achieving (the vision of) a good life (salutogenic), rather than focusing primarily on avoiding damage (pathogenic).

Box 30.1   Needs and satisfiers

Distinguishing needs from satisfiers is at the core of a scholarly discourse about quality of life that is based on the notion of needs. A well-known proponent of this difference is Manfred Max-Neef. Satisfiers are the means that are used to satisfy needs. Satisfiers are what people do and what they make use of to live a good life, and can be both material and immaterial (i.e., legislation). Products, services, practices, and infrastructures are satisfiers. While needs are assumed to be universal and stable, satisfiers are not; they are assumed to change over the course of time but also depend on historical and societal context.

While a considerable part of the public discussion on human well-being revolves around satisfiers and confuses them with needs, the concept of CC draws attention to clearly distinguishing needs from satisfiers. It seeks to change and replace satisfiers that lead to unsustainable patterns of consumption. It also suggests initiating a societal debate about satisfiers and about how societies organize to satisfy people's needs.

DOI: 10.4324/9781003584056-33

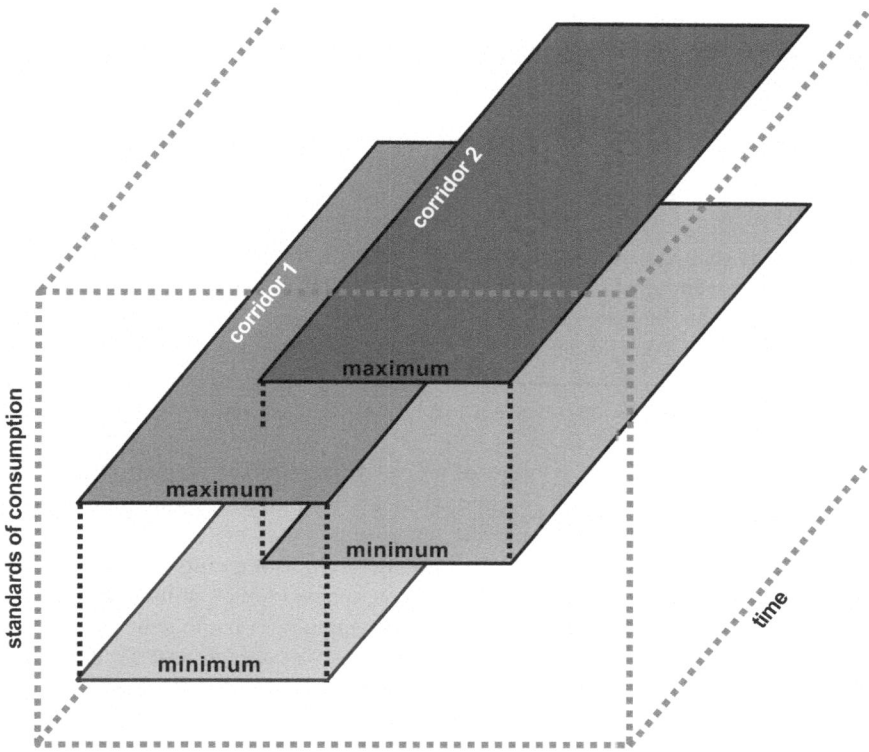

Figure 30.1 The concept of Consumption Corridors suggests defining minima and maxima of consumption that are both justified by quality of life and justice. The corridors have to be adjusted over time

Source: Di Giulio, A., & Fuchs, D. (2014). Sustainable consumption corridors: Concept, objections, and responses. *GAIA – Ecological Perspectives for Science and Society*, 23(S1), 184–192. https://doi.org/10.14512/gaia.23.S1.6

History

The concept of CC emerged as one result of an inter- and transdisciplinary research program (2008–2013) funded by the German Federal Ministry of Education and Research (BMBF). Some 150 academic and experience-based experts, covering more than 15 scientific disciplines and a broad diversity of areas of practice, investigated different aspects of sustainable consumption and engaged in integrating their results. One of the integrated results were eight recommendations ("Konsum-Botschaften": consumption messages), each presenting a pathway toward sustainable consumption. The drafts of these messages were discussed and validated with 70 representatives from the arenas of politics, administration, the economy, and civil society organizations. In one of these messages, the concept of CC is introduced ("Korridor-Botschaft", https://doi.org/10.48350/49106).

Since its launch in 2013 in Germany, the concept of CC has been introduced to a broad international audience and gained growing attention. It is used as guidance for exploring a broad spectrum of concrete fields of consumption, such as land use, urban mobility, park spaces, fashion, laundry, and meat consumption (see, for instance, the examples in Collection SSPP [2021]). Using the case of Switzerland, empirical work has been conducted to show how the concept would be received in one society (Defila & Di Giulio, 2020). On a regional level, the concept of CC has recently been taken up by the European Environment Agency (EEA, 2023). With a view to going beyond single fields of consumption, it has been discussed as a new paradigm to approach **sufficiency** (Lombardi & Cembalo, 2022) and as a promising concept to navigate the global land squeeze (Erb et al., 2024).

Different Perspectives

The concept of CC is related to sufficiency, and aligned with the principles of a **well-being economy**, as well as the concept of limitarianism developed by Ingrid Robeyns.

There are also concepts that the concept of CC is sometimes mixed up with. One of these is **Doughnut Economics** (DE) developed by Kate Raworth, which draws on the concept of Planetary Boundaries (PB), promoted by Johan Rockström and colleagues. Another more recent one is **Fair Consumption Space** (FCS). DE operates with inner and outer boundaries, FCS with underconsumption and overconsumption.

In contrast to these concepts, the concept of CC adopts a salutogenic approach and is deeply informed by quality of life and justice. Although acknowledging the vulnerability of (natural and societal) resources, both the lower and the upper boundaries of CC are justified by human well-being and justice. The concept of CC integrates the notions of quality of life, individual freedom, vulnerability of resources, and justice with a view to developing public policies.

Both DE and PB primarily aim at averting damage, pursuing a pathogenic approach. Neither DE nor PB ingrain the notions of a good life and justice as overarching criteria, they do not draw on the needs-satisfier distinction, and they do not explicitly deal with the question of individual freedom and diversity. According to PB, the destabilization of the natural conditions that mark the Holocene epoch must be averted. In DE, PB are used to define the "ecological ceiling", and the purpose of this ceiling is averting "planetary degradation". Similarly, FCS defines overconsumption as consumption that harms planetary systems. The other damage that has to be averted, according to DE, is "critical human deprivation", and this leads to the doughnut's inner boundary. To define this boundary, it suggests drawing on the social priorities defined by the Sustainable Development Goals.

Another difference concerns the question of how to determine the limits. While in DE the boundaries are defined by either natural scientists (outer boundaries) or by (international) political bodies (inner boundaries), the concept of CC posits that the minima and maxima of consumption have to be societally negotiated by adopting a participative approach. Determining them has to be based on an inter- and transdisciplinary collaboration that integrates a broad diversity of academic and non-academic perspectives and knowledge systems.

Application

The concept of CC challenges satisfiers, it does not challenge human needs. In addition to being a framework for governance, it is an invitation to reflect upon how quality of life is achieved and discussed in societies (see Box 30.2). It emphasizes the importance of the social contract within and across nations, and it posits that not damaging others is not enough. It points out the necessity for limits while recognizing that there will always be some inequality and diversity in how needs are satisfied in a given society. The approach thus acknowledges that there is no absolute level of sustainable consumption, but always a relative one.

The concept of CC offers a promising way forward that resonates with a broad diversity of people, but with a view to its implementation, some crucial questions remain to be explored. One

Box 30.2   The Theory of Protected Needs (PN) – a concept of quality of life that is a suitable fundament for defining consumption corridors

The concept of CC presupposes, first of all, that a universal definition of human well-being is possible and that it should not be reduced to basic needs. Such a definition must provide solid ground to proceed from in determining the satisfiers (and resources) that every individual must be provided with (minima) and that must not be put at risk (maxima). The Theory of PN (Di Giulio & Defila, 2020) provides a theory of well-being that serves this purpose.

The Theory of PN provides a list of universal (and not hierarchical) needs. These needs are called protected because they claim to be (1) needs that deserve special protection within and across societies since they are crucial to human well-being, and they claim to be, at the same time, (2) needs for which special societal protection is possible since they are needs that a government or community can reasonably be made responsible for. The Theory of PN proposes nine universal needs. Each is specified by a thick description. The needs denote what individuals must be allowed to want (Figure 30.2), and the thick descriptions describe the possibilities individuals should be provided with. Concurring with a needs approach, the nine needs are ends in themselves. That is, they cannot be further reduced and they are non-substitutable. The nine needs are context-sensitive despite being universal: the thick descriptions serve as a starting point for their cultural and historical adaptation. Empirical evidence shows that these needs resonate in different cultural contexts.

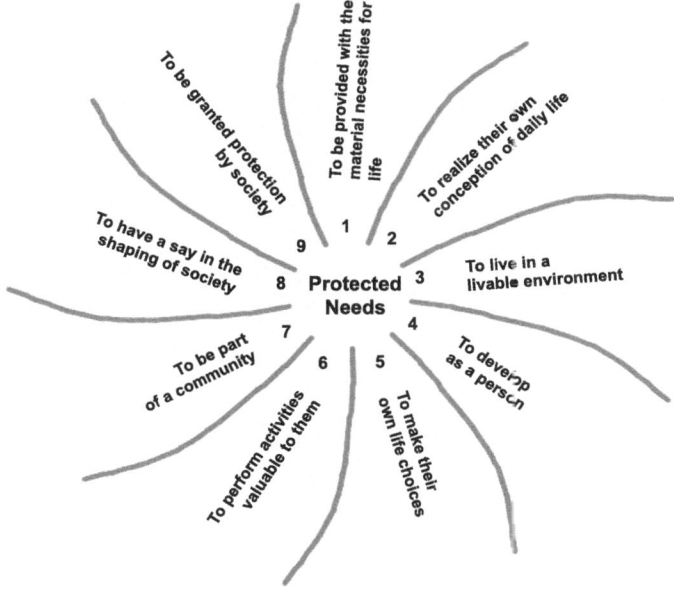

Figure 30.2 Protected Needs: what individuals must be allowed to want. The Theory of Protected Needs provides a comprehensive, thick, and salutogenic definition of quality of life

Source: Di Giulio, A., & Defila, R. (2020). The 'good life' and Protected Needs. In A. Kalfagianni, D. Fuchs, & A. Hayden (Eds.), *The Routledge handbook of global sustainability governance*, pp. 100–114. London. https://doi.org/10.4324/9781315170237-9

is the question of how to bridge the divide between a comprehensive approach to quality of life, on the one hand, and approaches that focus on the natural environment, on the other, in ways that provide a firm foundation on which to build when determining minima and maxima of consumption. Another, addressed by a number of recent contributions to the CC literature, is the question of how to initiate such a deep transformation and where to start.

Further Reading

Collection SSPP. (2021). Consumption corridors: Advancing the concept and exploring its implications. In M. Sahakian et al. (Eds.), *Sustainability: Science, practice and policy (SSPP)*. Available at: https://www.tandfonline.com/journals/tsus20/collections/consumption-corridors (accessed: 8 January 2025).

Defila, R., & Di Giulio, A. (2020). The concept of "consumption corridors" meets society – How an idea for fundamental changes in consumption is received. *Journal of Consumer Policy*, 43, 315–344. https://doi.org/10.1007/s10603-019-09437-w.

EEA (European Environment Agency). (2023). *Conditions and pathways for sustainable and circular consumption in Europe*. Briefing No. 11/2023. https://doi.org/10.2800/137584.

Erb, K.-H., Matej, S., Haberl, H., & Gingrich, S. (2024). Sustainable land systems in the Anthropocene: Navigating the global land squeeze. *One Earth*, 7(7), 1170–1186. https://doi.org/10.1016/j.oneear.2024.06.011.

Lombardi, A., & Cembalo, L. (2022). Consumption corridors as a new paradigm of sustainability. *Resources, Conservation & Recycling*, 184, 106423. https://doi.org/10.1016/j.resconrec.2022.106423.

31 Fair Consumption Space

Thomas S.J. Smith and Lewis Akenji

Definition

A fair consumption space has been defined as "an ecologically healthy perimeter that supports within it an equitable distribution of resources and opportunities for individuals and societies to fulfill their needs and achieve wellbeing" (Akenji et al., 2021: 13). Recognizing that there are power imbalances in society, especially between rich and poor, and growing competition over resources to meet people's needs, the approach suggests parameters for ensuring well-being for everyone within ecological limits. Given that household consumption currently drives about two-thirds of greenhouse gas emissions, high importance has been placed on identifying low-carbon lifestyles within an ecologically safe band or area, between an environmental ceiling (or maxima) and a social floor (or minima) (see Figure 31.1). This framework serves to focus on planning and practical decision-making regarding the range of consumption choices that exist across key lifestyle domains such as housing, transport, services, food, leisure, and consumer goods (see Box 31.1).

The approach is based on three key principles:

1. *Limits*: Referring to the need to stay within ecological boundaries and carrying capacity, this principle is implemented, for instance, by developing a carbon or resource budget determined in dialogue with the latest environmental science (see e.g., **Personal Carbon Allowance** and **Ecosocial Contract**).
2. *Equity*: This principle ensures that access to resources and opportunities in a climate- and resource-constrained world is fair and enables dignified lives, both now and into the future. This is particularly important given that the worst impacts of ecological devastation are felt by those least responsible for it, often far from where it has been caused (see **Climate Justice**).
3. *Well-being*: This means that the transformations required to live within the fair consumption space are guided by the well-being of individuals, society, and wider ecologies, rather than other metrics such as economic growth (see e.g., **Well-being Economy**).

The fair consumption space framework is part of a group of concepts that define the upper and lower boundaries for consumption and economic activity that provide the essential quality of life for humans in the present, while not degrading it for future generations and nature. **Consumption Corridors, Doughnut Economics,** and Earth System Boundaries are other concepts in that class. The fair consumption space framework particularly ensures that currently vast inequities in environmental impact – resulting from extremes of *both* overconsumption and underconsumption – are taken into account. Consumption beyond one's fair consumption space deprives others of their own livelihoods, leading to ecological disequilibrium and social tensions. Given that the emissions of the richest 10% of people account for 36–49% of the global total, while the poorest 50%

DOI: 10.4324/9781003584056-34

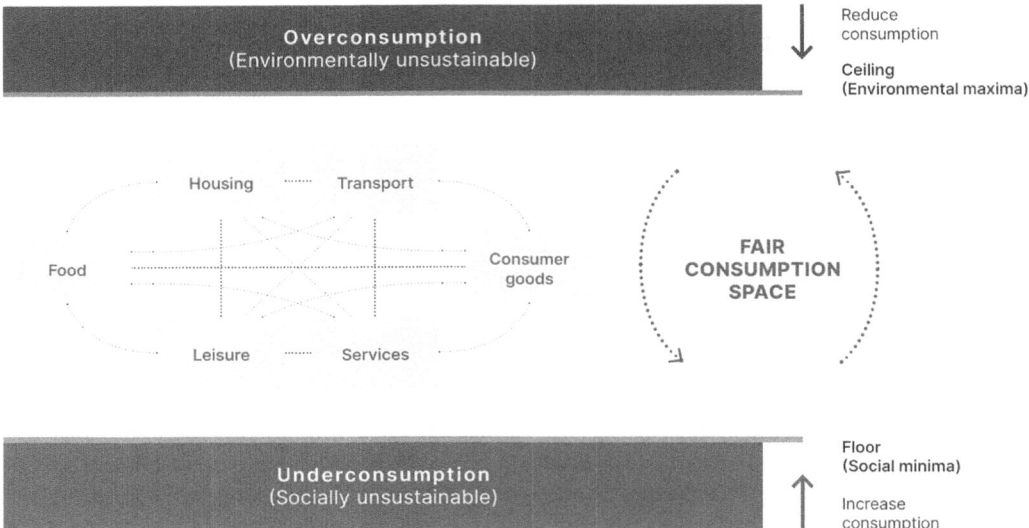

Figure 31.1 A fair consumption space for sustainable lifestyles

Source: Akenji et al. (2021)

account for just 7–15%, necessary changes in lifestyles and consumption will differ drastically between social groups (Akenji et al., 2021). There is, for instance, far more scope and necessity for change among those with high-impact lifestyles undertaking frivolous consumption (such as mega-mansions and private yachts and jets).

History

This concept emerged in the wake of decades of unsuccessful attempts to stem ecological impacts through techno-fixes, market-based approaches, and cycles of intergovernmental negotiation. The term itself was first introduced by Lewis Akenji in 2017 at an event organized by the United Nations Environment Programme (UNEP) examining how ecological and social crises can be jointly addressed. While coming into use in the early 2020s, the approach has a longer historical precedent, being adapted from the idea of "environmental space" developed by ecological economist Hans Opschoor in the late 1980s. Opschoor and colleagues focused primarily on the limits of natural resource use, describing external limits to – and the need to drastically reduce the use of – things like oil, copper, natural gas, and biomass. This work foreshadowed later definitions and ideas identifying a "safe operating space" for humanity emerging in the 2000s, such as the "planetary guardrails", "planetary boundaries", and more recently, **doughnut economics** and **consumption corridors**.

Different Perspectives

Fair consumption space takes a **political economy** approach and accounts for power disparities among different actors in society – for example, protecting the need for increased consumption by disadvantaged groups to meet fundamental needs for well-being. A clearly articulated space provides a balance in which wants should not overpower needs, economic demands stay within environmental limits, and political platforms and policies should not exacerbate social disparities and ecological deficits.

The potential of changes in lifestyle and demand has often been underappreciated or overlooked in favor of technological changes or upstream innovations in production. This bias has its roots in skepticism of individualistic consumption shifts, most famously recognized through practices such as taking shorter showers and using paper straws. The fair consumption space challenges the false dichotomy between individual **behavior change** and system change. Lifestyles are instead understood to include a complex web of topics such as care, voluntary and community work, and social and political activism (see **Social Movements, Consumer-Citizen,** or **Boycott and Buycott**). It therefore factors in broader patterns and social practices, which are often intimately related to social systems and societal provisioning (see **Social Practice Theory**).

A second site of tension is that the focus of the fair consumption space approach to date has largely foregrounded climate change, identifying lower-carbon lifestyles that keep us on track to stay below a 1.5-degree rise in global temperatures. This focus on carbon emissions alone – or what has been called "carbon reductionism" – can exclude a more holistic treatment of challenges like resource exhaustion, habitat destruction, biodiversity loss, and environmental pollution. For this reason, a more holistic environmental footprint or resource-focused approach is sometimes used.

Thirdly, the approach faces uphill struggles in implementation as it constitutes a direct challenge to business-as-usual politics and economics. By explicitly acknowledging resource limits and the need to radically transform social systems for an ecologically constrained future, the idea of a fair consumption space challenges mainstream economic assumptions of infinite economic growth through innovation and resource substitution. The inequities it addresses are likely to provoke political resistance, particularly from elites seeking to protect the most high-impact lifestyles, even if the global majority and most marginalized would instead benefit from more equitable social relations.

Application

Studies have shown vast potential for demand-side climate mitigation, reducing sectoral emissions by 40–80% (Creutzig et al., 2022). A fair consumption space allows global and national targets around more sustainable lifestyles to be formulated, and progress toward them monitored. As currently implemented, it is used to identify what have been called "**1.5-degree lifestyles**" – that is, lifestyles that are sustainable within the thresholds, which would curb drastic climate change, while providing for a good life for all.

Rather than relying on unproven technologies (such as carbon capture and storage) to avert climate change, three parallel efforts are called for in this approach: *reducing* high-impact consumption (such as phasing out fossil fuel-powered cars); *shifting* toward more sustainable options (such as getting people to walk, cycle or take public transport, rather than driving); and *improving efficiency* through technologies, where appropriate (such as shifting to electric cars where there is no alternative to individual mobility) (see Box 31.1).

Box 31.1   Applying the fair consumption space in Norway

Through close empirical analysis of lifestyles in particular countries, the fair consumption space has been applied to identify national-level scenarios and highlight particularly carbon-intensive lifestyle patterns. Bengtsson et al. (2024) apply the approach in Norway, finding that nearly two-thirds of the average lifestyle carbon footprint (LCF) is related to just two lifestyle domains: personal transport (36%) and nutrition (27%). From this, calculations are made to identify high- and medium-impact options that could tackle the most damaging lifestyle patterns.

High-impact options in this case are defined as those estimated to achieve reductions of between 500 and 1,200 kgCO2e/capita/year. In the case of Norway, these include:

- Diets with less meat, especially red meat (640–1,200 kgCO2e). This is particularly aided by adopting a completely plant-based diet, but shifting to a vegetarian diet also results in substantial reductions (see **Protein Shift**).
- Reduced international flying by traveling less or shifting to train (750–1,130 kgCO2e).
- Climate-smart personal car travel, by switching from fossil fuel-based cars to electric vehicles (560 kgCO2e).
- Reduced purchasing of new consumer goods, together with deep decarbonization of the production system (690 kgCO2e).

These figures are based on the average, and so it is crucial to also look at the particularly high-impact lifestyles undertaken by economic elites and the ultra-wealthy. It is also important to acknowledge, as the authors of Bengtsson et al. (2024) do, that the "focus on lifestyles does not imply that individuals alone can make the necessary changes, simply by shifting their preferences and habits. Our lifestyles are part of the cultures we grow up in and live in. They are to a high degree shaped by **social norms** and expectations, by the economic system and the incentives it provides, and by the technical infrastructure surrounding us."

Proponents of the fair consumption space concept have also identified potential in three policy approaches:

1. *Choice editing*: This relates to both editing out harmful consumption options (such as banning short-haul flights where trains could be taken) and editing in more sustainable alternatives (such as increased access and quality of public transport). Key questions remain here, of course, about the democratic legitimacy of choice editing criteria and who does the editing.
2. *Setting limits*: Given that offsetting and market-based approaches have failed, capping emissions will be essential to setting the environmental ceiling. This can happen through forms of rationing, though this is controversial and its implementation must be democratically legitimate (see **Freedom of Choice**).
3. *Sufficiency*: Rather than seeking unbridled economic growth and consumer lifestyles, a needs orientation is required to focus on what is really needed to live a good life within limits. One form this could take is to develop a social guarantee for **Universal Basic Services** (UBS), facilitating equitable and sufficient access to necessities like healthcare, transport, housing, childcare, and education.

Further Reading

Akenji, L., Bengtsson, M., Toivio, V., Lettenmeier, M., Fawcett, T., Parag, T., Saheb, Y., Coote, A., Spangenberg, J.H., Capstick, S., Gore, T., Coscierme, L., Wackernagel, M., & Kenner, D. (2021). *1.5-degree lifestyles: Towards a fair consumption space for all*. Berlin: Hot or Cool Institute. Available at: https://

hotorcool.org/resources/1–5-degree-lifestyles-towards-a-fair-consumption-space-for-all/   (accessed: 8 January 2025).

Bengtsson, M., Latva-Hakuni, E., Toivio, V., & Akenji, L. (2024). Towards a fair consumption space for all: Options for reducing lifestyle emissions in Norway. *The Future in Our Hands*, Oslo. Available at: https://www.framtiden.no/filer/dokumenter/Rapporter/2024/Online_Lifestyle-Emissions-in-Norway_v2-2.pdf (accessed: 8 January 2024).

Creutzig, F., Niamir, L., Bai, X., Callaghan, M., Cullen, J., Díaz-José, J., Figueroa, M., Grubler, A., Lamb, W.F., Leip, A., Masanet, E., Mata, É., Mattauch, L., Minx, J.C., Mirasgedis, S., Mulugetta, Y., Nugroho, S.B., Pathak, M., Perkins, P., & Ürge-Vorsatz, D. (2022). Demand-side solutions to climate change mitigation consistent with high levels of well-being. *Nature Climate Change*, 12(1), 36–46. https://doi.org/10.1038/s41558-021-01219-y.

Opschoor, H. (2010). Sustainable development and a dwindling carbon space. *Environmental and Resource Economics*, 45(1), 3–23. https://doi.org/10.1007/s10640-009-9332-2.

United Nations Environment Programme. (2022). *Enabling sustainable lifestyles in a climate emergency*. Nairobi: UNEP.

32 Social Tipping Points

Alessandro Tavoni and Veronica Pizziol

Definition

Social tipping points refer to critical thresholds in socioeconomic systems where small changes can trigger significant, rapid, and difficult-to-reverse shifts in collective behaviors and norms. These tipping points may occur spontaneously through self-reinforcing contagion and feedback loops, leading to the swift and widespread adoption of new behaviors and norms. Tipping points may also be triggered by committed minorities – sometimes called trendsetters, change instigators, or early adopters. Although the direct influence of these minority groups may be limited, their growing impact can inspire broader societal change, even when existing norms are deeply entrenched.

Social tipping points are increasingly important for understanding how societies experience rapid transitions. They hold promise for addressing global challenges such as climate change, where the shift might involve moving from a high-carbon status quo to a new societal state characterized by lower emissions. While social tipping points mark the transition from one social state to another, social tipping interventions can actively provoke this transition by disrupting the status quo (see Figure 32.1).

History

Concepts akin to tipping originated in the natural sciences, with early references in 19th-century chemistry and mathematics describing systemic qualitative changes. Although the term tipping point was not used, studies on the equilibrium of systems (i.e., rotating fluid masses) showed how small perturbations could lead to new equilibrium states, illustrating the idea of critical transitions in systems.

The term gained widespread use in its ecological meaning among Earth and climate scientists in the early 2000s to describe the critical thresholds at which abrupt, rapid, and non-linear changes in Earth's climate systems take place (see Box 32.1). Examples include the dieback of the Amazon rainforests and the melting of the Greenland and Antarctic ice sheets, with their tipping potentially significantly affecting the global climate, ecosystems, and human societies.

Box 32.1   Introducing tipping points as a concept

Malcolm Gladwell's book *The Tipping Point* widely popularized the concept both across the sciences and among the general public. He illustrated how small changes in policy, behavior, or social norms can trigger large-scale transformations. For instance, Gladwell described

---

DOI: 10.4324/9781003584056-35

how the sudden rise in popularity of Hush Puppies shoes in the 1990s was driven by a small group of trendsetters in New York City who began wearing them, ultimately leading to a broader cultural phenomenon. He also discussed the steep drop in New York City's crime rate after 1990, attributing it to a policing strategy that cracked down on minor offenses, dramatically shifting the public perception of safety and encouraging law enforcement and residents to change their behaviors, in a cycle of increased trust and cooperation.

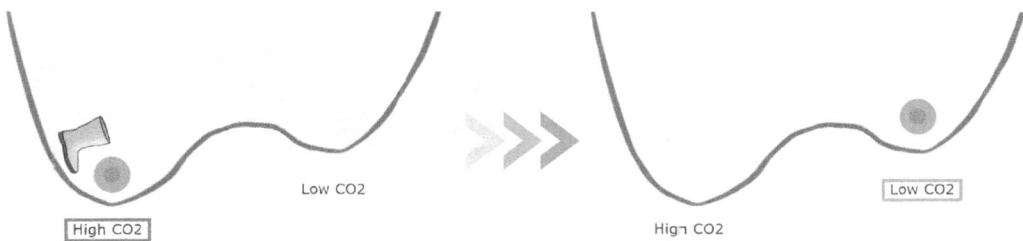

Figure 32.1 An example of a social tipping intervention (the "kick") disrupting the existing status quo (High CO_2) and guiding the system toward a new stable equilibrium with low CO_2 levels

Source: Adapted from Pizziol and Tavoni (2024)

See Figure 32.2 for an illustration of the concept of tipping points across the natural and social sciences.

The notion of social tipping has become increasingly relevant in recent years, especially in the context of climate change mitigation. Researchers have explored how interventions designed to shift **social norms** and behaviors could accelerate the transition to a low-carbon economy. For instance, recent research looks at contagion in the household adoption of rooftop solar panels, electric vehicles, and other renewable energy technologies. Another often-cited example of social contagion is the rise of the *Fridays for Future* campaign from a protest with a sole participant, Greta Thunberg, to a **social movement** of six million people in one year, from August 2018. Some research points to the movement's role in triggering rapid and widespread **behavioral change** to lower one's carbon footprint among citizens worldwide.

The term tipping point has also gained attention in media and public discourse, where journalists and commentators use it to describe various situations, such as the fight against the COVID-19 pandemic and critical moments in political movements like Black Lives Matter.

Different Perspectives

Due to its recent emergence and surge of interest across disciplines, the concept of social tipping has yet to acquire a commonly agreed-upon meaning. As a result, it suffers from frequent misuse, possibly due to misunderstandings about key concepts such as emergence, irreversibility, and non-linearity, which may undermine its applicability.

A further complication has to do with the fact that different scholars attach either a positive or normative view to social tipping points. While some focus on identifying the occurrence of social tipping points, others aim to find ways to trigger cascading change through interventions (see

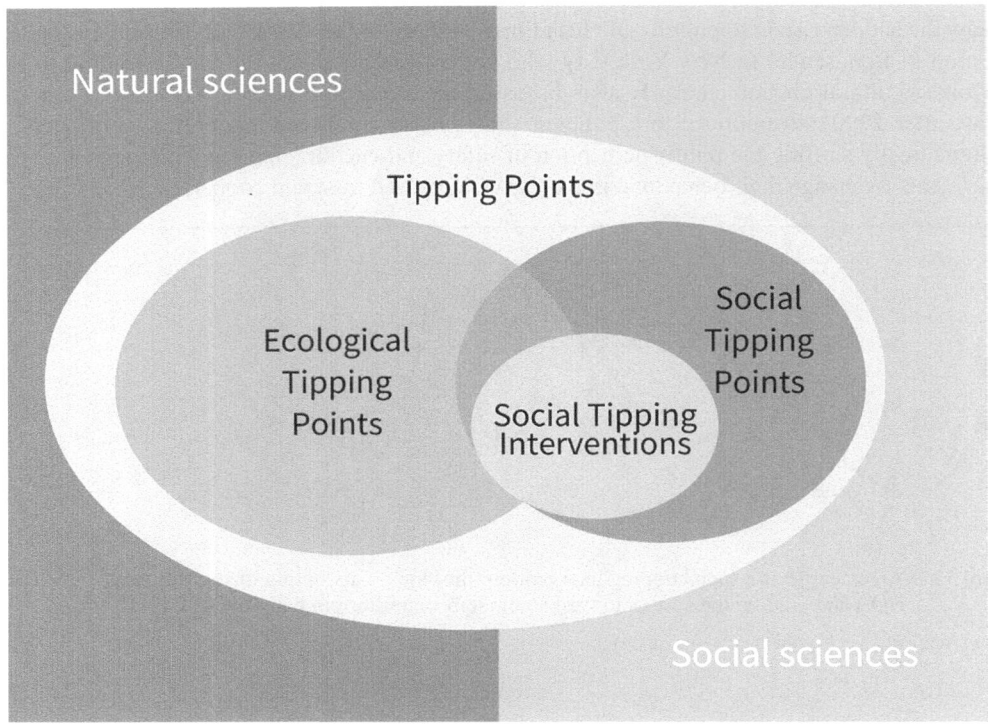

Figure 32.2 Overview of tipping point concepts across natural and social science

Source: Pizziol and Tavoni (2024)

Living Labs). However, the cost-effectiveness of such interventions remains an open question, as efforts to trigger social tipping may require significant financial resources as well as overcoming social and political resistance. Besides the confusion arising from the different meanings attached to the concept, it is worth stressing that the underlying social systems are inherently complex, making it hard to predict when a threshold has been reached, let alone intentionally design interventions to achieve it. This is because, unlike natural systems, where tipping points are tied to more measurable factors, social systems are influenced by a mix of cultural values, political institutions, environmental responses, and individual behaviors. Disentangling every parameter of the system remains a challenge for science and policy.

A related critical question regards thresholds for the proportion of the population that needs to adopt the new norms and behaviors to reach a tipping point for social change. Depending on the situation, research suggests that the share ranges between 10% and 40% of the population. These early adopters of new behaviors, or committed minorities taking the lead in abandoning entrenched norms, play a crucial role by demanding change and pushing for alternative lifestyles. The rest of the population can be represented as followers or conformists with little intrinsic motivation to abandon the old ways.

However, in light of the important role of **social norms** in steering societies toward one of multiple possible equilibria, the literature has yet to shed light on what a likely outcome is when conflicting minorities push in different directions. Put differently, possible outcomes of the interaction of different groups with differing views and objectives with respect to change include not only the status quo and positive tipping to a preferable equilibrium (e.g., widespread decarbonization) but also negative tipping toward a socially undesirable one (e.g., an even higher-carbon pathway).

This is particularly relevant for problems such as tackling climate change, whose solutions entail reconfiguring economic systems toward renewables and away from fossil fuels. Lobby groups coalescing around concentrated special interests, for example, preserving the reliance on oil and other hydrocarbons to power our economies, will inevitably push to preserve the status quo (see **Political Economy of Consumerism**). One such tool is the spread of misinformation campaigns on social networks. Such efforts can be thought of as reinforcing the status quo (deepening the left valley in Figure 32.1) and increasing the barrier for tipping.

Relatedly, another impediment to social tipping is the presence of contrasting social identities in polarized groups. The case of the lower willingness to act to mitigate climate change between Republicans and Democrats in the United States, as a result of the topic's politicization, is a potential barrier to change.

Applications

The spread and pace of social change have also been linked to the topology of the network characterizing the social interactions. Loosely speaking, individuals' connectedness and position in the network, as well as the content of what is being transmitted (e.g., simple information vs. complex **behavioral change**), are determinants of whether social contagion takes place. A recent publication points to a key role played by the periphery of the network in kicking off **social movements** and even revolutions.

Examples of rather abrupt social tipping include support for same-sex marriage in the United States, smoking cessation, and abandonment of formerly normative practices such as female foot binding and female genital mutilation. In the domain of sustainability, recent evidence of tipping pertains to the adoption of rooftop solar in US neighborhoods and the transition from cars to bikes in Copenhagen. The perception of change is, to some extent, influenced by the timescale of observation; whether a shift appears gradual or abrupt often depends on the level of zoom in one's perspective.

In other domains, consensus may be already shifting, but triggering a tipping point often requires external interventions to accelerate and amplify these changes across social networks. Looking ahead, there are promising areas, such as environmentally friendly food labeling (see **Ecolabeling**), which could contribute to shifting consumer behavior toward products with a lower environmental impact, especially if coupled with carefully designed incentives. As demand for low-carbon food grows, companies would likely compete to improve their sustainability practices, creating a positive feedback loop, possibly culminating in a state where sustainable food choices are the norm. Similarly, a combination of subsidies, preferential access to lanes, charging infrastructure, and information provision may dramatically speed up the adoption of electric vehicles, even in the absence of a ban on combustion engines.

While some highlight the relatively higher acceptability of social tipping interventions compared to traditional policy instruments like taxes, their cost-effectiveness remains uncertain – an important open empirical question, as is the case with many policy instruments (see **Co-Benefits of Climate Policy**). Although all policy approaches can face challenges such as unintended consequences, scaling difficulties, and governance complexities, social tipping interventions may encounter unique obstacles, including resistance to change, implementation hurdles, and the speed required to achieve meaningful change.

Further Reading

Gladwell, M. (2000). *The tipping point*. New York: Little, Brown.

Lenton, T.M., Mckay, D.I.A., Loriani, S., Abrams, J.F., Lade, S.J., Donges, J.F., Buxton, J.E., Milkoreit, M., Powell, T., Smith, S.R., & Constantino, S. (2023). *The global tipping points report 2023*. University of Exeter. Available at: https://pure.iiasa.ac.at/19228.

Otto, I.M., Donges, J.F., Cremades, R., Bhowmik, A., Hewitt, R.J., Lucht, W., Rockström, J., Allerberger, F., McCaffrey, M., Doe, S.S., & Lenferna, A. (2020). Social tipping dynamics for stabilizing Earth's climate by 2050. *Proceedings of the National Academy of Sciences*, 117(5), 2354–2365. https://doi.org/10.1073/pnas.1900577117.

Pizziol, V., & Tavoni, A. (2024). From niches to norms: The promise of social tipping interventions to scale climate action. *npj Climate Action*, 3(1), 46. https://doi.org/10.1038/s44168-024-00048-3.

Segerson, K., Polasky, S., Scheffer, M., Sumaila, U.R., Cárdenas, J.C., Nyborg, K., Fenichel, E.P., Anderies, J.M., Barrett, S., Bennett, E.M., Carpenter, S.R., Crona, B., Daily, G., de Zeeuw, A., Fischer, J., Folke, C., Kautsky, N., Kremen, C., Levin, S.A., & Weber, E.U. (2024). A cautious approach to subsidies for environmental sustainability. *Science,* 386(6717), 28–30. https://doi.org/10.1126/science.ado261.

Cluster III
Political Economy

33 Political Economy of Consumerism

Manu V. Mathai and Soumyajit Bhar

Definition

The Cambridge English Dictionary states that consumerism is a state of affairs typical of "an advanced industrial society in which a lot of goods are bought and sold". A second definition it provides observes that it is "the situation in which too much attention is given to buying and owning things". The use of "too much" in the second definition serves as a point of departure for this entry. Consumerism is defined here as the buying of commodities at a scale that is socially and ecologically undesirable, untenable, and driven by a value system that bases personal and social evaluation on it (see **Consumerism** and **Conspicuous/Positional Consumption**). The term "political economy" indicates that our focus is on the historical and contemporary political, social, and economic factors that brought about and maintain this undesirable scale. This shifts research and policy attention away from individual choice **behavior change** alone to understand consumerism and extends it to consider structural explanations. For example, is choosing to drive a personal car an entirely individual choice? Or is it also a value, norm, and practice produced by social, political, and economic processes that have failed to create dignified and efficient public transport systems? The political economy of consumerism concerns itself more with the latter form of inquiry (see also **Consumer Scapegoatism**).

History

Consumerism as seen today is a new conceptual category in human history. It represents the mid-20th-century culmination of great historical transformations of thought and action that also produced capitalism and the industrial revolution. These changes that accrued over centuries produced entirely new concepts and practices, such as the commodity form, capital, capital accumulation, consistent technological innovation, conditions for mass production, international trade, and money as wealth and capital. Writers like Peter Stearns point out that accompanying these changes were transformations in values, the weakening of social hierarchies, the rise of secularism and individual liberty, and the resulting question of individual identity as powerful cultural drivers. However, the preoccupation of the political economy even until the 18th century was not consumerism, but inquiring into the "nature and causes of the wealth of nations" under these new conditions. For example, references to consumption – not consumerism – are tangential in Adam Smith's famous treatise on the "wealth of nations". His chapter discussing "taxes upon consumables" offers an insight into how consumption was then understood. Smith distinguishes commodities that are "necessaries" from those that are "luxuries". The former was necessary for the "support of life" "which the established rules of decency have rendered necessary to the lowest rank of people". All others were included as "luxuries". The latter was not necessary, for "custom nowhere renders it

DOI: 10.4324/9781003584056-37

indecent to live without them". Yet, Smith qualifies that by "luxuries" he did not wish to "throw the smallest degree of reproach upon the temperate use of them". Earlier in the book, Smith offers that it is when "society is advancing to the further acquisition . . . of riches" that the "great body of people, seem to be the happiest and the most comfortable . . . The stationary state is dull; the declining melancholy". Discouraging consumption, even of luxuries, under these circumstances would be logically inconsistent. Mainstream social and economic policy continues to lean on this linkage of production, consumption, and social welfare (see also **Well-being Versus Income**).

Different Perspectives

Many other early political economists remained preoccupied with understanding and enabling the "causes of the wealth of nations". Writers like David Ricardo emphasized technology and international trade in this context, and Karl Marx highlighted the social contradictions of the capitalist industrial arrangement. However, more than a century after the *Wealth of Nations,* we notice that consumption, not production, came into focus in political economic analysis. A key text in this regard was Thorstein Veblen's *The Theory of the Leisure Class* published in 1899. It introduced the ideas of conspicuous consumption, conspicuous leisure, and pecuniary culture, to political economy. These insights emerged from the book's focus on the corruption of capitalism such that a "leisure class" in pursuit of profit-making, and status signaling instead of usefulness, became the drivers of consumption of unessential goods, replacing the "modes of productivity and the pride of workmanship that furnish a sound society with its true necessities".

Other writers such as W.S. Jevons offered that the "problem of economics" was not one of contest and negotiation between social classes over the allocation of capital, labor, and profits. He argued instead that it is a "calculus of pleasure and pain" where rational individuals interacting via free and competitive markets maximize their individual utility. These diverging orientations toward social structure seen in Veblen, and individual behavior and utility inaugurated by Jevons, remain a fault line between economics and political economy. Veblen worried that the lower classes would emulate the "leisure class" instead of resisting their consumption culture. This concern played out by the mid-20th century in the industrialized world and continues today globally, now as consumerism.

Application

The modern environmental movement, since the 1960s, has emerged as the key driver of critical interest in consumerism. The attribution of causality for the environmental crisis is an application of the broad fault line in the political economy identified above. Building on the utilitarian framework of writers like W.S. Jevons and Alfred Marshall, mainstream economists and policymakers concluded that the crisis resulted from the failure of prices to capture the full costs of producing, consuming, and disposing of commodities, leading to irrational decisions laden with externalities. Fields such as marketing, consumer behavior theory, and social psychology were used to produce a policy emphasis on capturing the externalities or social costs of consumerism in the prices of commodities (e.g., carbon tax to check the consumption of fossil fuels) and other efforts such as the provision of information and signals (e.g., labeling and branding) to nudge individual behavior (see **Green Nudging** and **Ecolabeling**). Such thinking, however, has mostly failed to dent consumerism and the resulting material footprint of nations. This recognition has brought attention to frameworks that engage consumerism in relational and evolutionary terms. Prominent here is the human tendency to distinguish oneself in relation to social hierarchies (as seen above in Veblen and in Bourdieu's *Distinction*). Applying these insights, efforts to address consumerism now include proposals such as the solidarity economy and the **sharing economy**. In both instances, the idea is to emphasize values of care and sharing as opposed to individual utility maximization as drivers of consumption decisions.

The political economy of consumerism also brings attention to the fact that, as noted by Moore (2015), the latter is enabled by a production system in which labor and nature are appropriated on the "cheap" and even via a "supply chain of violence" to enable mass production of affordable, often low-quality commodities (e.g., **fast fashion**, fast food) or those designed for rapid obsolescence (e.g., textiles, consumer electronics) (see also **Ecodesign** and **Extended Producer Responsibility**). Power, politics, and justice emerge as variables of central analytical importance and interest here – an insight that is now also recognized by Working Group III of the IPCC in its Sixth Assessment Report. It led Schandl et al. (2018: 834) to recognize that "the level of well-being achieved in wealthy industrial countries cannot be generalized globally based on the same system of production and consumption" (see **Carbon Inequality**, **Climate Justice**). This realization calls the mainstream development discourse and its promise of liberalism and national or state sovereignty built on open-ended economic growth and technological innovation into question, as recognized by international development scholar Sachs (1992) over three decades ago. It critiques the development path of the "developed world", even as it questions the direction of the "developing world".

In sum, the political economy of consumerism is a necessary corrective engagement with development studies, which offers a robust literature, policy, and practice that appears to have developed thus far in parallel to, and without much apparent interest from, the literature on sustainable consumption.

Further Reading

Guha, R. (2006). How much should a person consume? Environmentalism in India and the United States. University of California Press.

Mathai, M.V., Isenhour, C., Stevis, D., Vergragt, P., Bengtsson, M., Lorek, S., Mortensen, L.F., Coscieme, L., Scott, D., Waheed, A., & Alfredsson, E. (2021). The political economy of (un)sustainable production and consumption: A multidisciplinary synthesis for research and action. *Resources, Conservation and Recycling*, 105265. https://doi.org/10.1016/j.resconrec.2020.105265.

Moore, J.W. (2015). Capitalism in the web of life: Ecology and the accumulation of capital. 1st ed. Verso.

Sachs, W. (Ed.). (1992). The development dictionary: A guide to knowledge as power. Zed Books Ltd.

Schandl, H., Fischer-Kowalski, M., West, J., Giljum, S., Dittrich, M., Eisenmenger, N., Geschke, A., Lieber, M., Wieland, H., Schaffartzik, A., Krausmann, F., Gierlinger, S., Hosking, K., Lenzen, M., Tanikawa, H., Miatto, A., & Fishman, T. (2018). Global material flows and resource productivity: Forty years of evidence. *Journal of Industrial Ecology*, 22(4), 827–838. https://doi.org/10.1111/jiec.12626.

34 Consumer Scapegoatism

Aitor Marcos

Definition

The defining characteristic of consumer scapegoatism is the tendency to assign responsibility for the sustainability transition to individual consumers and citizens, without adequately considering whether they can significantly influence the outcome or the relevant players in the system. A paradigmatic case of consumer scapegoatism is the systematic use of pro-environment claims to encourage the consumption of goods and services, a practice known as green consumerism. The promotion of green consumerism by governments and markets makes consumers responsible for both sustaining economic growth and pushing the socio-economic system toward ecological sustainability – despite consumers' limited ability to do so.

Consumer scapegoatism arises whenever consumers' agency is not properly weighted against the urgency, magnitude, and scale of the problem of Earth's ecological imbalance. Therefore, a policy can be considered a form of consumer scapegoatism if it exaggerates the importance of individual **behavior change** while obscuring the – more important – role of broader socio-technical context and power systems that ultimately shape ways of living. This kind of reasoning tends to align with market-based regulatory measures and favors efficiency over **sufficiency**, thereby not challenging the paradigm of continued economic growth.

An example of consumer scapegoatism is the belief that a tax on plastic bags will effectively address the issue of plastic waste. However, this economic incentive fails to tackle the root causes of plastic proliferation. Such market-based measures simply force consumers to adjust their behavior, while failing to force companies to reconsider the economic logic that makes producing vast quantities of cheap, disposable plastic bags financially viable to the point of distributing them for free.

History

The concept of consumer scapegoatism, as it relates to sustainable consumption, was introduced in 2014 by Lewis Akenji in an article titled *Consumer Scapegoatism and Limits to Green Consumerism*. The concept embodies a critique that gained momentum in the early 2000s denouncing how sustainable consumption, with its focus on increasing "ecoefficiency", has been poorly used as a policy concept since the Rio Earth Summit in 1992. Thereafter, from the World Summit for Sustainable Development in 2002 to the establishment of Sustainable Development Goal 12 in 2015, sustainable consumption only entered the policy agenda by being increasingly appropriated by corporate marketing. It deliberately conflated sustainable consumption with green consumerism, which refers to the consumption of products and services that claim to be environmentally friendly.

The primary intent and theoretical contribution of the concept of consumer scapegoatism is to clarify that green consumerism and sustainable consumption are different approaches to

DOI: 10.4324/9781003584056-38

sustainability (see Table 34.1), despite current policies and mainstream thinking often failing to distinguish between the two.

Akenji argued that promoting green consumerism is a form of consumer scapegoatism because it misses the core idea of sustainable consumption and provides an illusion of progress that distracts from the urgent structural changes needed for the sustainability transition. This timely theoretical distinction between green consumerism and sustainable consumption was welcomed as a response to governments and companies that expected individual consumers to drive the shift toward sustainable consumption and production patterns through their purchasing choices alone.

The uptake of the concept of consumer scapegoatism directly challenges the market-driven axiom that if consumers were aware of the environmental impact of their choices, they would fix unsustainable consumption and production simply by buying differently and pressuring companies with their market power. However, the reality is that the burden of complex choices and everyday decisions further reduces the already little influence of consumers over powerful actors in large corporations and government agencies.

Different Perspectives

Consumer scapegoatism is a useful concept for scholars who want to convey the difference between weak and strong sustainability approaches. In broad terms, weak sustainability is similar to green consumerism, while strong sustainability relates to sustainable living. Weak sustainability involves buying energy-saving, recycled, or recyclable products out of a sense of moral duty. It emphasizes efficiency and new technologies, failing to question an economic growth model fueled by an ever-increasing production of green goods and services. In contrast, strong sustainability aims to understand the root causes of why we consume, emphasizing **sufficiency**. From this perspective, an excessive focus on promoting weak sustainability and not intervening at a preventive level is consumer scapegoatism.

As Akenji and Bengtsson suggest in their "Triple I framework", consumers might not be the most influential stakeholder in the value chain – consumers just happen to be the most visible actor. The market and policy emphasis on promoting green consumerism ignores that most consumption decisions are influenced by infrastructure, technology, **social norms**, and power dynamics within the production-consumption nexus. However, this conceptual flaw persists in popular models like the "ABC framework" and the "i-frame" (see Box 34.1).

Box 34.1 Social science frameworks highlight the lack of consumer agency

The "Triple I framework" in sustainable consumption

Akenji and Bengtsson's Triple I framework helps to identify consumption drivers and power distribution among stakeholders in a value chain, and points for policy interventions. The "I"s stand for each stakeholder group's Interest, their Influence upon one another, and the Instruments they use to exert power. The Triple I framework can be used to assign responsibility based on each stakeholder's capacity, rather than their visibility. This framework assumes that if the most influential stakeholder is targeted by policy, it can drive positive changes throughout the value chain due to its influence.

Moving beyond the "ABC framework" in the social sciences

The "ABC framework" is a dominant way of thinking in the social sciences that focuses on influencing individual behaviors to combat climate change by targeting Attitudes, Behaviors, and Choices. The emphasis on educating people, incentivizing ecofriendly actions, and promoting voluntary behavior changes often favors the shift of responsibilities away from governments and avoids addressing broader structural changes. Elizabeth Shove calls for moving beyond the ABC framework to critically consider **social practice theories**, recognizing that household consumption depends not only on consumer attitudes but also on the broader socio-technical context.

The "i-frame" and "s-frame" in behavioral science

Behavioral science has biased public policy toward addressing societal problems at the individual level without system changes. The "i-frame" approach, focusing on individual behavior, has shown modest results and diverted attention from systemic solutions. Chater and Loewenstein (2023) recommend designing systemic solutions (adopting an "s-frame") for more effective public policy, which can be applicable to a wide range of issues, from climate change to obesity.

Table 34.1 Comparison of green consumerism and sustainable consumption from the lens of consumer scapegoatism

	Green consumerism	*Sustainable consumption/living*
Definition	Production, promotion, and preferential consumption of goods and services on basis of their pro-environment claims	Equitable consumption and lifestyles that contribute to well-being within ecological limits
Indicators/key principles	Ecoefficiency, market options, and **freedom of choice**	Living within ecological limits, equitable consumption, and well-being of individuals and society
Perspective on the consumer	Individuated buyers of goods and services, making rational choices	Individuals and collectives enabled or constrained by institutions and systems of provision
Intervention node	Micro level – individual and household product shopping, use and waste disposal/ recycling	Micro, meso, and macro levels – individual and household behaviour, entire product value chains, provisioning systems, and related institutions
Sample policy instruments	Ecolabeling schemes for products and services, public awareness campaigns, ecoefficient production standards, recycling	Ecological and progressive charges for services and taxes, choice editing, promotion of out-of-market and non-economic opportunities
Key advocates	Businesses, governments, consumers	Researchers, communities, civil society groups, governments

Source: Akenji (2019)

The concept of consumer scapegoatism also resonates with ecological economists, particularly those who challenge the idea of green growth (i.e., progressively decoupling economic growth from environmental impacts). These voices argue that with an overemphasis on individual action, consumers are put in a delicate place: responsible for sustaining economic growth while burdened with steering the system toward sustainability. **Degrowth** scholarship, for instance, argues that expecting consumers to solely drive the sustainability transition through the market and better technologies oversimplifies the need for systemic changes. This exaggeration of consumer agency is fundamental to consumer scapegoatism. Moreover, consumer scapegoatism complements **ecological economics'** critique of the economic growth paradigm by identifying consumers as powerless actors, yet responsible for maintaining economic growth through their ever-increasing consumption.

Application

Consumer scapegoatism provides a blueprint for the type of reasoning to avoid when designing sustainable consumption policy. The Attitudes-Facilitators-Infrastructure (AFI) framework was developed to assist policymakers in steering clear of consumer scapegoatism when promoting sustainable consumption (see Box 34.2).

Box 34.2   The Attitudes-Facilitators-Infrastructure (AFI) framework

The three elements of the AFI framework enable a systemic approach to sustainable consumption policy design while avoiding consumer scapegoatism:

- Fostering pro-environmental attitudes.
- Establishing facilitators of access to sustainable options while restricting unsustainable ones.
- Developing the necessary infrastructure and product options for living sustainably.

By integrating these three elements in policy design, policymakers can simultaneously address drivers of consumer behavior, consider and influence other relevant stakeholders, and promote products and services to meet both individual and societal needs.
Policymakers following the AFI framework are better equipped to tackle issues such as:

- The **attitude-behavior gap** in consumer choice.
- **Behavior change** being restrained by lock-in to prevailing systems and infrastructure.
- Macro-level social and physical factors influencing consumption patterns.

The AFI framework can help guide a more holistic approach to sustainable consumption policy by redirecting policymakers' attention to reforming the systems and infrastructure, predetermining consumers' degree of flexibility in adopting sustainable lifestyles. This holistic approach avoids fully blaming consumers for a consumption process not fully in their control. For instance, the promotion of recycling can perpetuate the myth that minor post-consumption lifestyle changes and educated purchases of recyclable items are enough for sustainable living. In reality, recycling is a resource-intensive process that, without the required infrastructure, may significantly increase pollution.

Lastly, the notion of consumer scapegoatism serves as a warning that sustainable consumption policies risk being biased toward efficiency and market-based solutions. For example, incentivizing the adoption of a new generation of energy-efficient appliances, such as refrigerators or air conditioners, may initially seem like a sound policy, as these products typically consume less energy. However, rewarding the exchange of previously purchased appliances with discounts on similar products ultimately promotes further consumption, while failing to encourage the extended use of older, still functional appliances. Although consumers may benefit financially through discounts and energy savings, the potential ecological benefits are undermined by the ongoing cycle of consumption.

Further Reading

Akenji, L. (2014). Consumer scapegoatism and limits to green consumerism. *Journal of Cleaner Production*, 63, 13–23. https://doi.org/10.1016/j.jclepro.2013.05.022.

Akenji, L. (2019). *Avoiding consumer scapegoatism: Towards a political economy of sustainable living*. Doctoral dissertation, University of Helsinki. Available at: https://researchportal.helsinki.fi/en/publications/avoiding-consumer-scapegoatism-towards-a-political-economy-of-sus (accessed: 8 January 2025).

Akenji, L., & Bengtsson, M. (2010). Is the customer really king? Stakeholder analysis for sustainable consumption and production using the example of the packaging value chain. *Sustainable consumption and production in the Asia-pacific region: Effective responses in a resource constrained World*, 3, 23–46. Available at: https://www.iges.or.jp/en/pub/customer-really-king-stakeholder-analysis/en (accessed: 8 January 2025).

Chater, N., & Loewenstein, G. (2023). The i-frame and the s-frame: How focusing on individual-level solutions has led behavioral public policy astray. *Behavioral and Brain Sciences*, 46(e147), 1–84. https://doi.org/10.1017/S0140525X22002023.

Shove, E. (2010). Beyond the ABC: Climate change policy and theories of social change. *Environment and Planning A: Economy and Space*, 42(6), 1273–1285. https://doi.org/10.1068/a42282.

35 Energy Overshoot

William E. Rees

Definition

"Energy" is traditionally defined as the ability to do work; it represents the physical capacity to achieve anything. Indeed, nothing can happen without the irreversible expenditure of energy. It follows that adequate affordable energy is essential to the maintenance of civilization. Society should therefore be aware of the implications of the Second Law of Thermodynamics (the entropy law), which is the primary natural law governing biophysical processes, including energy use.

The Second Law states that every real process increases global "entropy" or disorder. Economic production and consumption *necessarily* consume and dissipate usable energy and other material resources and pollute the ecosphere no matter how efficient or circular the economy is – *there are no exemptions from the Second Law* (see **Circular Economy and Society** and **Steady-State Economy**).

History

Modern techno-industrial (MTI) society is the product of abundant cheap energy, mainly fossil fuels (coal, oil, and natural gas) (see **Energy Consumption Behavior**). Fossil fuels (FF) are the means by which an industrializing society could produce the food and all other resources needed to sustain the growth of the human enterprise. Since 1820, the human population has ballooned from one billion to eight billion and the world has experienced an unprecedented 100-fold expansion of real gross world product – all propelled by a 917-fold expansion of FF combustion (Table 35.1).

As expected, the MTI society continues to rely on FF for approximately 81% of its energy needs. Virtually every agricultural product, from artichokes to zucchinis, is grown using FF, and most industrial processes and products – including the service sectors – are powered mostly by FF. Even the recent explosion of seemingly material-free artificial intelligence applications has been accompanied by a massive increase in electricity consumption – and 60% of the world's electricity is still generated by FF.

There is an ominous consequence of the second thermodynamic law: Modern society's fossil-powered expansion has propelled humanity far into ecological overshoot. Overshoot drastically accelerates entropy, with humans consuming resources and polluting at rates that exceed the ecosphere's regenerative and assimilative capacities. Anthropogenic climate change, tropical deforestation, plunging biodiversity, ocean acidification, land and soil degradation, microplastic contamination, the pollution of everything – indeed, virtually all so-called environmental problems – are entropic co-symptoms of overshoot. MTI society is consequentially depleting and dis-ordering the biophysical basis of its own existence (Rees, 2023).

DOI: 10.4324/9781003584056-39

Table 35.1 Growth in energy consumption, population, and gross world product since 1820

	1820	*2023*	*Expansion factor*
Total Primary Energy +	6,264 TWh*	183,230 TWh	29
Fossil Fuel Consumption (coal, oil, natural gas)	152 TWh	140,231 TWh	917
Global Population	1 billion	8 billion	8
Energy Use Per Capita	21 Gigajoules	80 GJ	3.7
Fossil Energy Per Capita	.55 GJ	64 GJ	115
Gross World Product	1.4 trillion (2017 prices)	140 trillion (2017 prices)	100

Note: + Includes hydro and nuclear electricity, Terrawatt hours (TWh), *(98% traditional biomass: wood, crop waste, animal dung)

The best-known symptom of overshoot is climate change. 75% of climate change/global heating is attributable to increasing carbon dioxide (CO_2) emissions, *the major entropic waste output by the weight of industrial economies*. Industrialization has elevated atmospheric CO_2 by 50%, from a preindustrial 280 parts per million in 1800 to about 420 ppm in 2024. Other greenhouse gases have increased by even greater percentages. Consequently, the world is recording temperatures unprecedented in the past 24,000 years, with 2023 as the warmest year in the instrumental record so far. The International Panel on Climate Change's (IPCC) strong target of limiting the global temperature increase to 1.5°C compared to pre-industrial times is getting out of reach, as current government policies would cause an average global temperature increase of 2.7°C by 2100. This is sufficient to destroy agriculture in many regions and render large areas of Earth uninhabitable. There will likely be mass human migrations, complicated by global food shortages, regional famines, and inevitable geopolitical chaos. By definition, overshoot is ultimately a terminal condition.

Humans tend to respond to problems in simplistic, reductionist ways. Is anthropogenic global heating a concern? And is it caused by carbon dioxide emissions? No problem – without systems thinking or an ecological context (and ignoring overshoot), the solution seems simple: just replace fossil fuels with a clean, green alternative. Thus, a whole new industrial sector is born, encouraged by the widely publicized fantasy of 100% renewable green energy by 2050 (Jacobson et al., 2017). Indeed, according to industry hype, the modern renewable energy (RE) transition is well underway, led by wind and solar electricity generation.

The supply of RE has grown by approximately 2% annually since 1990 and global investment in energy transition technologies exceeded $1 trillion for the first time in 2022. However, wind and solar power – where most investment is going – still provided only 14.3% of global electricity in 2023 (compared to about 60% by FF). In short, wind and solar power only cover about 2.8% of the world's final energy consumption (see EI, 2024). It would be necessary to install *additional* capacity four times the current cumulative global stock of wind and solar infrastructure to fully displace fossil fuels from electricity generation alone – and the 80% of the global energy mix, which is non-electric would have yet to be addressed.

The above scenario assumes *no increase in demand*, even though demand *is* increasing – in 2023, total primary energy consumption grew by about 12.3 exajoules (EJ). FF accounted for 7.4 EJ, while the increase from wind and solar was about 5.4 EJ in FF equivalent terms. In short, the growth in RE electricity generation lags behind total demand growth. FF combustion and CO_2

emissions therefore continue their inexorable increase – at least until recoverable reserves are depleted/dissipated.

Different Perspectives

There are several additional sources of entropic disordering. The International Energy Agency projects the total mineral (e.g., copper, nickel, or cobalt) demand for clean energy technologies to double under existing policies and quadruple in their sustainable development scenario (SDS). Mineral demand associated with electric vehicles and batteries would grow approximately 30-fold under the SDS by 2040. Lithium consumption grows fastest at over 40 (possibly 50) times. The SDS projects a 40% increase in demand for copper and 60–70% for nickel and cobalt. It is further relevant to consider the massive increase in demand for steel, concrete, and glass – among the most carbon- and resource-intensive modern building materials – needed to build out RE infrastructure. All this portends an unprecedented explosion in pernicious mining, smelting, and manufacturing activities whose impact will likely worsen as the quality of ore grades declines over time.

And what about the competition for space – including agricultural land – represented by the expansion of wind and solar power. For example, given the average energy density of US solar power arrays (Bolinger & Bolinger, 2022) and the power consumption of electric vehicles, it would take as much as 4,400 sq miles of solar panels (almost the area of Connecticut) just to keep the electric equivalent of four million mostly diesel 500-miles-per-day Class 8 trucks on US roads – and this represents a small fraction of total US energy demand.

In short, the mining, transportation, refining, manufacturing, installation, and maintenance activities – mostly still fossil fuel dependent – associated with the growth of RE, combined with the loss of land productivity, would impose a formidable *additional,* often toxic, entropic burden on Earth's waters, land and atmosphere. All this is on an ecosphere already reeling from the excesses of overshoot (Fletcher et al., 2024).

Application

While the RE transition is barely underway, its implications for the planet and people are already alarming. It augments existing energy supplies and thus accelerates the entropic disordering of the planet. This comes as no surprise, as RE and supportive technologies are actually intended to maintain the growth-based *status quo* by alternative means. In this light, MTI society's simplistic approach to combating climate change is a delusional distraction from facing overshoot as a crucial meta-crisis.

What the leaders – and many citizens – of our "world-in-overshoot" refuse to acknowledge is that even *today's egregiously unequal levels of production and consumption are unsustainable.* Sustainable production and consumption mean absolutely *less* production and consumption combined with greater social equality (see **Sufficiency**, **Fair Consumption Space**, **Degrowth**).

The good news is that it is at least technically possible to increase the energy and resource productivity of industrial economies by up to 80%. Indeed, the measures to realize this productivity increase will almost certainly be demanded. It is ironically inevitable that renewable energy will power any future society, but much of it will be low-tech energy – think "oxen" and "water-wheels" – serving much smaller local economies and many fewer people living in inter-connected but more nearly self-reliant bio-regions.

The crucial question is whether the rulers of the MTI world will act in time to organize an orderly transition to "one-Earth living" or whether the planet's responses will impose their own solution.

Further Reading

Bolinger, M., & Bolinger, G. (2022). Land requirements for utility-scale PV: An empirical update on power and energy density. *IEEE Journal of Photovoltaics*, 12(2) (March), 589–594. https://doi.org/10.1109/JPHOTOV.2021.3136805.

EI. (2024). *Statistical review of world energy. 2024*. Energy Institute. Available at: https://www.energyinst.org/__data/assets/pdf_file/0006/1542714/EI_Stats_Review_2024.pdf (accessed: 8 January 2025).

Fletcher, C., Ripple, W.J., Newsome, T., Barnard, P., Beamer, K., Behl, A., Bowen, J., Cooney, M., Crist, E., Field, C., Hiser, K., Karl, D.M., King, D.A., Mann, M.E., McGregor, D.P., Mora, C., Oreskes, N., & Wilson, M. (2024). Earth at risk: An urgent call to end the age of destruction and forge a just and sustainable future. *PNAS Nexus*, 3(4), 106. https://doi.org/10.1093/pnasnexus/pgae106.

Jacobson, M.Z., Delucchi, M.A., Bauer, Z.A.F., Goodman, S.C., Chapman, W.E., Cameron, M.A., Bozonnat, C., Chobadi, L., Clonts, H.A., Enevoldsen, P., Erwin, J.R., Fobi, S.N., Goldstrom, O.K., Hennessy, E.M., Liu, J., Lo, J., Meyer, C.B., Morris, S.B., Moy, K.R., & Yachanin, A.S. (2017). 100% clean and renewable wind, water, and sunlight all-sector energy roadmaps for 139 countries of the world. *Joule*, 1(1). https://doi.org/10.1016/j.joule.2017.07.005.

Rees, W.E. (2023). The human ecology of overshoot: Why a major 'population correction' is inevitable. *World*, 4(3), 509–527. https://doi.org/10.3390/world4030032.

36 Carbon Inequality

Beatriz Barros and Kuishuang Feng

Definition

Carbon inequality concerns the disproportional distribution of carbon emissions across countries, populations, and time. Historically, high-income countries and individuals have been responsible for much higher carbon emissions (and the associated greenhouse gases responsible for climate change) than low-income countries and individuals. Despite contributing less to the problem, low-income countries and communities are more vulnerable to the destructive effects of climate change, such as extreme weather events, and have fewer resources to adapt and mitigate those effects (see **Climate Justice**). Carbon inequality also manifests itself across time, as past and present generations are exclusively responsible for the carbon emissions driving climate change, jeopardizing the well-being of future generations. Carbon inequality is a multifaceted concept that highlights the complexities of carbon footprint accounting, its intricate implications, and the need to address it in tandem with socio-economic inequality.

History

Although the exact origin of the concept of carbon inequality is impossible to pinpoint, it has been evolving and increasingly used since the beginning of the 21st century. The origin of the concept is linked to the rising awareness of the complexity of causes and effects of broader environmental degradation, which started in the later decades of the 20th century with the environmental justice movement. As with environmental justice, carbon inequality is concerned with equity, and any approach to tackle it needs to take into account related aspects of socio-economic inequality. Scientific research and policy discussions, at both the national and international levels, have become progressively cognizant of this interrelatedness, and, in consequence, carbon inequality has become a frequently used concept in those contexts.

In the first stage, scientific research and policy discussions around carbon inequality focused more on disparities between countries. Industrialized countries have, for years, relied on fossil fuels to meet their energetic demands, drive their development, and achieve high-income status. As a result, these countries have been responsible for extremely high carbon emissions that have been driving global warming and exacerbating climate change. Most Global South countries, in contrast, display much lower total and per capita emissions (a metric that divides a country's total carbon emissions by its population).

However, focusing solely on these metrics presents a distorted reality, since not everyone is contributing to carbon emissions at the same level within the same country. Because of this, scientific research and policy discussions concerning carbon inequality have started to consider, at a second stage, how different income groups, within and across countries, were disproportionately

DOI: 10.4324/9781003584056-40

responsible for carbon emissions (see **Household Income Versus Carbon Footprint**). Estimates show that the top 10% wealthiest individuals globally are responsible for approximately 48% of carbon emissions in recent years.

In addition to these two aspects – inequality between countries and income groups – carbon inequality as a concept has been progressively incorporating an intergenerational perspective. Given that past and older living generations are responsible for the bulk of carbon emissions driving climate change, it becomes a matter of equity to include the input of younger generations and consider the impact on future generations' welfare in policy discussions.

Different Perspectives

Carbon inequality is increasingly debated. As a result, the concept keeps expanding in complexity. Concerning the debate around carbon inequality between countries, critics have noted that focusing on per capita emissions is simplistic and might misguide policy by putting too much emphasis on individual behavior and not enough on infrastructural change (see **Consumer Scapegoatism**). In addition, focusing too much on per capita emissions might evade the impacts of globalization and ignore yet another facet of carbon inequality: offshoring or outsourcing production chains. Driven by environmental and labor regulations, as well as reduced costs, corporations in the Global North have moved all or parts of their production chains and associated carbon emissions to the Global South. Most of the resulting products and services are, however, ultimately consumed in the Global North.

To address this drawback, researchers have proposed consumption-based emissions accounting, which shifts the focus from production to consumption, attributing carbon emissions to end-users rather than producers (see **Consumption-Based Accounting**). It suggests that wealthier nations should take responsibility for the emissions embedded in the products they import and consume. Consumption-based emissions accounting also highlights carbon inequality through the lens of income disparities within and between countries. High-income individuals tend to have larger carbon footprints due to their carbon-intensive lifestyles, including frequent air travel and large homes, and also due to their investment choices.

Finally, some critics of carbon inequality as a working concept believe that it should not be considered at all and that policy efforts to address it might even be counterproductive. These critics are not concerned with reducing carbon emissions but with carbon capture and storage technology, arguing that governmental intervention might hinder entrepreneurship and the technological development needed to achieve carbon removal at a scale that might slow global warming.

Application

Carbon inequality illustrates how socio-economic disparities and the impacts of environmental degradation are intertwined. It can inform discussions at various levels and be applied to different sets of approaches aiming to address those disparities. Carbon inequality-based approaches can, for example, be used to identify high-emissions sectors and regions within and across countries. In understanding the distribution of carbon emissions, governments can design targeted interventions, such as progressive carbon taxes that impose a higher tax on those responsible for larger carbon emissions – often wealthier individuals and corporations – while protecting low-income households from disproportionate financial impacts.

In addition, carbon inequality may inform redistribution mechanisms. Revenue from carbon taxes or cap-and-trade systems can be redistributed to lower-income groups through rebates or direct payments, helping to mitigate the economic inequality exacerbated by climate change. It is important to raise awareness and develop strategies to communicate the effects of carbon inequality to all levels of society, from policymakers to organizations and individual consumers. This can potentially help harness support for the application of carbon taxes and associated measures to counteract carbon inequality and contribute to influencing consumers' behaviors (see **Behavior Change**).

From a sustainable supply chain perspective, focusing on carbon inequality can foster transparency and help identify instances of carbon outsourcing. Governments should establish carbon border adjustment mechanisms to prevent carbon outsourcing and push businesses to reduce carbon emissions across their supply chains and ensure that products are sustainable from production to consumption. Furthermore, transparency in corporate carbon emissions enables stakeholders to hold companies accountable for their environmental impacts (see e.g., *EU taxonomy for sustainable activities*), which may encourage further sustainability efforts and reduce greenwashing practices.

In discussions about allocating international climate finance, carbon inequality has proven a crucial concept. Countries with lower emissions but higher vulnerability to climate change should be prioritized for funding to support their mitigation and adaptation efforts. Wealthier nations with higher historical emissions should be responsible for higher climate financial assistance and should also contribute with technological support to developing countries, facilitating their transition to low-carbon economies.

In sum, carbon inequality is a crucial aspect of addressing climate change and should be used to inform decision-making on future policies and investments, ensuring that efforts to reduce emissions are aligned with social and economic equity goals.

Further Reading

Caney, S. (2014). Climate change, intergenerational equity and the social discount rate. *Politics, Philosophy & Economics*, 13(4), 320–342. https://doi.org/10.1177/1470594X14542566.

Chancel, L. (2022). Global carbon inequality over 1990–2019. *Nature Sustainability*, 5, 931–938. https://doi.org/10.1038/s41893-022-00955-z.

Jones, M.W., Peters, G.P., Gasser, T., Andrew, R.M., Schwingshackl, C., Gütschow, J., Houghton, R.A., Friedlingstein, P., Pongratz, J., & Le Quéré, C. (2023). National contributions to climate change due to historical emissions of carbon dioxide, methane, and nitrous oxide since 1850. *Scientific Data*, 10(155). https://doi.org/10.1038/s41597-023-02041-1.

Kenner, D. (2019). *Carbon inequality: The role of the richest in climate change*. New York: Routledge.

Khalfan, A., Nilsson Lewis, A., Aguilar, C., Persson, J., Lawson, M., Dabi, N., Jayoussi, S., & Acharya, S. (2023). Climate equality: A planet for the 99%. *Oxfam International*. https://doi.org/10.21201/2023.000001.

37 The Role of Business

Patrick Elf, Amy Isham, and Oksana Mont

Definition

A rapid reorientation away from perpetuating unsustainable lifestyles and toward facilitating sustainable consumption would require businesses to utilize their far-reaching influence as a positive force for sustainability. This will have to go beyond **greenwashing** practices and prioritization of short-term profit orientations that too often perpetuate unsustainable consumption. Here, we suggest that the role of businesses in promoting sustainable consumption lies not simply in changing product offerings, but also in leveraging their influence on consumers' lifestyle practices more broadly, ensuring that these align with long-term sustainability ambitions and, importantly, planetary and social carrying capacities (see **Doughnut Economy**).

In practice, businesses need to transcend short-termism to promote sustainable consumption and shift from mere ecoefficiency to comprehensive ecosufficiency, engaging customers in co-creative processes that encourage sustainable lifestyles. We underscore the necessity for businesses to move beyond superficial changes to products and processes to foster sustainable consumption and **sufficiency**. This will require businesses to redefine their roles to become enablers of sustainable living and drivers of socio-ecological change toward sustainability.

History

In the 19th and early 20th centuries, a belief emerged that business activities could drive economic development, which would automatically lead to societal improvement. Prominent business thinkers such as Peter Drucker championed this view, establishing it as the foundation for businesses' license to operate. During the pre-World War II period and early Cold War years, economists like Friedrich Hayek and Milton Friedman redefined the role of business in society. Advocating for free markets as the most efficient way to allocate resources and improve society, they promoted limited state intervention and unrestrained self-interest as an engine of economic progress, laying the foundation for neoliberalism. They asserted that the primary responsibility of businesses was the pursuit of profits, suggesting that broader social benefits would result from the "trickle-down" effect of economic activity. Profits, in turn, were generated by continuously selling more products and services to consumers.

Consequently, most businesses have perpetuated *un*sustainable consumption levels among their customers to ensure they comply with capitalist growth demands, with far-reaching implications for both people and the planet. Indeed, businesses have tactically employed product design, **advertising**, and marketing techniques to influence individuals to consume more and ensure short-term gains. These approaches have been instrumental in generating new "needs" (often conflated with "wants"), keeping customers in a state of perpetual dissatisfaction to maintain constant spending

DOI: 10.4324/9781003584056-41

and *un*sustainable consumption (see **Hedonic Treadmill**). Here, the still-dominant growth-driven practices present significant obstacles for businesses that promote sustainable consumption and sufficiency, as they must navigate a landscape where desires are constantly being reshaped and elevated.

With greater consumer awareness regarding social and environmental sustainability and simultaneous changes in legislation, the 21st century has seen a shift in business perspectives and practices. Businesses have increasingly tried to balance their unsustainable practices (e.g., via corporate philanthropy and Corporate Social Responsibility) and communicate their environmental and social impacts with greater transparency through standards and reporting initiatives (e.g., ISO standards and Environmental and Social Governance approaches). Alongside these efforts, diverse organizational forms are emerging – such as cooperatives, ecosocial enterprises, and B Corporations – that aim to balance profit with purpose and are committed to creating a positive impact for stakeholders, communities, and the environment (see Box 37.1).

Box 37.1   Example of a business accreditation in the environmental domain

Introducing BCorps: In recent years, more profound discussions have emerged within society regarding the role of businesses in economies. One prime example is the emergence of BCorps that draw on a mix of different approaches. To become BCorp accredited, businesses need to demonstrate their performance in key areas such as ethical business practices and innovation of sustainable products. Their popularity reflects a growing shift toward considering the (longer-term) impact of business practices on all stakeholders, one that considers internal actions, consumer engagement and education as important in an attempt to promote the structural and social conditions necessary to encourage responsible consumer behavior and sustainability more widely.

Different Perspectives

In an influential article by Donaldson & Walsh (2015), the authors invite readers to reflect on the role of business with the prompt: "Law is to justice, as medicine is to health, as business is to _____". Unsurprisingly, responses differ widely. This remains the case when contemplating the role of business in sustainable consumption. Some (still) argue that it is not the responsibility of businesses to address why and how (much) people consume, while others suggest that businesses should merely focus on sustainable products and supply chain management following ecoefficiency principles. These approaches may change *what* people consume, but not necessarily *how much*. Hence, while valuable for changing consumption *patterns*, these approaches have so far failed to reduce consumption *levels*.

Addressing consumption *levels* is the emergent discussion on post-growth and **degrowth** approaches for business. These approaches have gained popularity since the early 2000s, driven by critiques of consumer societies in Europe and the lack of significant progress in environmental and social sustainability. They advocate for "living well with less" (i.e., sufficiency). In a degrowth economy, businesses are in a state of contraction and may need to downsize. In contrast, in a post-growth economy, some businesses may find themselves in a steady state or are simply not expected to grow in size or efficiency, among others (see **Steady-State Economy**). While a more detailed introduction to either concept goes beyond the purpose of this entry, it is worth noting that it is an emerging field and neither approach rejects growth outright, usually distinguishing between quantitative and qualitative growth.

In this context, businesses are exploring alternative ways with a more holistic, long-term focus to create value for consumers *without* exceeding planetary boundaries, overconsuming natural resources, or depleting ecosystem services. A promising example are sufficiency-oriented businesses that have emerged, questioning the necessity of perpetual growth and the sustainability of expansionary business logic. These businesses are driven by environmental concerns, aspirations for better **work-life balance**, and a desire to conduct business based on sustainable (i.e., socio-ecological) values with sufficiency in mind rather than the values of investors or venture capitalists (Soulis & Mont, 2024). While creating tensions for conventional growth-oriented businesses, the role of social and environmental justice and equity (rather than just profit) becomes increasingly important in the post-growth discourse when limited resources must be used to satisfy the (consumption) needs of a growing population (see **Climate Justice**).

Application

The different perspectives on the role of business in the transformation to sustainable consumption are reflected in both the various roles that consumers play and in how businesses apply distinct communication and marketing approaches to engage consumers in *(eco)efficiency*, *shift,* and *sufficiency* strategies.

The (eco)efficiency approach is a management strategy that aims to create more goods and services while using fewer resources and reducing waste and pollution. Moreover, customer communication centers on educating consumers about the environmental impacts of the products they purchase thus helping consumers make more environmentally sound choices without necessarily altering their overall consumption levels.

In the shift approach, the focus moves beyond mere information and seeks to encourage changes in consumer behavior (see **Behavior Change**). Here, businesses shift to alternative business models such as reuse, renting, leasing, and sharing, which help increase product durability (see **Sharing Economy**). Communication strategies emphasize how products can be used and maintained over long periods, for example, repaired rather than replaced (see **Repair**). The shift approach fosters lifestyle changes and a wider, cultural change where consumers value longevity, quality, and the continued use of products in an attempt to move away from a "throwaway mentality".

The sufficiency approach is the most transformative, as it directly challenges existing norms of consumption (see **Social Norms**). Here, business marketing and communication aim to question the need for high levels of consumption and to encourage consumers to critically assess their purchasing habits. Businesses following a sufficiency approach seek to de-legitimize mass production, promoting the idea that less is more. This involves moderating sales tactics, avoiding discounts and seasonal promotions, and refraining from aggressive marketing campaigns that creatively invent new consumer wants (see **Advertising**). The communication focuses on demand reduction strategies. Consumers are encouraged to prioritize needs over desires, fundamentally shifting the way they think about products and engage with consumption.

The varying levels of consumer engagement across these strategies require businesses to develop new capacities and skills. In the efficiency approach, businesses must become adept at providing relevant information and transparency about their products. In the shift approach, they need to create systems that support repair, reuse, and thus overall circularity, and longer product life cycles, promoting alternative modes of provision. For businesses pursuing a sufficiency strategy, the challenge might be even greater, as they must not only shift their marketing approach but also radically transform their business models. This requires decoupling profit from both the environmental impacts and the volume of products sold, transitioning to a model that values sustainability and reduced consumption over perpetual growth. As an approach that aims to make a large-scale, meaningful difference, it will need to involve the active influence of the customer to consume better through transparent, co-creational processes. Take the UK outdoor brand Finisterre's *Wetsuit Project*, which developed a biodegradable wetsuit in collaboration with their local community and customers, aiming to promote circularity while supporting social causes.

In essence, businesses need to become *sustainable market-driving, not market-driven* (cf. Mont et al., 2025). Not to creatively invent new consumer wants but to satisfy actual environmental and societal needs. By leveraging their unique capabilities to influence people's lifestyles and underlying behaviors, businesses can drive the much-needed transition toward sustainable consumption and broader sustainability.

Further Reading

Donaldson, T., & Walsh, J.P. (2015). Toward a theory of business. *Research in Organizational Behavior*, 35, 181–207. https://doi.org/10.1016/j.riob.2015.10.002.

Elf, P., Isham, A., & Gatersleben, B. (2020). Above and beyond? How businesses can drive sustainable development by promoting lasting pro-environmental behaviour change: An examination of the IKEA Live Lagom project. *Business Strategy and the Environment*, 30, 1037–1050. https://doi.org/10.1002/bse.2668.

Elf, P., Werner, A., & Black, S. (2022). Advancing the circular economy through dynamic capabilities and extended customer engagement: Insights from small sustainable fashion enterprises in the UK. *Business Strategy and the Environment*, 31(6), 2682–2699. https://doi.org/0.1002/bse.2999

Mont, O., Elf, P., & Isham, A. (2025). Marketing sustainable lifestyles. In K. Peattie, R. De Angelis, & N. Koenig-Lewis (Eds.), *The Routledge Companion to Marketing and Sustainability*. Abingdon: Routledge. ISBN: 9781003412397.

Soulis, M., & Mont, O. (2024). Towards sustainable fashion: Exploring sufficiency-oriented business models in French fashion SMEs. *Sufficiency in Business*, 22 pages. Transcript, June. Available at: https://www.transcript-publishing.com/978-3-8376-6910-7/sufficiency-in-business/ (accessed: 8 January 2024).

38 Money

*Fatemeh Jouzi, Jarkko Levänen, Mirja Mikkilä,
and Lassi Linnanen*

Definition

Money can be broadly defined as anything that is widely accepted as a medium of exchange, a unit of account, and a store of value. It can be used to buy goods and services, measure their value, and retain value over time.

Money also facilitates consumption. People can satisfy their immediate needs, for instance, by borrowing money. When an individual spends less than they earn, they save money. Savings can then be used for future consumption or to buy financial assets like stocks. In everyday language, buying stocks or bonds is often called investment, but in economics, investment means spending on capital goods like machinery or buildings (see **Sustainable Finance**).

In the current system, most of the money supply is made up of bank-issued money, which is created through the lending process. This process is regulated by the state and the central bank. When a bank grants a loan, it credits the borrower's account with a deposit, effectively creating new money. Most of the money in the economy is created this way, rather than being issued by central banks.

Money fundamentally shapes sustainable consumption and lifestyles by influencing consumer behavior, driving resource allocation and market demands, and perpetuating unsustainable growth. It can thus both catalyze and hamper sustainability transitions through financial (dis)incentives and distribution logic and mechanisms.

History

Before barter systems, small communities relied on reciprocity and social bonds. In societies where everyone knew each other, trust was inherent, and exchanges were based on mutual aid and social obligations. As communities grew and trade between strangers increased, barter systems, where goods and services were directly exchanged, developed. To further facilitate trade, especially in larger and more complex economies, popular commodities such as grain, livestock, and eventually metal coins were used as mediums of exchange.

The introduction of paper money added another layer of trust, now dependent on state authority, and brought new features to money, including easier transport and storage. In the 20th century, the use of credit cards revolutionized transactions, allowing for faster global exchanges and enabling consumers to increasingly spread their spending over time. Later, with the expansion of online shops, micro-financing allowed consumers to buy immediately and pay later in installments. Gradually, the government allowed the role of banks to become more prominent, and the share of money legally issued by central banks considerably decreased relative to the money created by banks through their lending activities.

DOI: 10.4324/9781003584056-42

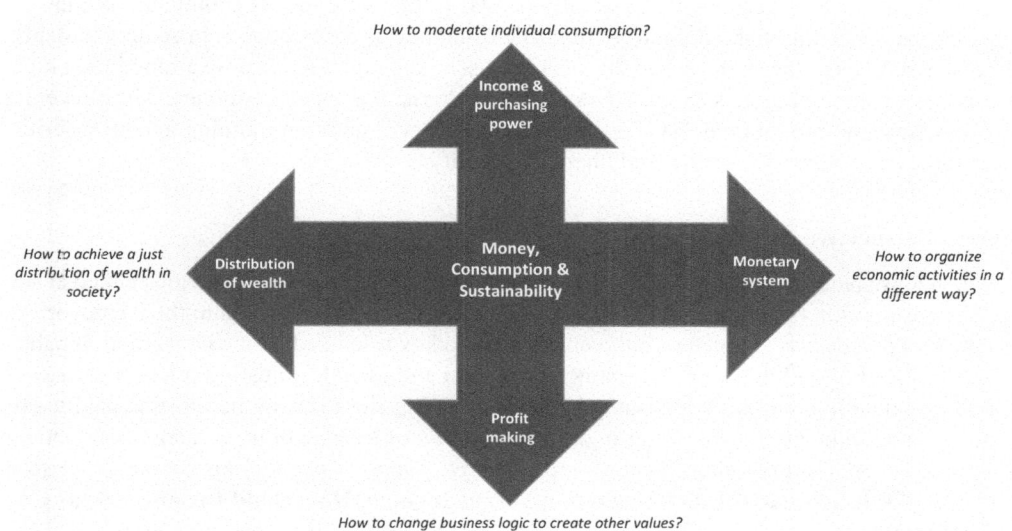

Figure 38.1 Different aspects of money are relevant to sustainable consumption and dimensions of the solutions within the economic system

Source: Adapted from Jouzi et al. (2024)

The concept of money as a store of wealth differs fundamentally from traditional forms of wealth. Traditional wealth has intrinsic value but can depreciate over time due to changes in owner preferences, wear and tear, and technological advancements. As an abstract representation of wealth within a monetary system, money does not suffer from obsolescence. However, the modern monetary system introduced the potential for inflation, which directly affects the purchasing power of each unit of currency and the consumption behavior of individuals.

Over time, money has gained various other social functions (see Box 38.1). It influences decision-making at multiple levels: individuals decide whether to purchase products and services, corporations evaluate costs and profits in a competitive market, and authorities plan policies to achieve environmental, social, and economic goals within their budget constraints. The economic system has become the foundation for most societal interactions, increasingly defined by monetary values. In modern societies, salary and income are often seen as measures of individual success, while businesses focus on maximizing profit, and states assess their performance based on measurable economic growth.

Box 38.1   Money-consumption interconnections

The relationship between money, consumption, and sustainability is complex and multifaceted. Consumption and production are interdependent processes in the economic system. At the business level, companies are obliged to make profits by producing goods and services. At the societal level, the just distribution of wealth is an ideal of strong sustainability, which contrasts with the competitive structure of the current system. Although at the individual level, money is linked to consumption as purchasing power, the interconnection of money and consumption cannot be understood without considering their embeddedness within the greater context of the socio-economic system.

Since the early 20th century, when environmental crises were not as prominent as they are today, scholars have highlighted the troubling role of money as an institution in society and critically discussed it. By the mid- to late 20th century, scholars and advocates examined the consequent crises of the monetary system, advocating for alternative forms of money. More recently, terms like "**sustainable finance**" have appeared in monetary discussions, aiming to address critics and promote environmental, social, and financial stability.

Different Perspectives

There is inconsistency in the approaches to using money for implementing strategies aimed at sustainable consumption (see Figure 38.1). These perspectives can be grouped into three categories.

In the first perspective, monetary intervention plays a key role in adjusting consumption habits. These solutions fall within the current monetary system and include adjusting prices and taxes in favor of sustainable choices for consumption and investing the extra money to financially support sustainable innovations. Investing in different aspects of sustainability is a common tactic in this viewpoint. At the individual level, this perspective supports decoupling consumption from income (see **Well-being and Life Satisfaction Versus Income, Household Income Versus Carbon Footprint**).

The second perspective is revolutionary and criticizes the current monetary system, often suggesting solutions based on avoiding money. Unlike the first viewpoint, monetary intervention is only accepted as a short-term solution and is not seen as effective enough to compensate for the harms caused by overproduction and overconsumption. This view emphasizes promoting non-monetary values and social activities, avoiding markets, and encouraging unpaid work.

The third perspective is a reformist view that calls for restructuring the monetary system while retaining the concept of money as the medium of transaction. This perspective includes reducing the size of the financial system, as seen in local economies, and using local and other alternative currencies. Alternative currencies are forms of money that exist outside of traditional government-issued currencies. Cryptocurrency is a virtual form of currency that allows for transactions over the internet, with Bitcoin being the most well-known example. Two other examples of alternative currencies are those backed by carbon credits to represent a reduction in greenhouse gas emissions, and time-based currencies where hours of labor are the exchange medium.

Economic growth generates money to repay loans, driven by interest, which harms the environment through increased waste and resource use. Proponents of decoupling believe emissions can be separated from growth, aligning with ecological modernization. Opponents argue that continuous growth leads to ever-increasing production and consumption (see **Degrowth**). Lowering consumption challenges growth expectations and the role of money (see **Sufficiency, Steady-State Economy** and **Foundational Economy**). Proposed solutions like green growth, the **circular economy**, and green consumerism have proven slow, ineffective, and insufficient to keep human activities within the planet's carrying capacity (see **Doughnut Economics**)

Application

Monetary intervention: At the state level, adjusting taxes and subsidies to control emission levels and promote sustainable consumption is a common solution. Proponents of decoupling suggest using the extra revenue from taxing environmentally harmful activities or products to support sustainable investments and innovations through grants and funds, or to make sustainable choices more affordable by adjusting prices. This includes the idea of life cycle pricing for products and campaigns to avoid cheap products that do not reflect their true costs. Other proposed monetary interventions include guaranteed income and unconditional basic income for all, with the goal of fair wealth distribution and social equality (see **Universal Basic Services**). The revolutionary perspective acknowledges the necessity of these actions only as temporary measures. It highlights the shortcomings of solutions involving monetary transactions and promotes solutions that are independent of money.

Behavior-income dependency: The wealthiest 25% of the global population is responsible for 74% of excess energy and material consumption. Increased income leads to larger living spaces, more air travel, and luxury goods, contributing to higher emissions and waste. While some argue that affluent individuals and high-income celebrities can promote sustainable behaviors, it is undeniable that investment is a primary income source for affluent individuals. This reliance on investment may cause resistance to post-growth solutions, thereby hindering changes in traditional business practices and the current economic system.

Money rebound effect: In the context of money-consumption dependency, the **rebound effect** of money refers to unintended increases in consumption from reduced consumption elsewhere. For instance, if affluent groups reduce consumption, prices may drop, leading to increased consumption by lower-income groups. This shift between groups of consumers can be justified by equality aims in sustainable societies. Supporters of the monetary intervention perspective debate if revenue from consumption or income taxes will increase consumption elsewhere unless the saved money or tax income is invested in sustainable practices and investments.

Income-well-being opposition: Many sacrifice well-being for money, working excessively, and missing family time. Reducing working hours is a practical solution, but it needs support from other social, cultural, and economic initiatives, due to the current system's emphasis on monetary values. Social ecological economists advocate restructuring economic activities to focus on meeting society's needs and provisioning, rather than monetary evaluation (see **Ecological Economics**). In such a system, individual life or a successful business is not dependent merely on salary and profit.

Further Reading

Alcott, B. (2008). The sufficiency strategy: Would rich-world frugality lower environmental impact? *Ecological Economics*, 64, 770–786. https://doi.org/10.1016/j.ecolecon.2007.04.015.

Core Economy. (2017). *The economy (eBook). Unit 10: Banks, money, and the credit market.* Available at: https://www.core-econ.org/the-economy/book/text/10.html (accessed: 29 October 2024).

Jouzi, F., Levänen, J., Mikkilä, M., & Linnanen, L. (2024). To spend or to avoid? A critical review on the role of money in aiming for sufficiency. *Ecological Economics*, 220. Elsevier B.V. https://doi.org/10.1016/j.ecolecon.2024.108190.

Nielsen, K.S., Nicholas, K.A., Creutzig, F., Dietz, T., & Stern, P.C. (2021). The role of high-socioeconomic-status people in locking in or rapidly reducing energy-driven greenhouse gas emissions. *Nature Energy*, 6, 1011–1016. https://doi.org/10.1038/s41560-021-00900-y.

Read, R. (2009). Towards a green philosophy of money. In L. Leonard & J. Barry (Eds.), *Special edition: Financial crisis – environmental crisis: What is the link? Advances in ecopolitics*, Vol. 3, pp. 3–26. Leeds: Emerald Group Publishing Limited. https://doi.org/10.1108/S2041.

39 Climate Justice

Jennie C. Stephens

Definition

Climate justice is a term used to describe a growing social movement and a transformative approach to climate action that centers equity and justice. Climate justice acknowledges the extremely uneven and inequitable impacts of climate change; it emphasizes the injustice that those who are contributing the most to worsening climate instability are among the least vulnerable and they are powerful actors resisting change (see **Household Income Versus Carbon Footprint**). The people, communities, and households who are suffering the most from climate disruptions are contributing the least and are often already marginalized (see **Carbon Inequality**). A climate justice approach to climate action, climate policy (see **Co-Benefits of Climate Policy**), and sustainable consumption goes beyond the mainstream technological emphasis on decarbonization and the goal of reducing greenhouse gas emissions. It prioritizes social, economic, and institutional innovations that advance societal transformation toward greater economic justice, global solidarity, and an inclusive vision of a hopeful future for all.

The term has been defined by diverse communities of scholars, activists, and change-makers in different contexts around the world. For some, climate justice is about linking human rights and development to achieve a human-centered approach that safeguards the rights of the most vulnerable. Climate justice is also about distributing and sharing the burdens and benefits of climate change and its resolution equitably and fairly. The term is often used to describe resistance to international climate inequities among and within different countries and global regions; it is also used to characterize local disparities in climate impacts and vulnerabilities. Acknowledging the colonial legacy of the uneven distribution of climate suffering, climate justice resists continued economic extraction and exploitation and prioritizes a vision of an alternative economic system that prioritizes human well-being and planetary health (see **Well-being Economy**). Advocates point out that climate action that ignores justice perpetuates the continued concentration of wealth and power and exacerbates climate inequities.

History

Since climate change became increasingly acknowledged as a major threat to humanity in the 1980s and 1990s, the injustices of its colonial legacy and the inequities in climate vulnerabilities have been increasingly evident. In response to the growing crisis in both global injustices and climate, the climate justice movement converged from international environmental, nongovernmental organizations and grassroots activism, focusing on local vulnerabilities and demands for community voice and sovereignty (see **Social Movements**, **Grassroots Innovation**).

Throughout the 2010s and 2020s, a global climate justice movement based on global solidarity and feminist principles has been steadily expanding, in response to a growing sense of

DOI: 10.4324/9781003584056-43

dispossession, disconnection, and disruption. This movement connects struggles for liberation, justice, and peace with resistance to capitalism, extraction of fossil fuels, and exploitation of people and communities. Within the climate justice movement, transformative proposals for new ways of structuring societies are gaining traction. Many climate justice activists are advocating for alternative economic systems that would promote more sustainable consumption patterns and more ecologically healthy lifestyles, including the solidarity economy, the **well-being economy**, **doughnut economics**, the caring economy, the post-growth economy, and the **sufficiency** economy.

Different Perspectives

Climate justice is both an aspirational vision of a more equitable, stable, healthy future and a process or approach to reorienting climate action to connect with other global struggles. Within academic literature, climate justice includes ideal conceptions and normative arguments of justice theory related to reparative justice, distributive justice, and procedural justice. Among activists standing up for marginalized and vulnerable people and communities, climate justice includes resistance, protest, and other forms of advocacy. For many, climate justice focuses on integrating Global South priorities in how humanity responds to the multiple intersecting crises. The focus within climate justice on empowering and supporting people and places that are most vulnerable and precarious is coupled with a focus on reducing the power of the most wealthy and influential people and places. Climate justice highlights the fact that structural economic and financial changes are required to reduce the consumption and greenhouse gas emissions of the top earners, rich countries, and other elite people and organizations (see **Fair Consumption Space**).

Given the disproportionate suffering from the impacts of climate change on Indigenous communities, many argue that Indigenous leadership is essential for climate justice to be achieved. Faced with the ineffectiveness of patriarchal, male-dominated technocratic approaches to climate action, many also argue that feminist principles are essential.

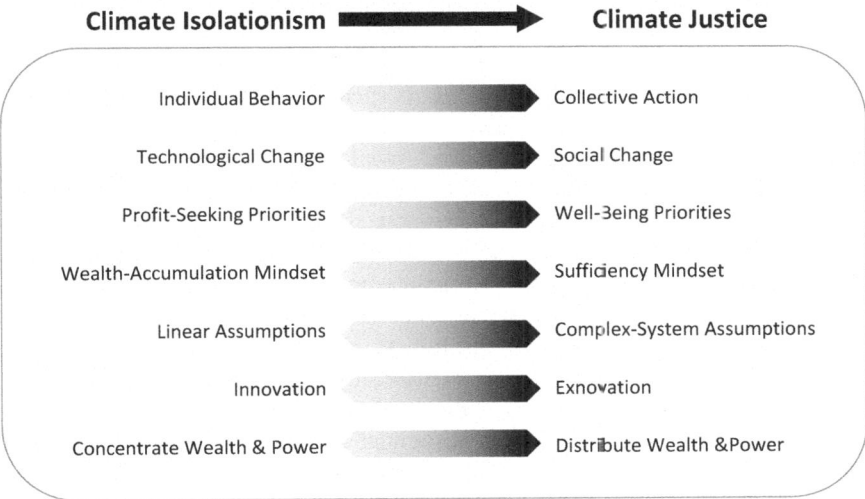

Figure 39.1 Climate policy approaches: from climate isolationism to climate justice

Source: Reproduced with permission from Jennie C Stephens in *Climate Justice and the University* (Johns Hopkins University Press, 2024)

As the term climate justice becomes more mainstream, it is increasingly being used performatively, without a strong commitment to transformative social and economic change. For example, some universities claim to advance climate justice while their climate action plans continue to focus narrowly on technological approaches to reducing greenhouse gas emissions. Other organizations also use the term without acknowledging how many so-called climate policies (including subsidies for electric vehicles and incentives for solar panels) disproportionately benefit economically privileged households and communities, which perpetuates economic inequities and exacerbates climate vulnerabilities.

As calls for climate justice grow louder in communities around the world, resistance to climate justice is also expanding. Many powerful actors and organizations, including the fossil fuel industry and those representing other corporate interests profiting from expanding unsustainable consumption, are threatened by the climate justice call for transformative change. The prospect of structural and systemic economic change focused on equity, justice, and meeting the needs of the most vulnerable people and communities is incompatible with many corporate priorities. Thus, efforts to delegitimize and dismiss climate justice advocacy are growing. For example, some fossil fuel companies have been attempting to undermine democratic processes in disadvantaged communities and to deflect blame from the devastating impacts on human and ecological health in climate-vulnerable places (see **Consumer Scapegoatism**).

Application

Climate justice is a concept applied at multiple scales. It is used internationally to address the global inequities of climate instability. Within the United Nations Framework Convention on Climate Change (UNFCCC) Conference of the Parties (COP) international climate negotiations, climate justice became a focus. At the international level, the devastating vulnerabilities of a country like Bangladesh demonstrate the urgent need for climate justice. Bangladesh has historically contributed only a fraction of the world's greenhouse gas emissions. Yet the low-lying delta nation is suffering from extreme floods and sea-level rise from climate change. The high-emitting countries responsible for intensifying the climate impacts in Bangladesh are not yet taking responsibility for the harm they are causing.

Climate justice is also applied to climate actions designed to address inequities and disparities of climate vulnerabilities within cities and regions. As such, it is used for advocating infrastructure improvements in the favelas of Rio de Janeiro to reduce the climate vulnerabilities of millions living in precarious conditions. In the United States, local, state, and federal policies to, for example, retrofit housing stocks, replace local fossil fuel distribution infrastructure, or improve public transit, sometimes include subsidies for low-income residents and so-called environmental justice communities. Similarly, the construction of subsidized housing for low-income residents is increasingly required to have high-energy performance and low greenhouse gas emissions.

Climate justice is increasingly recognized as a necessary paradigm shift, given the ineffective and insufficient mainstream approach to climate policy, often pushing technological solutions that are non-transformative and benefiting those who are already privileged. It breaks from approaching climate policy through a narrow "climate isolationism" lens that focuses on individual behavior, technological change, profit-seeking, a wealth-accumulation mindset based on linear assumptions, and innovation that concentrates wealth and power (see Figure 39.1)

Instead, climate justice encourages transformative structural change based on collective action, social change, well-being priorities, and **sufficiency** (see also **Social Norms**). It is based on complex system assumptions, the distribution of wealth and power, and exnovation, that is, an intentional move away from practices, systems, and technologies that are known to cause ecological destruction and harm (see **Sustainable Finance**, **Political Economy of Consumerism**).

Further Reading

Feminist Action Nexus. (2022). *Concept and key demands for economic and climate justice.* https://wedo.org/wp-content/uploads/2022/2005/ActionNexus_KeyDemands_EN.pdf (accessed: 8 January 2024).

Newell, P., Srivastava, S., Naess, L.O., Torres Contreras, G.A., & Price, R. (2021). Toward transformative climate justice: An emerging research agenda. *WIREs Climate Change,* 12(6), e733. https://doi.org/10.1002/wcc.733.

Rice, J.L., Long, J., & Levanda, A. (Eds.). (2023). *Urban climate justice: theory, praxis, resistance.* University of Georgia Press. Available at: https://ugapress.org/book/9780820363769/urban-climate-justice/ (accessed: 8 January 2024).

Stephens, J.C. (2024). *Climate justice and the university: Shaping a hopeful future for all.* Johns Hopkins University Press.

Sultana, F. (Ed.). (2024). *Confronting climate coloniality: Decolonizing pathways for climate justice.* Edward Elgar Publishing.

40 Ecosocial Contract

Saamah Abdallah

Definition

The United Nations Research Institute for Social Development (UNRISD) defines the social contract as "the explicit or implicit agreements between state and citizens defining rights and obligations to ensure legitimacy, security, rule of law and social justice". According to European think tanks IDDRI and the Hot or Cool Institute, it encompasses the "rights we enjoy, the duties we agree to, the responsibilities incumbent on institutions and the narratives we believe in". There is no agreed definition of what an *eco*social contract (also called a "natural social contract") is, but UNRISD argues that it "must recognize that humans are part of a global ecosystem. It must protect essential ecological processes, life support systems and the diversity of life forms, and pursue harmony with nature".

History

The concept of a social contract emerged in the 17th century in the United Kingdom and France as a way to understand political authority without relying on increasingly tenuous claims of "divine right". The argument is that citizens cede some of their freedoms and allow the state to control aspects of their lives because this grants them security or better living standards. For Hobbes, this was a way to justify the monarch's absolute power. Locke and Rousseau had more democratic objectives in their visions. To some degree, the social contract is a story or a metaphor – a set of beliefs and expectations. Where there are formal agreements and rules, for example, constitutions, these are typically ratified by elected representatives rather than citizens themselves.

The social contract has evolved over the centuries, with multiple pacts emerging. Hobbes originally focused on a security pact – where citizens concede the state has a monopoly on the legitimate use of violence in exchange for physical security. Later, the American and French revolutions provided scope for a democracy pact – where citizens forfeited a direct political voice in exchange for the possibility to elect representatives, and for political equality. In Western countries such as the United Kingdom and France, the 19th century saw the emergence of a working pact – with increasing protection and welfare rights for workers in exchange for an acceptance of Fordist production models and greater hierarchy. In the 20th century, a consumption pact emerged, whereby citizens had greater and greater expectations in terms of the freedom and possibility to consume while accepting greater alienation at work, competitive consumerist pressures, and increasing inequality. Meanwhile, the security pact continues to evolve, with further aspects of life falling under a similar logic of centralized power and rule-making to ensure greater safety – this logic applies for example to food, health, and more recently environmental considerations.

DOI: 10.4324/9781003584056-44

The idea of an *eco*social contract emerged in the 2020s, motivated by two arguments. On the one hand, there is the perception that our current social contract, shaped by individualism, materialism, short-termism, and the free market, is not compatible with the ecological challenges we face. On the other hand, there is the argument that, by failing to address climate change adequately, states are failing citizens on their most basic obligation – security

UNRISD has initiated a research program and a research network on the idea of an ecosocial contract, together with the Green Economy Coalition. Meanwhile, IDDRI and Hot or Cool have stepped into the discussion by trying to better understand the state of our *current* (non-eco)social contract. By doing so, these sustainability-minded organizations have tried to ensure that calls for a new social contract coming from more influential actors (e.g., the UN Secretary-General, or the International Trade Union Congress) are imbued with an ecological perspective.

Different Perspectives

The debate on the ecosocial contract is still nascent, but some differences in perspective can be observed, for example, regarding the framing of the problem to be addressed. For some, the only solution to the environmental crises we face is for nature (including the rights of nature) to be explicitly front and center of a new social contract. Others – while mindful of the need for greater environmental protection – see the main purpose of a new social contract as being about addressing the political backlash against pro-environmental policies.

From this perspective, many of the solutions to the sustainability challenge already exist – the problem is that there has been much less thought on *how* to ensure these solutions have political and public support. Discussing the social contract presents an approach whereby new societal rules can be co-created without being imposed upon society (see Box 40.1). In that sense, the ecosocial contract agenda overlaps with the citizen participation agenda, stressing the need for the public to be involved in the process of defining a new social contract through citizens' assemblies and other participatory approaches.

Box 40.1   Envisioning an ecosocial contract

What could an ecosocial contract look like?

- Explicit commitment by citizens to reduce their personal environmental impact in exchange for the commitment of the state to do everything in its power to minimize the threats of climate change (through mitigation and adaptation).
- Collectively abandoning the belief that individuals have the right to consume as much as their income allows (see **Personal Carbon Allowance**).
- An evolution of the work ethic existing in some countries. Working hard would not be seen as a virtue in itself, but rather virtuous to the extent that it contributes to society.
- Aside from the protection of those who are out of work, the state expects to ensure that those who are at work can attain a decent living standard, perhaps through the provision of **universal basic services**.

These are just speculative suggestions. For those who stress the importance of citizen participation, it makes little sense to already define what the social contract should be without involving citizens in the co-creation process.

Other authors take politically salient notions such as the Just Transition and Green New Deal as their starting point but argue that neither is sufficient to truly address the sustainability challenge. In effect, these authors are arguing that a focus on production and the welfare state is not enough to reduce emissions. Rather, over-consumption in the Global North needs to be addressed directly, and a new social contract can help toward that.

Applications

Three national-level examples have included some features that can be understood as elements of an ecosocial contract. Between 2007 and 2008, a Constituent Assembly was set up in Ecuador to draft a new constitution. The elected assembly included politicians and also academics, and representatives of NGOs and indigenous communities. The final constitution was then approved in a referendum by citizens. The constitution was partly inspired by indigenous philosophy, in particular the concept of Sumak kawsay (translated into Spanish as **buen vivir** or living well, but also sometimes as "living in harmony"). This was reflected in the inclusion of the rights of nature into the constitution, which has helped environmental movements in Ecuador, for example, to restrict oil drilling in the Amazon. Despite several changes of government, the constitution still holds in Ecuador today, and the clauses on the rights of nature have recently been used to protect the river through the capital city Quito.

Less successful was the attempt to introduce a new constitution in Chile in 2022. As with the case in Ecuador, the Convention charged with drafting the new constitution was elected and included a large spread of society, with many independent candidates. The proposed constitution also included elements of the rights of nature, as well as the right to a clean and healthy environment. In this case, however, the proposed constitution was rejected in a referendum for being too "left-leaning".

Another example, which has been framed as an ecosocial contract, is the 2015 constitution in Nepal. This constitution was focused on ensuring social rights and also included rich elements of environmental protection.

Further work is being conducted by the Green Economy Coalition, Hot or Cool, and IDDRI to explore how to better integrate the opinions of representative, marginalized, and "affected" groups (affected in the sense that they are likely to be most affected by societal or economic changes associated with the sustainability transformation) to create a new ecosocial contract.

Further Reading

Huntjens, P. (2021). *Towards a natural social contract: Transformative social-ecological innovation for a sustainable, healthy and just society.* https://doi.org/10.1007/978-3-030-67130-3.

Kempf, I., Hujo, K., & Ponte, R. (Eds.). (2023). *Global study on new eco-social contracts.* Geneva: United Nations Research Institute for Social Development.

Norton, A., & Greenfield, O. (2023). *Eco-social contracts for the polycrisis. Participatory mechanisms, Green Deals and a new architecture for just economic transformation.* Green Economy Coalition.

Saujot, M., Bet, M., Abdallah, S., Bengtsson, M., & Rogers, C. (2024). *Towards a 21st Century Social Contract – How did we get here? A short history of 19th and 20th century social contracts in France and the UK.* IDDRI & Hot or Cool.

Willis, R. (2020). A social contract for the climate crisis. *IPPR Progressive Review,* 27(2), 156–164. https://doi.org/10.1111/newe.12202

41 Ecological Economics

Lucia A. Reisch

Definition

Ecological economics focuses on rebalancing human-nature interactions within a finite system of planetary boundaries. Its main field of research is the environment and sustainable development, rather than the traditional economic concepts of efficiency, market equilibria, and growth. In contrast to the latter, it views the economy as embedded in natural processes, such as biological, chemical, and physical transformations. It highlights the limitations and opportunities of an economy and society that intentionally stays within the planet's carrying capacity.

Ecological economics regards the human economy as one subsystem embedded in a larger ecosystem, closely connected to and dependent on other systems through physical and chemical material flows. This contrasts with orthodox neoclassical economics, which focuses on markets and financial flows and views the market economy as the dominant system with a license to externalize environmental costs and exploit nature. While environmental and resource economics apply neoclassical economics to environmental problems, ecological economics uses a broader view and includes the material consequences of inputs, outputs, and waste in the analysis. In ecological economics, nature has an intrinsic value and is more than just a resource and sink for consumption and production purposes.

One key indicator ecological economics uses is the "ecological footprint", which measures the ecological impacts of everyday activities and practices of individuals or even whole industries. Based on local empirical evidence, policymakers can decide which consumption and production activities must be curbed – and which should be supported. A second set of important indicators are instances of violations of "environmental justice", such as the health and climate impacts of factories or mining on neighboring communities, or the fatal effects of large-scale deforestation on biodiversity as well as animal and human welfare (see **Climate Justice**). Intra- and intergenerational equity and fairness worldwide are further defining principles.

Ecological economics advocates for stable, **steady-state economies** to avoid ecological overshoot beyond the planetary boundaries. It addresses global challenges such as climate change, deforestation, overfishing, and biodiversity loss. With its focus on the interactions between human decision-making and the environment, ecological economics is a field that thrives on diverse perspectives and expertise converging in interdisciplinary and transdisciplinary work. It is a space where pluralistic frameworks and theories come together to study sustainable consumption. This inclusive and collaborative approach, which assembles researchers, students, policymakers, and **prosumers** who want to solve environmental problems, is at the core of the field's mission and scope.

DOI: 10.4324/9781003584056-45

History

Drawing on the work of economist Nicholas Georgescu-Roegen, Herman Daly masterminded the field's foundational models in the early 1970s. Ecological economics was institutionalized as a field of study by the end of the 1980s with the establishment of the *International Society for Ecological Economics* in 1988, the society's journal *Ecological Economics* in 1989, and its first international conference in 1990.

Environmentally motivated critiques of consumption and consumer society emerged in the mid-1960s, reflecting the challenges of the early environmental movement that was also formative for ecological economics. Early voices – such as Herman Daly (1968) and Fred Hirsch (1976) – warned of the environmental and social costs of the dominant paradigm of maximum economic growth and the societal expectation of ever-increasing consumption. Ecological economics developed an alternative, holistic, and long-term view of what "progress" means, using a systems perspective of the individual being embedded in a natural and social environment. However, the "double dividend" promises that reducing consumption would make us better off remained a niche belief.

The Brundtland Report and the Rio Conference – the first COP – officially acclaimed the relevance of consumption for sustainable development in 1992. A decade later, UNEP launched its 10-Year Framework Programme on Sustainable Consumption and Production Patterns, which laid the base for the Sustainable Development Goals (SDG) in 2015. SDG 12 sets concrete goals for more sustainable consumption patterns by ending water, energy, material, and food wastage, staying within the planetary boundaries, and reducing the ecological footprint. Since then, science and policy increasingly embraced the view of the economy as embedded in a larger ecosystem. Climate change and extreme weather events have made our dependency on a planet in balance ever more salient. Consumers, as one major contributor to the ecological crisis, are guided toward reducing waste, adopting circular approaches (such as **repairing**, reusing instead of buying, repurposing, recycling, or composting), increasing material efficiency by, for example, car sharing and swapping clothes and tools, and learning how to fulfill their wishes with more sufficient, less material and energy intensive options (such as swapping beef for plant-based diets or stopping using planes) (see **Sharing Economy**, **Protein Shift**, **Sustainable Mobility**, **Energy Consumption Behavior**).

Different Perspectives

Ecological economics has primarily developed from a critique of neoclassical, environmental, and resource economics. Its beginnings were fuelled by a deep frustration with the dominant economic growth paradigm that maximized consumption measured in GDP. For some voices, however, ecological economics does not distance itself enough from the traditional approaches, since it adopts some of the same fundamental concepts and stays within the contested perspective of capitalism. For instance, some critics claim that by using economic terms such as "natural capital", the interpretations of value and cost of nature, and the definition of the optimal scale of economic activity, ecological economics reinforces rather than breaks with dominant capitalistic values. These critical voices demand and develop explicitly anti-capitalist solutions, some within, others beyond the ecological economics paradigm. Others, however, question the value of fuzzy complex concepts such as "natural capital" or "optimal scale", which are complex to assess and subject to value judgments, limiting their practical impact. These different perspectives reflect the usual and much-needed debates that shape an emergent field or discipline, rather than constituting a profound rejection of its key propositions.

Applications

Beyond the continuous development of the principles and models of ecological economics, the field has inspired practical applications that promote sustainable consumption and production, translating theory into practice. One example with a significant impact has been the idea of natural capital and "putting a price tag" on ecosystem services. Over the years, several Natural Capital Accounting and assessment methods have evolved. Being able to estimate the value of ecosystem services, such as pollination or carbon sequestration, has been a big step forward in convincing governments about the benefits of environmental policies in traditional cost-benefit exercises. Moreover, having a shadow price available allows landowners, farmers, forest owners, and so forth to be compensated for maintaining or restoring the biodiversity of an area. A less obvious application is making the costs and benefits of circular economy projects transparent, to support this critical approach of minimizing waste and maximizing material efficiency. Last, estimates of ecosystem services are the base of carbon offset markets that allow emitters – including private consumers – to offset their carbon impact, at least on the balance sheet.

The story of ecological economics exemplifies a trend toward the establishment of applied, transdisciplinary fields that use a systems approach, especially those crossing natural and social sciences. Considering today's "new normal" of multiple permacrises, a field that (i) thinks the economic, social, and ecosystems together, (ii) studies feedback loops, non-linear dynamics, and interconnectedness, (iii) harnesses the knowledge of natural and social science disciplines, and (iv) views the individual as more than just a market actor is well equipped to develop solutions and policy recommendations. Orthodox economics failed to deliver viable solutions for the climate and biodiversity crises beyond internalizing externalities. Ecological economics, in contrast, has spurred innovations in consumption and production practices, promotes circular concepts, and aims to create welfare for all within the planetary boundaries (see **Doughnut Economics**).

The good news is that today, economics as a discipline has become broader in scope and more inclusive. Most economists today see the potential of heterodox economic approaches such as ecological economics. The complexity and escalating urgency of moving toward sustainable consumption and production patterns and the concrete threats of planetary overshoot, with all its ramifications, call ever more for an ecological economics approach.

Further Reading

Costanza, R., D'Arge, R., de Groot, R., Farber, S., Grasso, M., Hannon, B., Limburg, K., Naeem, S., O'Neill, R.V., Paruelo, J., Raskin, R.G., Sutton, P., & van den Belt, M. (1997). The value of the world's ecosystem services and natural capital. *Nature*, 387(6630), 253–260. https://doi.org/10.1038/387253a0.

Daly, H. (1968). On economics as a life science. *Journal of Political Economy*, 76, 392–406. https://doi.org/10.1086/259412.

Hirsch, F. (1976). *Social limits to growth*. Routledge.

Røpke, I. (2004). The early history of modern ecological economics. *Ecological Economics*, 50, 293–314. https://doi.org/10.1016/j.ecolecon.2004.02.012.

Røpke, I., & Reisch, L.A. (2004). The place of consumption in ecological economics. In *The ecological economics of consumption*. https://doi.org/10.4337/9781845423568.00007.

42 Well-being Economy

Anders Hayden

Definition

One definition of a well-being economy (WE) consistent with a post-growth, post-consumerist vision comes from the Well-being Economy Alliance (WEAll), a global collaboration of groups and individuals seeking economic system transformation. WEAll (2024) describes a WE as "an economy designed to serve people and the planet, not the other way around. Rather than treating economic growth as an end in and of itself and pursuing it at all costs, a Wellbeing Economy puts our human and planetary needs at the center of its activities . . .". For WEAll, a WE delivers on the need for fairness, dignity, purpose, nature, and participation. However, readers should be aware that when they encounter the term, others may use it differently (see Different Perspectives below).

History

Various streams of thought have contributed to the concept of a WE. The New Economics Foundation's 2004 "Well-Being Manifesto for a Flourishing Society" provided one early call for a WE. The manifesto advocated going beyond gross domestic product (GDP) to "measure what matters" – an idea central to all major variations on the WE today. It also (i) highlighted the limits of further income and consumption growth in improving well-being in affluent nations (see **Well-being and Life Satisfaction Versus Income**), (ii) critiqued the negative impacts of "materialism" on humans and the environment (see **Consumerism**), and (iii) proposed giving people more free time as an alternative to ever-increasing consumption (see **Work-Life Balance**).

In 2012, Bhutan – known for its commitment to Gross National Happiness – hosted a high-level meeting at the United Nations on "Happiness and Well-Being: Defining a New Economic Paradigm". It subsequently convened an International Expert Working Group (IEWG), leading to a report to the UN calling for a new paradigm centered on seeking happiness within planetary boundaries to replace the "doctrine of limitless growth". Some IEWG members carried on their work via the Alliance for Sustainability and Prosperity, which contributed to WEAll's founding in 2018.

In Finland – a prominent WE proponent – the "economy of wellbeing" concept was introduced by health and social affairs NGOs in 2012, emphasizing the importance of all three aspects of sustainability (economic, social, and environmental) in delivering human well-being, but without directly challenging the pursuit of economic growth as a goal.

The WE has made considerable inroads into the political mainstream, notably the creation of the Wellbeing Economy Governments (WEGo) by Iceland, New Zealand, and Scotland in 2018. The following year, the Council of the European Union endorsed an economy of well-being, a

DOI: 10.4324/9781003584056-46

concept Finland promoted during its Council presidency. Both Finland and Wales joined WEGo in 2020. Iceland hosted the first annual WE Forum in 2023, leading some to imagine that it could one day rival the World Economic Forum in Davos and its focus on economic growth.

Variations on the concept include the World Health Organization's Geneva Charter for Well-being, which highlighted the "urgency of creating sustainable 'well-being societies,'" while the World Wellbeing Movement – a coalition of business, civil society, and academic leaders – speaks of a WE as a goal.

Different Perspectives

A WE has emerged as one main alternative to an economy geared toward ever-growing consumption with success measured by GDP. It has the potential to unite supporters of various causes such as addressing the climate and biodiversity crises; tackling inequality, poverty, and social marginalization; enhancing citizen participation; and generating good lives and sustainable consumption patterns based on an ethic of **sufficiency**. However, as WE understandings vary, efforts are required to ensure that the concept means more than minor reforms to business-as-usual.

The EU Wellbeing Economy Coalition (2023), for example, describes a WE as one "in which everyone can thrive, within planetary boundaries" and "an economic system that is no longer structurally dependent on economic growth". This understanding of a WE has an affinity with other post-growth concepts such as **doughnut economics** and **ecological economists**' work on "prosperity without growth" (see also **Degrowth, Steady-State Economy**). Some WE formulations, however, do not directly challenge a growth- and consumption-centered economy.

Some observers celebrate the WE's capacity to bring post-growth ideas into the mainstream. However, research by Hayden and Dasilva (2022) and Mason and Büchs (2023) have shown that the WE vision that governments and major international organizations have embraced is not – at least so far – the radically transformative, post-growth approach that WEAll and some others have advocated.

The Organisation for Economic Cooperation and Development has been a significant source of mainstream backing for a WE. It emphasizes "a 'virtuous circle' in which individual well-being and long-term economic growth are mutually reinforcing". This idea has also been voiced by Finland and the Council of the European Union; it is a WE understanding with much in common with by-now conventional ideas of "sustainable and inclusive" growth.

Application

Hayden and Dasilva (2022) characterize the practice of WEGo nations as a "weak post-growth" approach. These nations all remain committed to economic growth either as a goal in itself or as a means to achieve well-being. Indeed, they still depend on growth for purposes such as generating revenue for well-being-enhancing social spending. Yet WEGos have taken small steps in line with post-growth thinking, including some downplaying of economic growth as the overarching societal objective, moving beyond GDP alone toward more comprehensive economic and social performance indicators, and starting to orient policy toward achieving well-being. Examples of the latter include New Zealand's "wellbeing budgets" and Iceland's well-being priorities to guide budget allocations. WEGo nations have also introduced some **sufficiency**-oriented policies that limit particularly damaging forms of consumption and encourage alternatives, with notable examples in Wales (see Box 42.1).

Box 42.1   Practice insight: well-being economy in Wales

Wales has arguably gone the furthest in pursuing a WE in a way that creates possibilities for transformative change. Its Wellbeing of Future Generations Act of 2015 established seven national well-being goals that emerged from a national conversation on the "Wales We Want": a prosperous, resilient, healthier, and more equal Wales, a Wales of cohesive communities, a Wales of vibrant culture and thriving Welsh language, and a globally responsible Wales. The Act requires public bodies to act in ways that are long-term, integrative, collaborative, preventative, and involve citizens in decision-making. It also requires the government to introduce national indicators to measure progress toward the well-being goals and "national milestones" to assess that progress. Importantly, the Act created the office of the Future Generations Commissioner to protect the interests of future generations and monitor the Act's implementation.

The Future Generations Commissioner has played a key role in curbing the growth of the country's road network and encouraging more sustainable mobility. The first step involved stopping a planned 13-mile-long relief road through environmentally sensitive wetlands to relieve motorway congestion at a cost exceeding £1.4 billion. The Commissioner, Sophie Howe, told the government it should spend its money differently as the plan was not in line with the Future Generations Act. In addition to scrapping the relief road, the government introduced a "new path" transport policy that prioritizes reducing the need to travel and shifting to more sustainable transport modes such as active travel and public transit. In February 2023, Wales took the groundbreaking step of canceling all major road-building projects, with strict criteria for any new construction.

Other **sufficiency**-oriented policies in Wales include establishing a default 20-mile-per-hour urban speed limit, discouraging second-home ownership, and encouraging re-use, repair, and sharing of goods (see **Repair** and **Sharing Economy**). While the Welsh government still pursues economic growth, a less growth- and consumption-centered vision of prosperity is evident – supported by an understanding of well-being that highlights the needs of future generations.

One noteworthy element of a WE is a preventative approach that aims to avoid social and environmental problems rather than paying large sums to fix them later. For example, in healthcare, the emphasis shifts toward social and environmental determinants of health – and to factors such as a healthier **work-life balance**, less environmental pollution, and lower inequality that can help reduce the need for costly healthcare services and pharmaceuticals.

Reducing inequality is a key element of a WE. Evidence shows that countries with lower inequality and strong welfare states have better well-being outcomes on various indicators (see **Carbon Inequality**). Meanwhile, if well-being rather than economic growth is the ultimate goal, it strengthens the case for actions that can boost well-being in less consumption-intensive ways, such as work-time reduction, limits on **advertising** (which generates dissatisfaction with what we already have to get us to buy more), or creating sites where people can share and repair goods and strengthen community ties at the same time.

Alongside such possibilities, a danger exists of watering down the WE concept, allowing governments and other actors to co-opt it without significant change. Some WE advocates, such as Black (2023), warn of "wellbeing washing" – using well-being language without truly transforming policies and practices – while others propose criteria to distinguish a "genuine" WE from "window dressing". Much work remains to shape what is done in the name of a WE.

For a post-growth, post-consumerist WE to prevail, some major obstacles must be overcome. These include the predominance of conventional economic thinking and training, opposition from vested interests, and ultimately the growth dependency of contemporary societies (which requires substantial economic transformation). Some positive initial steps have nevertheless already been taken toward a WE that can generate good lives in a sustainable and equitable way.

Further Reading

Black, I. (2023). *Wellwashing: Why a superficial approach to wellbeing economics will fail.* Glasgow: Common Weal. Available at: https://commonweal.scot/wp-content/uploads/2023/10/Wellwashing.pdf (accessed: 5 February 2024).

EU Wellbeing Economy Coalition. (2023). *Discussion paper on EU wellbeing economy*, p. 8. EU Wellbeing Economy Coalition. Available at: https://ieep.eu/wp-content/uploads/2023/06/Discussion-Paper-on-EU-Wellbeing-Economy.pdf (accessed: 9 December 2023).

Hayden, A., & Dasilva, C. (2022). The wellbeing economy: Possibilities and limits in bringing sufficiency from the margins into the mainstream. *Frontiers in Sustainability*, 3. https://doi.org/10.3389/frsus.2022.966876.

Mason, N., & Büchs, M. (2023). Barriers to adopting wellbeing-economy narratives: Comparing the Wellbeing Economy Alliance and Wellbeing Economy Governments. *Sustainability: Science, Practice and Policy*, 19(1), 2222624. https://doi.org/10.1080/15487733.2023.2222624.

WEAll. (2024). *What is a wellbeing economy*. Wellbeing Economy Alliance. Available at: http://weall.org/what-is-wellbeing-economy (accessed: 14 January 2024).

43 Foundational Economy

Richard Bärnthaler

Definition

The foundational economy is both a concept designed to rethink the character and purpose of contemporary economies and an approach to socio-economic development that focuses on providing everyday universal basics within planetary limits (see **Universal Basic Services**). It emphasizes that there is no single "economy", but rather different economic zones, each with distinct rationales and forms of consumption (see Table 43.1). The foundational economy facilitates the often-collective consumption of daily essentials such as energy, electricity, water, and food, along with welfare-state services like health, education, and social housing. In addition to being essential for meeting human needs, foundational sectors differ from other sectors of the economy in two main ways: (i) consumption needs are predominantly delivered through infrastructures, networks, and branches, rather than through the purchase of individual commodities; and (ii) much is relatively sheltered from international competition.

Table 43.1 Zonal schema of the economy

Economic zones				
Unpaid	Activities valued and paid for in monetary terms in which *rents* can be extracted			
Core economy	*Foundational economy*		*Overlooked economy*	*World-market oriented, tradable, and competitive economy*
	"Material"	"Providential"		
Examples				
Unpaid care and housework	Food, utilities	Health, education, social housing	Gastronomy, hair salons, furniture	Cars, electronics, business consultancy
Form of consumption				
Daily essentials in households and communities	Daily essentials via infrastructure of networks and branches		Occasional purchases of mundane, cultural necessities	(Aspirational) private purchases
Dominant business model				
No charging or recovery of "cost"	*Used to be* low risk, low return, long time horizon for public and private providers		Small enterprises and SMEs vs. financialized corporates	High risk, high return, short time horizon

Source: Bärnthaler, based on various adaptations since 2013

DOI: 10.4324/9781003584056-47

History

The foundational economy approach has evolved over the past decade (2010s). In 2013, the Foundational Economy Manifesto identified a distinct part of the economy "that creates and distributes goods and services consumed by all (regardless of income and status) because they support everyday life". This led to the introduction of a zonal understanding of contemporary economies (see Table 43.1).

Each economic zone has unique principles, ways of working, and forms of consumption. The zonal schema critiques the dominant focus on growing GDP/capita (which adds heterogeneous forms of consumption according to a singular definition of worth), boosting international competitiveness, and viewing the market economy as the defining system (see **Ecological Economics**). Instead, it emphasizes the importance of everyday consumption for the quality of life. While income is portable, consumption in the foundational economy is place-based and depends on household location; deficiencies cannot be remedied by simply raising personal income for private consumption. Therefore, per capita disposable income poorly reflects living standards. The zonal schema also shows that applying a uniform market logic – including privatization, financialization, and short-termism – can threaten universal access to the consumption of daily essentials.

In 2018, the *Foundational Economy* book expanded the discussion from rethinking the economy to providing new framings for politics and policy. In terms of politics, it emphasizes that entitlement to foundational provision is the material basis for the actual exercise of citizenship, enabling social participation. In terms of policy, it began to explore how to restore the collective foundations of everyday life, including reinventing taxation and instituting social licensing (see Box 43.1).

In 2023, amidst the cost-of-living crisis, the book *When Nothing Works* introduced a three-pillar concept of liveability: residual income after covering essential household costs; access to foundational goods and services; and social infrastructures (e.g., parks, community hubs, public spaces). This highlights the need for policy to balance focus between residual-income-based private consumption (e.g., for occasional purchases in the overlooked economy) and infrastructure- and services-based collective consumption. Given that 70% of UK citizens live in multi-person households sharing expenditures, the book argues that the household, rather than the individual, should be the primary unit of analysis.

Different Perspectives

Initially developed in the United Kingdom, foundational thinking has evolved with contributions from across Europe over the past decade, sparking debates and variations within and between national groups. For example, internal and external critiques have pointed out its neglect of unpaid work and ecological concerns, leading to increased engagement with feminist theories and ecological limits as well as exploration of synergies with other concepts of alternative economies.

The foundational economy, **well-being economy**, and **doughnut economy** all advocate for high-income countries to move away from growth-focused strategies to ones that prioritize public services and redistribution. Thereby, foundational thinking is central to bridging the gap between the doughnut economy's heuristic recognition of social foundations and practical policymaking (see Box 43.1).

Critiques highlight that the foundational economy's lack of biophysical realism may hinder its ability to operate within ecological limits, emphasizing that the current provision of services such as food, mobility, housing, and even health are highly resource- and emissions-intense. Engaging with these critiques, some scholars have begun integrating **sufficiency** research into the foundational approach. Examples include addressing excessive residential floor space, implementing progressive pricing models with absolute consumption ceilings, and placing greater emphasis on prevention (see **Sustainable Mobility**, **Sustainable Housing**, **Energy Consumption Behavior**, and **Personal Carbon Allowance**).

Box 43.1   Foundational economy: policy recommendations and examples

The foundational policy objective is to improve access to essential goods and services, which depend on social investment and tax revenues. Compared to **Universal Basic Services**, this approach tends to be more sensitive to local context and systems and is agnostic on ownership models (e.g., private, public, communal). While all places face challenges related to foundational services (e.g., transportation, housing, and adult care), the specific conditions for change vary. Foundational policy differs from mainstream approaches by prioritizing issues such as housing costs, tenure, and redesigning service provision over jobs and growth. Foundational policy priorities include:

Social licensing is essential for regulating corporations in foundational sectors (see **Extended Producer Responsibility**). Businesses extracting cash from a territory via networks and branches should provide social benefits in return, regardless of ownership. For example, Austrian limited-profit housing developers face profit and rent restrictions and must reinvest profits into social housing. Development contracts can also require private developers to include social housing units and invest in public infrastructure. Supermarkets, with their territorial franchises, could be required to promote healthy diets and reduce single-use plastics.

Reinventing taxation, financing, and charging systems to support social investment is another priority. This could involve traditional tax mechanisms (e.g., wealth taxes) and capturing rising land and property values and redirecting them into the foundational economy to enable collective consumption, rather than boosting private investment. Progressive utility charging systems, such as for water, would ease financial burdens on poorer households, ensure long-term financing, and align with theories of production and **consumption corridors**.

Strengthening collective consumption, or what households cannot buy from private income (e.g., local parks, free healthcare), is crucial. Collective consumption can also be advanced through public procurement. For example, the Welsh government's Economic Action Plan, inspired by the foundational approach, emphasizes sustainable public-sector food procurement and free school meals to enhance public health, social justice, and ecological integrity.

Application

It is clear that foundational provisioning systems must undergo radical decarbonization (e.g., housing retrofit), conversion (e.g., focusing on sharing and repairing), transformation (e.g., reducing car dependency, diet reform), and extension (e.g., including the provision of clean air and public green spaces, and restoring ecosystems). Against this backdrop, the foundational economy offers a specific contribution to enabling sustainable consumption and lifestyles, given its (i) focus on collective provisioning for human needs; (ii) effect on inequalities; and (iii) potential role as a cornerstone for a radical social-ecological transformation beyond capitalism.

First, foundational thinking highlights that our capacity to act is always realized in systems we produce and reproduce. Having the capacity to be mobile or to dwell sustainably requires the provision of accessible, affordable, and sustainable mobility and housing systems. This shifts the focus from changing individual behaviors within existing systems (see **Consumer Scapegoatism**) toward the collective shaping of these systems as a primary purpose of politics. Collective in-kind provisioning generally leads to better universal need satisfaction and lower energy requirements than individualized consumption. Thus, improving access to infrastructure-based collective consumption – from local recreational areas and public kitchens to free healthcare and energy communities – contributes to sustainable consumption and lifestyles.

Second, foundational policies recognize that everyday essentials will not be accessible to all according to need if they are left to the realm of private consumption from individual income. This shifts public policy toward renewing collective foundational provisioning, grounded in social investment and a commitment to social priorities. Since, in OECD countries, low-income groups spend around two-thirds of their income on essential services, this approach reduces residual income inequality. As Kate Pickett and Richard Wilkinson have famously shown, reducing inequality benefits individuals, households, and society as a whole, as inequalities erode trust, increase anxiety and illness, undermine public support for climate policies, and encourage excessive consumption (see **Conspicuous/Positional Consumption**).

Third, universal access to the foundational economy serves as a cornerstone for **Degrowth/ Postgrowth**. Prioritizing the foundational economy in public policy requires shrinking economic zones and activities that directly undermine universal foundational provisioning, particularly those driven by rent extraction and redirecting productive capacities toward the universal provision of everyday basics and thus toward collective consumption. This counteracts the creation of artificial scarcities, which restrict people's access to daily essentials and have historically served as the engine of capitalist expansion, from the enclosure of the commons to the commodification of public services. As foundational policy aims to ensure universal access to the goods and services necessary for a good life without requiring high levels of income, it establishes the precondition for "radical abundance" beyond capitalism.

Further Reading

Bärnthaler, R., Novy, A., & Plank, L. (2021). The foundational economy as a cornerstone for a social–ecological transformation. *Sustainability*, 13, 10460. https://doi.org/10.3390/su131810460.

Calafati, L., Froud, J., Haslam, C., Johal, S., Williams, K. (2023). *When nothing works: From cost of living to foundational liveability*. Manchester University Press.

Foundational Economy Collective. (2022). *Foundational economy: The infrastructure of everyday life*. New ed. Manchester University Press.

Schafran, A., Smith, M.N., & Hall, S. (2020). *The spatial contract: A new politics of provision for an urbanized planet*. Manchester University Press.

Wahlund, M., & Hansen, T. (2022). Exploring alternative economic pathways: A comparison of foundational economy and Doughnut economics. *Sustainability: Science, Practice and Policy*, 18, 171–186. https://doi.org/10.1080/15487733.2022.2030280.

44   Steady-State Economy

Brett Dolter

Definition

The concept of a steady-state economy, as it is currently used, was introduced by Herman Daly. In a steady-state economy, physical quantities of people (world population) and our artifacts are held constant. These artifacts include homes, vehicles, fridges, stoves, factories, power plants, and other consumer and producer goods used by people.

Daly's idea was to constrain the stocks of population and artifacts to reduce the demands we put on the planet. Humans and our belongings are created using materials and energy and require a continual stream of materials and energy to function. When these stocks are growing, then the flow of energy and materials required to create and sustain them must also grow, leading to resource depletion and exhaustion (see Box 44.1 and Figure 44.1). Growing stocks of people and artifacts also lead to growing levels of waste and pollution that overwhelm Earth's assimilative capacity, creating problems such as climate change, acid rain, ozone depletion, and the eutrophication of water bodies. (See Figure 44.1 where a growing "full-world" economy takes up much more of the ecological carrying capacity of the planet).

Box 44.1   System dynamics differentiates between stocks and flows

Stocks change over time and are subject to flows. The level of water in a bathtub is a simple stock, subject to the inflows of water from the faucet and outflows of water down the drain. Similarly, the world population is a stock of people that changes subject to the inflows of births and the outflows of deaths. When births outnumber deaths in a given period, the world population increases. When the opposite is true, the world population decreases. If births and deaths are equal, the world population stays "steady", in dynamic equilibrium, at a given population level. This same kind of dynamic equilibrium can be achieved with stocks of consumer and producer goods if inflows of investment equal outflows of retirement and depreciation (see **Stocks Versus Flows**).

The stocks of consumer and producer goods that Daly suggests we constrain are not without benefit; they all provide services in our economy. The housing stock provides shelter (see **Sustainable Housing**). The stock of automobiles, buses, trains, planes, and bicycles offers the service of mobility (see **Sustainable Mobility**). If we constrain the number of homes and vehicles we produce on Earth, we risk creating a stagnant economy that does not meet the needs of the population.

To avoid this stagnation, Daly suggests we focus on constraining the energy and materials that we use as inputs to create and sustain the stocks. If we constrain our energy and resource inputs,

DOI: 10.4324/9781003584056-48

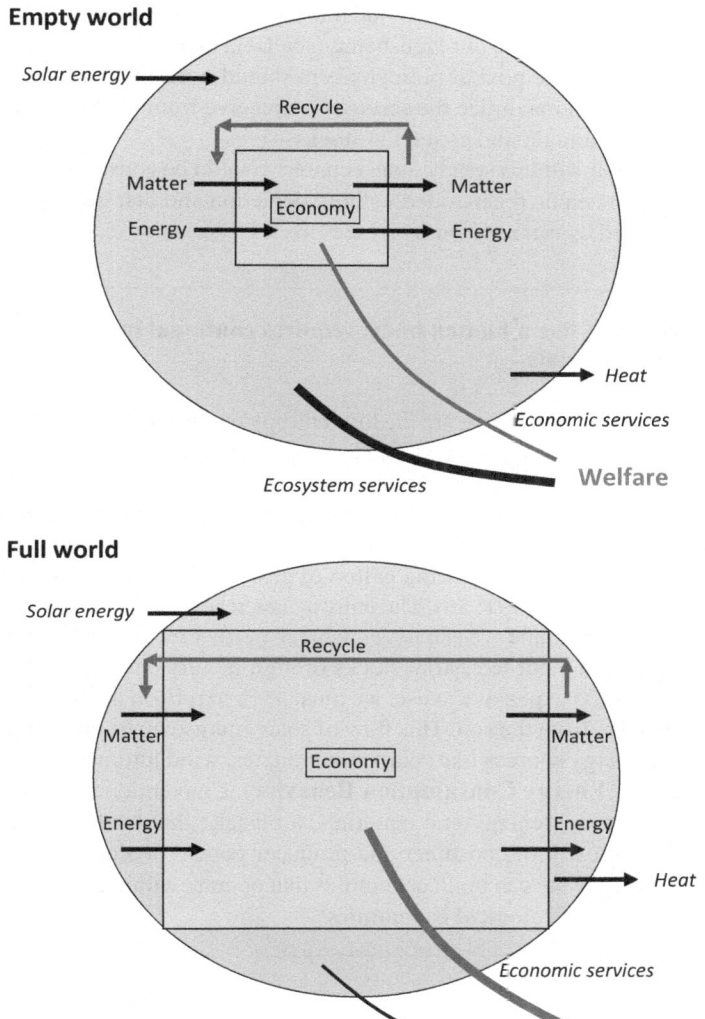

Figure 44.1 Welfare in a full versus empty world

Source: *Ecological Economics 2nd Edition*, by Herman E. Daly and Joshua Farley. Reproduced by permission of Island Press, Washington, DC

we can still create more efficient stocks that use less energy and materials, and that better meet our needs. This would be a process of qualitative improvement that Daly calls "economic development". Just like a new recipe can provide culinary delights with the same quantities of ingredients, chef's labor, and cooking time, economic development would allow a given level of stock and material throughput to generate greater levels of happiness (see **Well-being Economy**).

In contrast to this vision of sustainable economic development, Daly defines economic growth as a focus on growing the stock of consumer and producer goods, requiring increasing inputs of energy and materials. Bigger stocks might provide more services to humanity, but would also generate ecological costs that undermine our well-being (see **Degrowth**).

In a steady-state economy, the goal of policymakers should not be to maximize the size of our stocks of goods, but instead to maximize the services we receive from our stocks, and to minimize the throughput needed to maintain and create the stocks.

Sustainable consumption within a steady-state economy would be consumption that uses energy and materials at a rate that can be regenerated without depletion and that generates levels of waste that can be safely absorbed by our environment (see Box 44.2).

Box 44.2 An economy, like a human body, requires continual inputs of high-quality energy and materials

In thermodynamic terms, these inputs are the low-entropy energy and material throughput we need to drive the system. Entropy is key to understanding the logic of the steady-state economy. According to the first law of thermodynamics, energy cannot be created or destroyed. However, according to the second law of thermodynamics – the entropy law – the energy available to do work will decline in each transformation of energy. As Daly writes, "Were it not for entropy, we could burn the same gallon of gasoline over and over, and our capital stock would never wear out" (1991: 36). The entropy law teaches us that we can never create a perpetual motion machine, and we can never achieve perfect recycling of materials. As we burn fossil fuels, we use up our terrestrial stocks of high-quality energy. Once this "ancient sunlight" is used up, or too expensive to use, we must learn to rely on the flow of solar energy that we receive each day from the sun. This flow of solar energy can be converted to electricity using renewable energy sources like solar photovoltaics, wind turbines, and hydroelectric power production (see **Energy Consumption Behavior**). Conventional economics ignores our reliance on low-entropy energy and materials, a mistake that leads economists to pursue the never-ending growth of consumer and producer goods. Ecological economists like Herman Daly explore how we can build economies that operate within the flow of renewable energy and resources (see **Ecological Economics**)

History

Daly built his concept of a steady-state economy on the work of his PhD supervisor Nicholas Georgescu-Roegen, who identified entropy as the ultimate cost of economic activity. He was likewise influenced by the work of Kenneth Boulding and his conception of a "spaceman economy" that must operate within the finite limits of the Earth.

Herman Daly was not the first economist to discuss the possibility of a steady-state economy. Daly points back to John Stuart Mill (1848) and his writings on the subject in *The Principles of Political Economy*:

I cannot . . . regard the stationary state of capital and wealth with the unaffected aversion so generally manifested towards it by political economists of the old school. I am inclined to believe that it would be, on the whole, a very considerable improvement on our present condition.

Mill, like Daly, argues that growth involves increasing costs. Both Mill and Daly also argue that a steady (or stationary) state of population and artifacts would not mean the end of human progress. Mill wrote:

> It is scarcely necessary to remark that a stationary condition of capital and population implies no stationary state of human improvement. There would be as much scope as ever for all kinds of mental culture, and moral and social progress; as much room for improving the Art of Living, and much more likelihood of its being improved, when minds ceased to be engrossed by the art of getting on.

Different Perspectives

Theories of complexity and adaptive resilience challenge the steady-state economy proposed by Daly. In a steady-state economy, humanity works to find a level of throughput at which we avoid passing what would now be called "planetary boundaries", those critical thresholds after which great calamity befalls human civilization. Existing at this level would assume a balanced view of nature where we can stay within limits and maintain stability. However, ecosystems also experience random, stochastic change, and dynamic instability. A forest may grow to maturity, begin aging and decaying, and then experience a fire that burns its aging structures, leaving a meadow where the potential for new growth begins anew. This process of adaptive cycles was called *Panarchy* in a key text by Lance Gunderson and C.S. Holling, who envision our world as a dynamic, ever-changing place in which systems can be more or less resilient. Rather than working toward a steady-state economy, we might work to build a resilient economy, in which we plan for the unexpected, learn from our mistakes, and recognize the dynamic nature of Earth systems.

Application

To implement a steady-state economy, Daly proposes three institutions:

1. A system of "transferable birth licenses" would control human population growth. This institution was partially established in China with its one-child policy, although China's policy lacked a mechanism for transferring the right to have a child from one family to another. This policy proposal attracts criticism from conservative religious circles for encouraging birth control and from progressive circles for threatening women's ability to choose how many children they have.
2. A system of "depletion quotas", auctioned by governments, would control the rate of energy and material use and depletion. The idea of depletion quotas for key energy and materials has not been pursued widely, though cap-and-trade systems are used to restrict emissions of pollution that result from burning fossil fuels (see **Personal Carbon Allowance, Carbon Inequality**).
3. To ensure inequality does not worsen, Daly proposes minimum and maximum income limits and a maximum limit to personal wealth. In a steady-state economy, we cannot depend on a growing stock of economic artifacts to raise the living standards of the poor. When the pie is not growing larger, we must ensure a fair distribution of the pie. Minimum incomes have gained prominence with calls for a universal basic income. Maximum limits on income and wealth have not been as widely pursued (see **Household Income Versus Carbon Footprint, Life Satisfactions Versus Income**, and **Fair Consumption Space**).

Apart from Daly's suggestions, scholars such as Peter Victor have worked to detail the policies that would allow countries to enhance quality of life, while achieving sustainability goals like eliminating greenhouse gas emissions (see also **Well-being Economy**). Victor recommends policies such as a reduced workweek that allows more leisure time and reduces a key labor input into the economic production process (see **Work-Life Balance**). Tim Jackson makes similar policy proposals and Victor and Jackson collaborate on modeling policies that can achieve stable, non-growing economies. Kate Raworth's **Doughnut Economics** concept similarly helps conceptualize an economy that operates within planetary boundaries and that meets the needs of the world's population. The **Degrowth** movement advocates for simple lifestyles that require lower levels of consumer and producer goods.

Further Reading

Boulding, K.E. (1966). The economics of the coming spaceship earth. In H. Jarrett (Ed.), *Environmental quality in a growing economy*, pp. 3–14. Resources for the Future, Johns Hopkins University Press.
Daly, Herman E. (1991). *Steady-state economics*. 2nd ed. Island Press.
Jackson, T. (2017). *Prosperity without growth: Foundations for the economy of tomorrow*. 2nd ed. Routledge.
Raworth, K. (2017). *Doughnut economics: 7 ways to think like a 21st century economist*. Chelsea Green Publishing.
Victor, P.A. (2019). *Managing without growth: Slower by design, not disaster*. 2nd ed. Edward Elgar Publishing.

45 Doughnut Economics

Elena Dawkins, Angela Druckman, and Benedikt Schmid

Definition

The Doughnut Economy is a conceptual framing best described as a "safe and just living space for humanity" (see Figure 45.1), where the needs of all humanity are met within the physical boundaries of the planet. It combines social justice issues with environmental concerns.

The "sweet spot" is between the inner circle, which depicts the social foundation of human well-being that no one should fall below, and the outer circle, which represents the ecological ceiling of planetary pressure that humanity should not go beyond. Above the social foundation, peoples' needs for life's essentials, such as food, water, healthcare, housing, gender equality, and political voice are met. The ecological ceiling represents planetary boundaries, such as climate change, land conversion, and biodiversity loss, which should not be transgressed for the Earth to remain in a stable state within which humanity can thrive. This is part of a group of concepts that define the upper and lower boundaries for consumption and economic activity levels that provide the essential quality of life for humans in the present, while not degrading it for future generations and nature (see also **Consumption Corridors**, **Fair Consumption Space**).

Doughnut Economics aims to design a regenerative and distributive economy: an economy that is more equitable and that creates a net positive impact on the environment and society. It puts social and environmental principles at the center of policymaking, thus leading to an economy where the social foundation is met without overshooting the planetary boundaries. Doughnut Economics was first introduced by Kate Raworth. In her seminal 2017 book, Raworth posits that moving toward the Doughnut requires seven major economic transformations, including three fundamental shifts in conceptualizing the economy: (i) change the understanding of the goal of the economy, from chasing GDP growth to living within the Doughnut; (ii) recognize that the economy depends on society and the living world, rather than treating it as self-contained; and (ii) relinquish the baseless belief that economic growth, as measured by GDP, will eventually reduce social inequalities, and instead design a deliberately redistributive system. Doughnut Economics is growth-agnostic: while it advocates a move away from the myopic pursuit of economic growth, it does not take a position regarding the green growth/degrowth/post-growth debates. Instead, Doughnut Economics fore-grounds the material (and social) corridor within which human consumption needs to take place. It calls for a shift away from high mass consumption and the narrow view of humans as consumers rather than citizens (see **Consumer-Citizen**).

History

Doughnut Economics originates from Kate Raworth's work at Oxfam, a non-governmental organization (NGO) dedicated to alleviating global poverty. In her 2012 Oxfam Discussion Paper titled

DOI: 10.4324/9781003584056-49

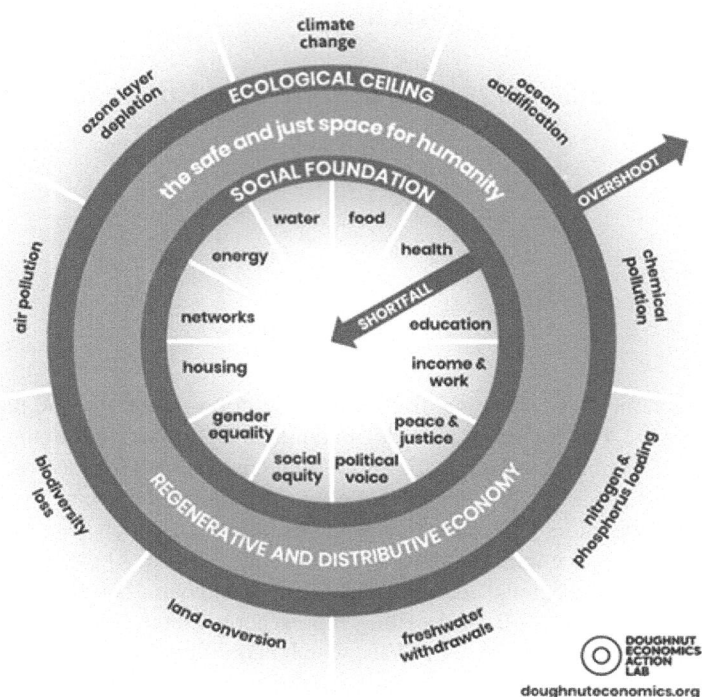

Figure 45.1 The Doughnut

Source: Doughnut Economics Action Lab

"*A Safe and Just Operating Space for Humanity*", Raworth introduces the Doughnut framework as a new model of prosperity.

Doughnut Economics builds on long-standing efforts and concepts of sustainable development, integrating the work of human rights advocates who champion every person's right to life's essentials with that of **ecological economists** who stress the importance of situating the economy within environmental limits. Related complementary concepts include, for example, the **Well-being Economy**, **Foundational Economy**, **Fair Consumption Space**, **Consumption Corridors**, and **1.5-Degree Lifestyles**.

The social foundation of this framework was initially based on United Nations discussions leading up to the Rio+20 conference in 2012. During these discussions, governments identified 11 social priorities to be achieved over the next decade, focusing on enhancing well-being, productivity, and empowerment. The 12 dimensions in the 2015 UN Sustainable Development Goals are now used as the social foundation.

The ecological ceiling of Doughnut Economics is based on the concept of planetary boundaries, established by 28 Earth system scientists in 2009. This concept proposes thresholds for nine critical Earth system processes, which, if exceeded, could lead to irreversible and, in some cases, abrupt environmental change, moving Earth out of a stable state. Led by the Stockholm Resilience Centre, the planetary boundaries framework has been revised multiple times since its inception, with the latest update in 2024 quantifying all boundaries and concluding that six of the nine have already been breached.

Raworth followed up on the 2012 discussion brief with a 2017 book titled *Doughnut Economics: Seven Ways to Think Like a 21st-Century Economist* that further explored the economic thinking required to bring humanity into the Doughnut. These seven principles are shown in Figure 45.2.

Different Perspectives

The Doughnut Economics framework has sparked various critiques, with the question of growth proving particularly contentious. Advocates of growth-based development challenge its proposed economic rethinking, relying on familiar arguments in favor of competition, market efficiency, and technological progress. In contrast, critics of growth as a societal priority argue that the agnostic position of Doughnut Economics' thinking fails to address the destructive aspects of GDP and allows for continued growth rather than promoting genuinely alternative development pathways (see **Degrowth**). Empirical studies on the implementation of the tools and concepts of Doughnut Economics suggest that growth, indeed, often reasserts itself – not unlike findings in related approaches like the **Well-being Economy** or **Circular Economy**. As a result, some critics frame Doughnut Economics as "watered down" and easily instrumentalized for "business as usual".

Such criticisms often overlook what proponents perceive as a key advantage of Doughnut Economics compared to more radical approaches: it has successfully prompted engagement by various actors and communities (see the Applications section) including institutional actors integrating the Doughnut framework into formal structures. Empirical evidence suggests that the model's clear communicability and ideological flexibility are central to its success. Raworth herself acknowledges ties to more radical approaches such as Degrowth, stating, "*It's not the intellectual position I have a problem with. It's the name*". The model's "growth agnostic" stance may, therefore, be seen as a strategic choice, enabling it to gain traction without being dismissed immediately by established stakeholder groups.

However, these critiques highlight an important gap in the approach's consideration of power dynamics, politics, and instrumentalization. Doughnut Economics' "go-where-the-energy-is" attitude tends to avoid directly confronting incumbent actors and institutions and instead focus on successes rather than shortcomings. While telling alternative stories of economic thriving is crucial, it is equally important to consider those who remain disadvantaged, despite – or even because of – initiatives inspired by Doughnut Economics (see **Climate Justice**).

Application

Initially designed with a global perspective, the Doughnut Economics framework has been adapted for national and subnational applications, exploring its potential to drive meaningful change at various levels. To support the practical application of the framework, the Doughnut Economics Action Lab (DEAL) was established in 2019. Launched as a Community Interest Company in 2020, DEAL aims to move from the ideas of the Doughnut framework to transformative action. The platform provides tools and resources for collecting locally relevant, data-led targets and indicators to create a "Portrait" of a place with local aspirations and global responsibilities for the social foundation and ecological ceiling.

A challenge that arises when scaling down the planetary boundaries for local applications is that the baseline remains global, and calculating national or local shares requires the application of a distributional mechanism. It could be the case that improving access to basic resources and services in some areas would increase the pressure on, and potentially contribute to breaching, the ecological ceiling. This would have to be addressed through extensive redistribution and reduction

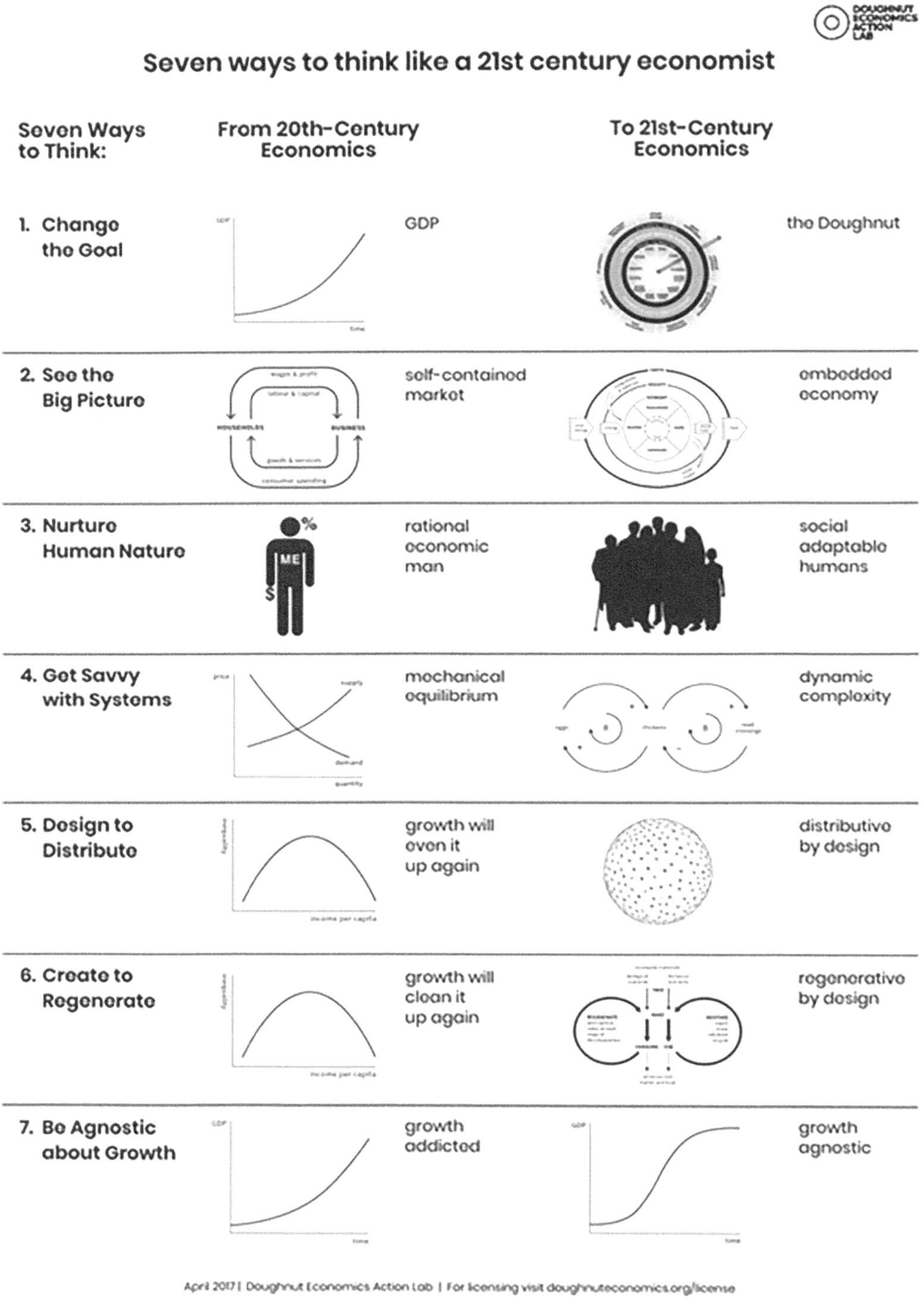

Seven ways to think like a 21st century economist

Seven Ways to Think:	From 20th-Century Economics		To 21st-Century Economics	
1. Change the Goal		GDP		the Doughnut
2. See the Big Picture		self-contained market		embedded economy
3. Nurture Human Nature		rational economic man		social adaptable humans
4. Get Savvy with Systems		mechanical equilibrium		dynamic complexity
5. Design to Distribute		growth will even it up again		distributive by design
6. Create to Regenerate		growth will clean it up again		regenerative by design
7. Be Agnostic about Growth		growth addicted		growth agnostic

April 2017 | Doughnut Economics Action Lab | For licensing visit doughnuteconomics.org/license

Figure 45.2 Seven ways to think like a 21st-century economist

Source: Doughnut Economics Action Lab

of consumption and pressures through, for example, technological and institutional innovations, to ensure that minimum needs are met for all (see **Sufficiency**).

DEAL also collates stories of application of the framework in different communities globally. Applications include engagement, communication, awareness raising, education, data-led evaluations of progress, and supporting local authority policy, planning, and decision-making (see examples in Box 45.1). The aims, outputs, and outcomes of applications vary according to the lead organization (e.g., local authority, community hub, education, or action group), resources and funding, data availability, and participation.

While the various applications of the Doughnut framework testify to its value as an analytical tool and a visual framework for communication, awareness-raising, and decision-making, its use encounters several challenges. These include general obstacles faced by transformative approaches, such as siloed decision-making, lack of higher-level political support, and difficulties in navigating power dynamics, inequalities, and trade-offs (see **Political Economy of Consumerism**). Furthermore, the downscaling of the model often comes along with a lack of data or indicators on regional and local levels and needs to acknowledge the specificities of place. Academic studies have focused on resolving technical and data-related issues and several practitioner-led approaches have been developed.

Box 45.1   Examples of applications of Doughnut Economics

Community groups

Numerous Dommunity groups, community hubs, and action organizations have explored the Doughnut framework for their locality, by creating what the DEAL calls a "Community Portrait of a Place" using participatory workshop approaches. This aims to provide a holistic picture with diverse inputs and perspectives that can be a starting point for action. A Community Portrait of a Place can be completed alongside a Data Portrait of a Place, which aims to build a quantifiable picture of a place in the Doughnut. Examples listed include a neighborhood in Birmingham (UK), Minato ward in Tokyo (Japan), and Bielsko-Biała (Poland).

Policy, planning, and decision-making

Several cities and regions are applying Doughnut Economics in planning and policymaking including Amsterdam, Melbourne, and the Brussels City Region. The most prominent example is Amsterdam, which was the first municipality to develop a "City Portrait": similar to a Community Portrait, this methodology downscales the Doughnut from global to city level. Other examples include the German municipality Bad Nauheim, which has launched a comprehensive public engagement initiative aimed at integrating Doughnut Economics principles at the local level. The Swedish municipality Tomelilla is using the Doughnut model for building a new school (Schmid, 2024).

Education

According to the DEAL website, children and young people understand Doughnut Economics faster than most adults. DEAL provides activities and lesson plans. Youth activities have been carried out in countries such as Greece, Slovakia, and the Netherlands.

Business

DEAL launched a tool for businesses to engage with Doughnut Economics in 2022. It provides guidance on how businesses can transform their "deep design", focusing on a company's purpose, networks, governance, ownership, and finance (see **The Role of Business**, **Sustainable Finance**). The aim is that following this, companies will be empowered to pursue strategies, practices, and business models to help move toward the Doughnut.

Further Reading

Doughnut Economics Action Lab. (n.d.). Available at: https://doughnuteconomics.org/ (accessed: 20 August 2024).

Fanning, A.L., O'Neill, D.W., Hickel, J., & Roux, N. (2022). The social shortfall and ecological overshoot of nations. *Nature Sustainability*, 5(1), 26–36. https://doi.org/10.1038/s41893-021-00799-z.

Raworth, K. (2017). *Doughnut economics: Seven ways to think like a 21st-century economist.* Chelsea Green Publishing.

Savini, F. (2024). Post-growth, degrowth, the doughnut, and circular economy: A short guide for policymakers. *Journal of City Climate Policy and Economy*, 2(2), 113–123. https://doi.org/10.3138/jccpe-2023-0004.

Schmid, B. (2024). *The spectre of growth in urban transformations: Insights from two Doughnut-oriented municipalities on the negotiation of local development pathways.* Urban Studies. https://doi.org/10.1177/00420980241305322.

46 Degrowth

Sam Bliss, John Mulrow, Megan Egler, and Lindsay Barbieri

Definition

Sustainable consumption decisions at the individual and community scale will fail if humanity consumes too much of the Earth overall. We will transgress critical thresholds of planetary sustainability, beyond which lurks a future global ecology that is likely inhospitable to many species, including our own. Thus, the overconsuming nations of the world must reduce the extraction and pollution their economies cause. This is degrowth.

Degrowth is an intentional, democratic downscaling of the global economy that also creates space for the realization of social equity. It is sustainable consumption applied to the level of the whole economy (see Figure 46.1). It entails shrinking rich societies' use of materials and energy in ways that prioritize justice and well-being for all (see **Ecosocial Contract**, **Well-being Economy**, **Doughnut Economics**).

History

The workers and dispossessed peoples on whose backs the global economy has grown have long criticized growth. Degrowth draws on historical resistance to the violence that economic growth requires: labor exploitation, industrial-scale extraction, and the poisoning of landscapes. But degrowth also comes from the dissatisfaction of growth's supposed beneficiaries, who have struggled against ills of modernity such as alienation and the acceleration of the pace of life (see **Work-Life Balance**, **Well-being and Life Satisfaction Versus Income**). Even scholars famous for their ideas about how economies achieve growth have imagined worlds beyond it. In 1848, John Stuart Mill described a culturally rich, post-growth "stationary state", while in 1932 John Maynard Keynes predicted a future where productivity increases would bring us a 15-hour work week: more leisure time rather than growing material wealth. In the 1960s and 1970s, academics and environmentalists began to warn that Earth's limits would impose the end of growth, as critical resources like oil ran out. Half a century later, the size of the global economy – gross world product – and the total mass of materials extracted for it have both tripled, growing hand-in-hand (see **Stocks Versus Flows**). As the climate heats and species go extinct at one thousand times the background rate, it is now obvious that what we are running out of is not a particular resource, but the functionality of the whole Earth.

Degrowth has emerged in the 21st century as a proposal for collectively limiting ourselves, as human societies, according to the world we want to inhabit. Its recent origins lie in the French scholar-activist movement for *décroissance*. Rather than waiting for the planet to impose limits on growth, degrowth is about stopping economic expansion on purpose. Policy proposals for degrowth draw on the work of academics who were worried about reaching the limits to growth decades ago,

DOI: 10.4324/9781003584056-50

such as the caps on the use of various critical resources that Herman Daly devised to maintain a **steady-state economy**. Today, degrowth is a diverse international movement and a familiar, if somewhat misunderstood, term to anybody who reads the news (see **Social Movements**).

Different Perspectives

Degrowth is not simply a matter of shrinking the economy in its current form. That would be a recession, and it would come with all the impoverishment and suffering that recessions bring. Instead, degrowth entails reorganizing society such that everybody's basic needs can be met even as the economy contracts (see **Sufficiency**). As the editors of a volume on degrowth put it, "The objective is not to make an elephant leaner, but to turn an elephant into a snail" (D'Alisa et al., 2015).

The opportunity to *reorganize* the economy opens degrowth to myriad perspectives on economic justice, political organization, and ways of living together. A great variety of vocabularies and disciplinary backgrounds meet and mingle in the field of degrowth. Here, we summarize three complementary perspectives:

- Biophysical perspective: For the last 10,000 or so years, the Earth provided humanity and the larger community of life with rather stable living conditions. Around these, human communities patterned their social development, migrations, and rituals. The planet's relative stability has been supported by a finite collection of biophysical resources. Humanity has overcommitted those resources to its own purposes, causing planetary instability. The sustainability literature is full of metrics for these limits, such as the nine planetary boundaries. Degrowth, then, is about steering the human ship back within the boundaries of its host. However, scholarship on degrowth acknowledges that if unsustainability is seen as simply a math problem, we can lose sight of social complexities and power relationships.
- Decolonial perspective: The colonial powers that have dominated economic decision-making and resource flows over the past several centuries have played an outsized role in our species' takeover of the Earth. Because these powers have mismanaged things to the point of disaster, degrowth calls for both reducing their power and uplifting voices and ways of living from outside the colonial frame. Today, resources still flow overwhelmingly from South to North (see **Consumption-Based Accounting**). Modifying the international architecture of power through reparations and redistribution is a key goal of degrowth. The consumption of the rich, and that of rich countries, must degrow dramatically not just to make ecological space for less-developed countries to grow if they choose to, but to give the rest of the world *conceptual* space and control over their own resources and destinies. With this, they can define the good life as they see fit and pursue it within their means (see **Buen Vivir and Buenos Convivires**, **Ubuntu** and **Food Sovereignty**).
- Anti-capitalist perspective: Degrowth theorists tend to blame capitalism for the drive toward endless economic growth. The process of capital accumulation generates a continuous need for consumption growth and an ongoing pursuit of new profitable frontiers for investing the profits from previous rounds of money-making activity. In practice, this entails extending the frontiers of resource extraction and pollution. Thus, to downscale the economy in a way that can be maintained socially, it is contended that nothing less than a political transformation beyond capitalism will suffice.

Advocates of degrowth emphasize these and other perspectives to varying degrees depending on their backgrounds and affinities. A key strength of the degrowth movement – though not without its challenges – lies in its ability to bring these diverse perspectives into dialogue, fostering mutual learning and, ultimately, integration. As such, degrowth is often described as a "movement of movements" aimed at achieving global social-ecological transformation.

Application

No nation has pursued degrowth explicitly, though many have implemented policies and programs that make progress toward creating an economy that neither grows nor collapses for lack of growth. These include reducing working hours, scaling down harmful sectors like the military, and providing universal access to necessities like housing and healthcare (see **Universal Basic Services, Foundational Economy**).

At a smaller scale, degrowth is reflected in small-scale experimental projects and traditional societies in which people live together convivially with low levels of material consumption. From poor neighborhoods to ancient civilizations, modern ecovillages to foraging societies, people have long been practicing the art of pursuing a frugal, shared good life without growth, sometimes under the banner of degrowth but mostly not (see **Eco-Communities, Alternative Consumer Cooperatives, Alternative Hedonism, Grassroots Innovation**). Many seek simple and peaceful cohabitation; others are explicitly developing low-carbon or low-consumption lifestyles (see **Quiet Sustainability, Voluntary Simplicity**).

Certainly, initiatives for reducing the environmental impact of consumption abound, many of which are discussed in this book. Such initiatives build readiness for degrowth. Indeed, degrowth demands sustainable lifestyles for all. However, as long as governments are officially and legally committed to growth, all our well-meaning efforts will not deliver planet-saving power at a scale that matters (see Box 46.1). Degrowth is thus sustainable consumption's willing – and inevitable – partner in pursuing just and lasting planetary health for all.

Box 46.1   Banking sustainable consumption's savings requires degrowth

Do an image search for "sustainable future" and you will find some very predictable visions: collages of wind turbines, solar panels, and people tending to gardens, interspersed with bike lanes and rail lines; hands holding some soil and a seedling; hands holding a globe, or a globe shaped like a light bulb. Such images are meant to be both enticing and informative. The future looks greener and cleaner and also consists of a distinct set of equipment and consumption choices. The implied logic of transitioning to an ecology-compatible future is this: install alternative equipment, cultivate sustainable habits, and the transition to a sustainable future will happen.

What is hardly ever depicted in these bright green visions of a sustainability transition is the *economic context*. It's easy to imply that biking instead of driving, or using solar power instead of coal, has a smaller ecological impact. However, the ecological resources saved through sustainable infrastructure and consumption decisions are not left unused; their fate is shaped by globe-spanning economic linkages, most of which prioritize economic growth. There are plenty of potential fossil fuel users lined up to claim those saved liters of gasoline. This partially explains why steady renewable energy growth has occurred alongside the growth of fossil fuels – and thus global carbon emissions. Renewables are being added to energy supplies, not replacing existing ones (see **Energy Overshoot, Rebound Effects**).

If we want to bank the ecological savings made possible by sustainability efforts, we'll need to set a limit – not just on carbon (see **Personal Carbon Allowance**), but on economic activity itself. This will allow alternative energy and ecological practices to claim an increasing share of limited economic space. By drawing the macroeconomic border around sustainability visions, degrowth clarifies the planetary context for delivering on the eco-promises being made by dedicated communities and caring individuals across the world.

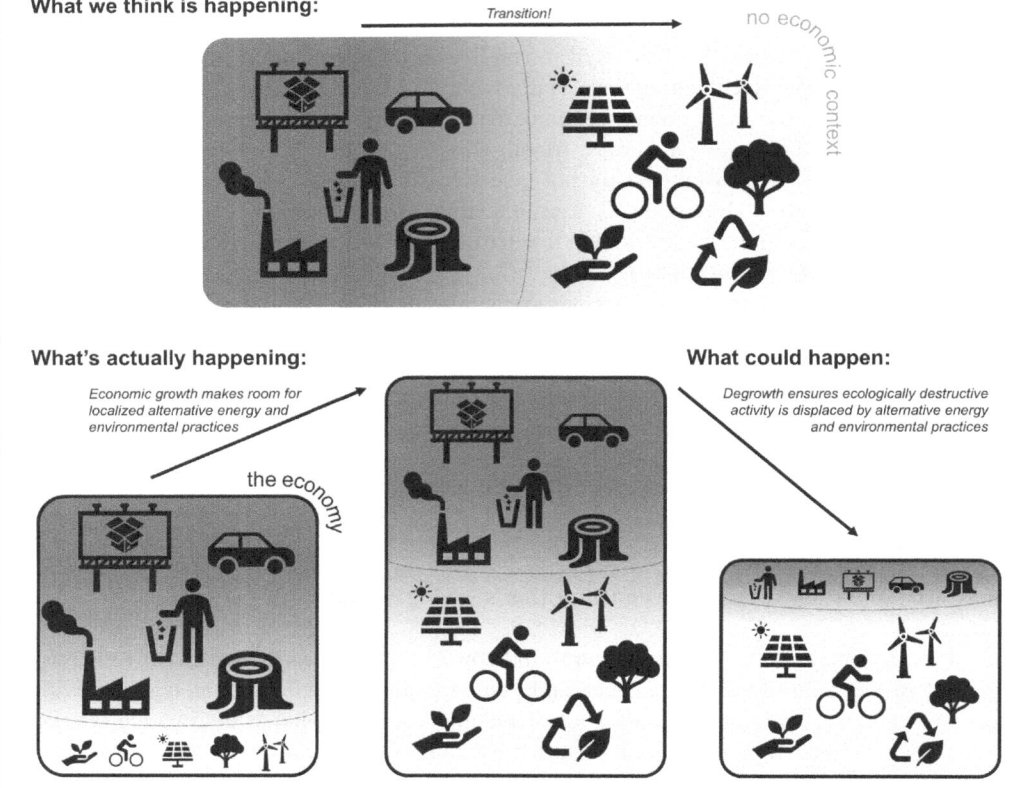

Figure 46.1 Images of a sustainable future

Source: John Mulrow and coauthors, 2025. Own illustration created for this publication

Further Reading

D'Alisa, G., Demaria, F., & Kallis, G. (Eds.). (2015). *Degrowth: A vocabulary for a new era*. London: Routledge.
Hickel, J. (2021). *Less is more: How degrowth will save the world*. London: Windmill.
Meadows, D.H., Meadows, D.L., Randers, J., & Behrens, W. (1972). *The limits to growth: A report for the club of Rome's project on the predicament of mankind*. New York: Universe Books.
Paulson, S. (2017). Degrowth: Culture, power and change. *Journal of Political Ecology*, 24(1), 425–448. https://doi.org/10.2458/v24i1.20882.
Schmelzer, M., Vetter, A., & Vansintjan, A. (2022). *The future is degrowth: A guide to a world beyond capitalism*. New York: Verso.

47 Sustainable Finance

Rens van Tilburg

Definition

Finance is the management of money through banking, asset management (investments), ministries of finance, central banks, and other financial instruments and institutions. Finance thus links people with a surplus of money with those who are in need of money on terms agreeable to both, like the interest paid on a loan or the share of the profit and loss for a more risky equity investment. Central and commercial banks have the ability to create money. Money is often defined by its three key functions: "Medium of Exchange", "Unit of Account", and "Store of Value" (see **Money**).

History

Throughout most of human history, people lived in groups of hunter-gatherers where each contributed according to ability and used according to need. It was only after humans settled that money came into use. The oldest known form of money can be found in clay tablets from Mesopotamia, dating back three thousand years, and used to record stored goods. Money is thus inextricably linked to the concept of debt. Owning money makes one a creditor, and hence someone else a debtor, who needs to do or give something to get even.

Since then, humanity has had an ambiguous relationship with money, debt, and the inequalities it gave rise to. Hence, for instance, the Jubilee Year developed as a periodic forgiveness in which all debts were forgiven. By the end of the second millennium BC, the Jubilee was scarcely practiced, hurting the general population and contributing to the "Bronze Age collapse". This illustrates the great importance of society's rules around money.

A Biblical prohibition of interest followed that lasted three thousand years. During the Enlightenment, the taboo on interest was lifted, giving way to the modern era of banking. For centuries, there have been various combinations of public and private money creation and rules for private money creation. Historically, from the 8th century BC when the first coinage was introduced, it was usually the sovereign who had the right to create this money. As coins were increasingly deposited at central and commercial banks, the paper proof of deposit also came into use as money. Because generally not all the coins were withdrawn at once, banks could issue more of this paper money. Thus banks also became money-creating institutions.

The current organization of the global banking sector arose after the financial crisis of 1929, in part caused by commercial banks creating too much money. The deep economic crisis that followed gave way to experiments and fundamental reform proposals (see Box 47.1). However, this was to no avail. Commercial banks even got an explicit guarantee from the state for their deposits, alongside an increase in regulation and supervision.

DOI: 10.4324/9781003584056-51

Box 47.1   The 1930s experiment and proposal that could have changed finance

In 1931, a historic monetary experiment took place in the Austrian village of Wörgl. Here, the municipal treasury was empty, as were the wallets of its inhabitants. The mayor initiated a system where the national currency, the Schilling, was exchanged for municipal vouchers with a negative interest rate of 1% per month. This stimulated the inhabitants to start spending their money, using the economic capacities of their neighbors who were lying idle. Suddenly there was plenty of economic activity again. And while unemployment continued to rise throughout Austria, it fell by 25% in Wörgl. The municipality was able to make additional expenditures: all houses were given a drinking water network connection, a new bridge was built, and even a ski jump. All this happened before the Austrian central bank ended the experiment (van Arkel, 2024).

On the other side of the Atlantic Ocean, in 1933, an equally bold plan was launched by some of the most prominent US economists of that time. The so-called Chicago Plan proposed to strip commercial banks of the possibility to create money through their lending. Such "full reserve" banks would essentially become asset managers that need to attract funds before they can be lent out again. Thus the public central bank would become the only money-creating institution, as it has been throughout a large part of history.

After the 2008 financial crisis, we witnessed a rerun of the discussion of the 1930s, with a similar outcome. Commercial banks are still the main money creators in the world economy (WRR, 2019). Next to banks, pension savings and growing inequality have led to a globally concentrated asset management industry that owns most companies. A handful of asset managers thus set the tone in the real economy, pushing relentlessly for the maximization of short-term profits.

Different Perspectives

Money and finance are relevant for sustainable consumption in two ways: first, as an enabler of the investments needed to make consumption sustainable, like in renewable energy capacity and the circular economy (see **Energy Consumption Behavior**, **Energy Overshoot**); and, secondly, finance can also play a less helpful role, providing a growth imperative, pressuring debtors to increase their income so they can pay the interest on their loans, or pressuring companies to increase their profits to satisfy their shareholders (see **The Role of Business**). This fuels the economic system to create new needs to be satisfied through economic growth.

Five thousand years of monetary history teaches us that finance can be organized in very different ways. The dominance of the current monetary and financial system by private financial institutions is historically an outlier. Whereas this system has been effective in growing financial capital (the value of the stock market is now a hundred times bigger than it was 50 years ago), this has come at too high a natural and social cost. It has become counterproductive, now that the economy needs to be brought within the planetary boundaries and work for all people (see **Doughnut Economics**, **Fair Consumption Space**). We seem stuck with a monetary system that is geared toward further growing financial capital, while the essential scarcities for human well-being are in natural and social capital (see **Well-being Economy**).

This is widely acknowledged. The main controversy seems to be whether reform of private finance will suffice, or whether public money creation is also needed.

Application

Sustainable consumption and lifestyles can be enabled through respective financing, for example, structural changes and complementary behavioral interventions. There is much that can be done along the lines of reform. Recent EU legislation on capital regulation (CRR/CRD), reporting (CSRD), and due diligence (CS3D) obliges banks to present transition plans that show how they will make their balance sheet aligned with the Paris Climate Accord. This way banks need to reduce their lending to harmful projects and increase their lending to households and companies for more sustainable consumption and production. Supervisors can also press asset owners, like pension funds and insurance companies, and their asset managers to draft transition plans (e.g., toward Socially Responsible Investing [SRI]). This will diminish the shareholder pressure for short-term profit at the expense of social and ecological values.

Sustainable finance, according to an EU definition, refers to the process of taking environmental, social, and governance (ESG) considerations into account for investment decisions so that these lead to more long-term investments in sustainable economic activities and projects. However, it is unlikely that this will be enough to bring the economy back within the planetary boundaries fast enough. For this, the unpriced externalities are too large and the wealth of a large part of the world is simply too small (see **Climate Justice**). Saving the world is simply not profitable.

This is where public finance needs to step in. All through history, central banks have created money if needed. Examples include the Bank of Venice being set up to finance the Crusade of Pope Urban the Second in 1157 and the Bank of England, founded in 1694 as a result of the deteriorating state of government coffers during the Nine Years War with France (1688–1697). In response to the Euro crisis, the ECB expanded its balance sheet, buying mostly government debt with no less than 6,000 billion newly created euros. Most recently, during the COVID crisis, the IMF created 650 billion US dollars worth of Special Drawing Rights (SDRs). New issuances of SDRs can pay for "nature and carbon coins" that also allow emerging economies to make the necessary investments (van Tilburg et al., 2023).

At the local level, local currencies can provide the means to start more sustainable local economies, with exchanges between neighbors and local small- and medium-sized enterprises. These can be designed not to push for ever more growth, for instance through negative interest rates (van Arkel, 2024).

Further Reading

Graeber, D. (2011). *Debt: The first 5000 years.* London: Melville House Publishing.

Schoenmaker, D., & Schramade, W. (2019). *Principles of sustainable finance.* Oxford: Oxford University Press.

van Arkel, H. (2024). *Boosting your local economy by making money virtuous.* Utrecht: Uitgeverij Jan van Arkel.

van Tilburg, R., Simić, A., & Murawski, S. (2023). The climate trillions we need; proposals for a new global financial architecture to end poverty and save the planet. *Sustainable Finance Lab*, November.

WRR. (2019). *Money and debt.* Den Haag: Netherlands Scientific Council for Government Policy.

48 Sharing Economy

Diana Ivanova and Tamar Makov

Definition

The sharing economy encompasses a diverse range of practices, activities, and business models where resources are borrowed, lent, shared, bartered, rented, and swapped, often in innovative ways revolutionized by digital technologies (see **Information and Communication Technology**). The monetized and non-monetized sharing of underutilized products or assets is a common practice within social networks. However, while applications of the sharing economy have become ubiquitous, the term's definition and the scope of businesses and practices it covers remain contested. Most agree that at its core, the sharing economy is about severing the link between ownership and access and that its novelty (vs. traditional sharing) is in reliance on digital technologies. Beyond that, however, there is much debate regarding what is truly "sharing" and whether a specific activity or business model falls within its scope.

The sharing economy cuts across various domains of consumption and lifestyles, including accommodation, transport, food, skills, spaces, and other goods and services. While there is much overlap between definitions around the sharing economy, there also remain important differences around whether sharing must happen on a temporary basis and in exchange for payment; involve a physical asset; be owned by an individual (also referred to as peer-to-peer sharing); or can also be owned by a company (also referred to as **product-service systems** or PSS). The positive connotation of "sharing" as a communal activity complicates these boundaries, as many companies (i.e., platforms) are eager to identify themselves as part of the sharing economy and gain the social approval and consumer trust associated with the term. More traditional non-profit actors promoting surplus food exchange, tool libraries, or service barters have also embraced the term (see **Grassroots Innovation**).

History

While humans have always shared within circles of families and communities, digital technologies have brought new possibilities around scale and speed (e.g., immediate and global), networks and relationships (e.g., peer-to-peer ratings, stranger sharing), legal frameworks and institutions (e.g., digital cooperatives and commons). The early 2010s were dominated by the so-called idealist discourse, which is perhaps best captured by the term "collaborative consumption" (see **Alternative Consumer Cooperatives**), see Figure 48.1. The promises of the sharing economy were numerous, and it was assumed that its benefits would be widely shared. Consumers would gain access to goods that they may otherwise be unable to afford. The shared assets would typically be under-utilized, and sharing them would reduce demand for new goods and their production lowering environmental pressures. Workers would earn additional income and enjoy flexible working arrangements on

DOI: 10.4324/9781003584056-52

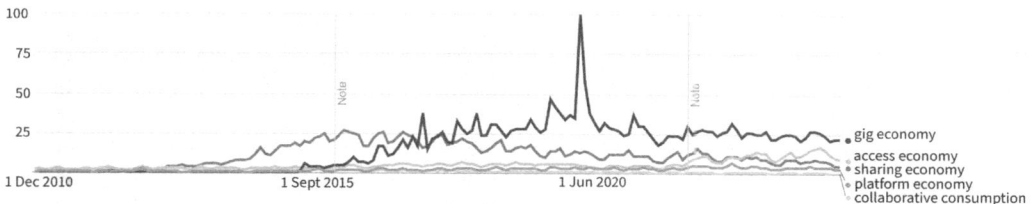

Figure 48.1 Global interest in the sharing economy and related terms over time on Google Trends. Numbers represent search interest relative to the highest point on the chart for the given region and time (2010–2024)

Source: The authors generated this figure based on Google Trends

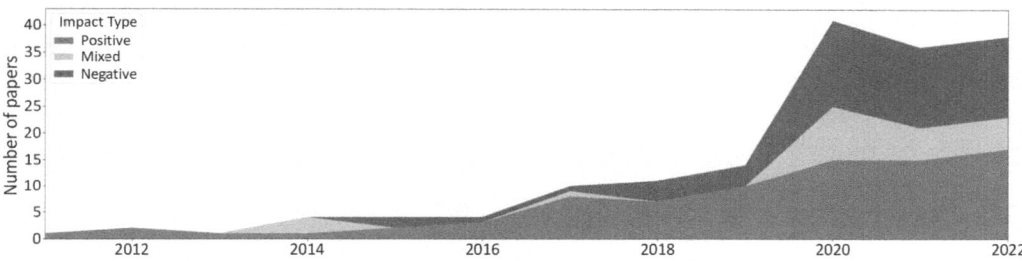

Figure 48.2 Environmental impact score of empirical peer-reviewed studies of the sharing economy over time. Results and labels were adapted from Meshulam et al. (2024). The figure presents the frequencies of papers included in the systematic literature review by Meshulam et al. (2024), by studies' publication year, and environmental impact findings

Source: Meshulam et al. (2024)

the platforms, which in turn would generate profits. Social networks would expand and people would build relationships around sharing activities.

While proponents of the sharing economy are still plentiful, empirical evidence over the past decade has cooled off the initial enthusiasm, painting a more complex and nuanced picture. The early promises of social and environmental benefits are increasingly put into question in the evolution of the sharing economy. Developments scrutinizing the environmental impacts (Figure 48.2), quality of relationships, and access associated with the sharing economy have been particularly prominent.

Different Perspectives

Environmental Impacts: Shared consumption at the household, neighborhood, or community level holds the potential to reduce environmental impacts; for example, as people share living space, rides to work, or appliances, the provisioning system requires fewer resources per person (see **Stocks Versus Flows**). Emerging research, however, suggests that such ecoefficiency gains are often eroded when sharing systems scale up. First, supportive logistics operations may increase (e.g., rebalancing bicycle stocks at the end of the day or more deadhead miles on Uber as geographic areas expand). Second, lower costs for sharing economy users, together with added income for providers, may trigger **rebound effects** leading to an overall increase in demand for products and services. Third, sharing can displace not only high-impact but also low-impact products or services. For example, shared scooters often replace walking and public transport rather than single occupancy car travel (see **Sustainable Mobility**). Finally, when such sharing economy alternatives become mainstream, economic incentives could change, leading to negative environmental and social spillovers. For example, as Airbnb grew in popularity, the incentives to purchase residential real estate that could serve as full-time Airbnb rentals increased, raising housing costs and lowering overall average occupancy. Critically, the sharing economy under capitalism is characterized by an incentive to scale up to make profits, which limits the potential for environmental benefits.

Social Impacts: The utopian rhetoric of the widely shared social benefits of the sharing economy and the establishment of strong bonds between strangers has also been disputed. Some have praised workers' freedom and autonomy, particularly those that rely on additional income. However, the sharing economy has also been criticized for its precarious labor conditions, defined by income instability, lack of security, discrimination, and exploitation. Class, race, and **gender** dynamics dominate for-profit and not-for-profit sharing economy enterprises, and the high prerequisites for participation (e.g., skill, technology, time) act as barriers to securing access to essentials among the most vulnerable. The sharing economy thus tends to reproduce wider inequalities across class, race, gender, and other social characteristics in the distribution of its benefits.

Applications

Today, the digital sharing economy is a $150 billion market where individuals share or gain temporary access to a broad range of assets, from tourist accommodation and transportation to clothing and food. Some of the earliest well-known examples of peer-to-peer sharing platforms include *Airbnb* and *Couchsurfing* in accommodation; *Uber*, *Lyft,* and *Didi* in ride-hailing; and *Taskrabbit* in freelance labor. Centralized ownership sharing (i.e., PSS) includes platforms such as *Car2Go*, *Lime*, and *Rent the Runway*. Yet, while sharing platforms can be found across most consumption domains, including prams (*Buggybooker*), tool libraries (*Tulu*), and food sharing (*Olio*), finding rules of thumb for optimizing the environmental or social benefits of sharing is challenging.

Research suggests that small-scale, local sharing of physical products or assets among peers (e.g., neighborhood sharing of books, seeds, and other items) can have environmental benefits if the potential for added operations (e.g., long-distance travel) is limited. In business-owned sharing, benefits are more likely in the employment of newer, more efficient stocks where use-phase impacts are high, but business models are not reliant on overproduction and overconsumption. Critically, sharing has recently expanded into luxury consumption domains such as yachts and even private flights. In such cases, sharing likely induces demand for services and goods that otherwise would not be consumed.

When considering social benefits, unmonetized peer-to-peer sharing may indeed foster stronger community ties (e.g., *Olio, HomeExchange*). Yet sharing does not seem to transcend social classes, as individuals tend to share mostly with others of similar socio-economic status and it often requires a degree of cultural or physical capital to participate (e.g., technological literacy and access). In addition, research suggests that social capital is likely a prerequisite for successful sharing.

As research increasingly shows, the sharing economy faces significant barriers to delivering improved environmental outcomes, access to essential goods and resources, and stronger social bonds. Such developments are unlikely to be realized through platforms and environments characterized by inequality, exploitation, commodification, ecological overshoot, and individualism. Promising alternatives have been proposed, including platform cooperatives (e.g., *Stocksy*) and solidarity networks (e.g., *Shareable*); policy innovations and partnerships (e.g., co-creating urban commons); and universal, democratic service provisioning (see **Universal Basic Services**). However, such alternatives would have to directly engage with profound economic, political, and socio-cultural challenges to contribute to a wider socio-ecological transformation.

Further Reading

Ivanova, D., & Buchs, M. (2023). Barriers and enablers around radical sharing. *The Lancet Planetary Health*, 7(9), E784–E792. https://doi.org/10.1016/S2542-5196(23)00168-7.

Makov, T., Shepon, A., Krones, J., Gupta, C., & Chertow, M. (2020). Social and environmental analysis of food waste abatement via the peer-to-peer sharing economy. *Nature Communications*, 11(1156). https://doi.org/10.1038/s41467-020-14899-5.

Meshulam, T., Goldberg, S., Ivanova, D., & Makov, T. (2024). The sharing economy is not always greener: A review and consolidation of empirical evidence. *Environmental Research Letters*, 19(013004). https://doi.org/10.1088/1748–9326/ad0f00.

Schor, J.B. (2020). *After the gig: How the sharing economy got hijacked and how to win it back*. Oakland, CA: University of California Press.

Schor, J.B., & Vallas, S.P. (2021). The sharing economy: Rhetoric and reality. *Annual Review of Sociology*, 47, 369–389. https://doi.org/10.1146/annurev-soc-082620–031411.

49 Circular Economy and Society

Martin Calisto Friant and Melanie Jaeger-Erben

Definition

In the last decade, the concept of the circular economy (CE) has emerged as a major discourse on sustainability in policy, business, NGO, and academic domains. While the concept of the circular economy (CE) continues to evolve, encompassing a plurality of approaches, the main idea is to optimize the use of the Earth's natural resources to meet human needs in a sustainable, resilient, and regenerative manner. CE thereby seeks to create a balanced and harmonious flow of socio-ecological resources between human and natural ecosystems so humanity may thrive within planetary boundaries. There are two broad conceptualizations of circularity: a CE, which only seeks to circulate material and energy resources sustainably, and a Circular Society (CS), which also seeks to circulate wealth, political power, care, and knowledge (including technologies) in a democratic and redistributive manner (see Figure 49.1).

History

The idea of a CE is nothing new. For the greatest part of humanity's presence on Earth, material and energy flows were in harmony with the Earth's regenerative capacities. It is only with the rise of capitalism and the Industrial Revolution that humanity permanently broke this balance through the increasing use of fossil fuels, colonization, and the global dominance of growth-dependent economic systems.

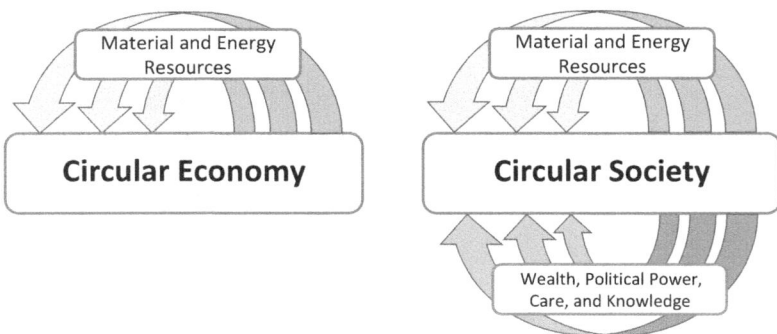

Figure 49.1 Conceptual differentiation between circular economy and circular society

Source: Adapted from Calisto Friant et al. (2020, 2024)

DOI: 10.4324/9781003584056-53

The CE concept arose as a response to these socio-ecological crises and had various stages and phases in its development, summarized in Table 49.1. Considering the diverse history and the variety of concepts related to CE, it is best understood as an "umbrella concept" that combines and embraces many key elements of sustainability thinking.

The 1960s and 1970s were the first significant moments in CE development when many of the modern precursors to the CE concept emerged, driven by a rising awareness of the impossibility of eternal economic growth on a finite planet (see **Degrowth**). A myriad of ideas were put forth for post-capitalist socio-economic systems that do not depend on economic growth to survive (e.g., ecoanarchism, **steady-state economics**, and Buddhist economics, see "precursors" in Table 49.1). They proposed a wholesale transformation of our production and consumption systems to create slower and more convivial ways of life centered on non-material aspirations. This coincided with the emergence of global environmental movements and the institutionalization of environmental protection (e.g., in the 1972 United Nations Conference on the Human Environment and the creation of UNEP).

During the 1990s to early 2000s, neoliberal thinking dominated global debates and centered on market-based approaches to CE, focusing on economic growth, technological innovation, and competitive markets as avenues to address social and environmental problems (see, e.g., **Extended Producer Responsibility**). However, particularly in the 1990s, little attention was paid to social justice issues or systemic transformations to reduce unsustainable production and consumption patterns.

In the 2000s, new CE concepts – with a more holistic and socially inclusive approach to consumption and production – were developed, such as Cradle-to-Cradle, the Performance Economy, **Doughnut Economics**, as well as **Degrowth**. In addition, several transformative concepts from the Global South emerged during this time, such as **Buen Vivir and Buenos Convivires** by Latin American Indigenous movements, Ecological Swaraj in India, and **Ubuntu** from South Africa. These movements stress the importance of global justice and decolonization for a just transition and place Mother Earth as an equal partner endowed with inalienable rights.

During the same time, many national governments, such as China, Japan, and the EU, as well as cities such as Paris, Amsterdam, and London, started incorporating CE into their policies. This is also the moment when many corporations around the globe integrated the CE concept as part of their sustainability and corporate social responsibility strategies, such as Unilever, IKEA, Patagonia, Renault, Fairphone, and Swapfiets.

Different Perspectives

A circularity discourse typology was developed to help navigate the rich history and diversity of CE concepts and ideas, dividing the discourses into two main criteria. First, it distinguishes segmented discourses, which focus on the technical and business components of circularity, from holistic discourses, which include social justice and political empowerment. Second, it divides optimist and skeptical perspectives regarding the possibility of decoupling environmental degradation from economic growth. Different combinations of these two criteria lead to four main circularity discourse types (Table 49.2).

Research on CE has found that the most dominant and widespread discourse type is currently the Technocentric Circular Economy. Over 80% of CE definitions fall in this discourse type, which is particularly widespread in business consultancies and corporate CE strategies. Governments and the EU tend to follow a more mixed approach with some holistic discursive elements that, however, often fail to translate into more sustainable and socially inclusive policies.

Table 49.1 Timeline of circularity concepts and ideas (adapted from Calisto Friant, M., Vermeulen, W.J., & Salomone, R. (2020). A typology of circular economy discourses: Navigating the diverse visions of a contested paradigm. *Resources, Conservation and Recycling*, 161. https://doi.org/10.1016/j.resconrec.2020.104917)

Precursors to circularity	Circularity 1.0 ana 2.0: lecnno-nxes to waste		Circularity 3.0: integrated socio-economic approaches to resources, consumption ana waste			
	Circularity 1.0: Dealing with Waste	Circularity 2.0: Connecting Input and Output in Strategies for Ecoefficiency	Circularity 3.1 Reformist views on Circularity		Circularity 3.2 Transformational views on Circularity and visions of the Global South	
Preamble Period	Excitement Period				Validity Challenge Period	
1945–1980	1980–2010				2010 to present	
			First holistic Circularity frameworks	New holistic Circularity views	Transformational views of Circularity	Non-western visions of Circularity
Gandhian economics (1945)	Waste-Water Treatment	Industrial Ecology	Rio Declaration on Environment and Development (1992)	Blue Economy (2010)	Transition Movement (Hopkins, 2008)	Buen Vivir/ Sumak Kawsay
The Economics of the Coming Spaceship Earth (1966)	Solid Waste Management and Recycling	Ecodesign/Design for environment	Regenerative design (1994)	Third Industrial Revolution (2013)	Degrowth	Ubuntu
The tragedy of the Commons (1968)	Bio-Digestion	Cyclic Economy (1993)	Natural Capitalism (1999)	Ecosystem Economy (2013)	Ecosocialism	Ecological Civilization
The Population Bomb (1968)	Energy Recovery	Industrial Metabolism (1994)	Cyclical Economy (2001)	Regenerative Capitalism (2015)	Low-Tech	Ecological Swaraj
Bioeconomics of Georgescu-Roegen (1971)		Cleaner Production	Materials Matter (2001)	Sharing Economy	Laudato si' (2015)	Suma Qamaña / Vivir Bien
The Closing Circle (1971)		Reverse Logistics	Cradle to Cradle (2002)	Doughnut Economics (2017)	Transition design (2015)	Buddhist, Confucian and Taoist ecology
Ecoanarchism (1971)		Ecoindustrial parks and networks	The Natural Step (2002)	Symbiotic Economy (2017)	Economy for the Common Good (2015)	Radical Pluralism/Pluriverse
The Limits to Growth (1972)		Biomimicry (1998)	The Performance Economy (2010)	Social Circular Economy (2017)	Post-growth	
Ecological Design (1972)		Product Service System	Sound Material-Cycle Society in Japan (2000)	Spiral Economy (2019)	Permacircular Economy (2018)	
United Nations Conference on the Human Environment (1972)		Extended Producer Responsibility	Circular Economy in China (2002)	Coviability (2019)	Voluntary Simplicity	
Buddhist economics (1973)		Industrial Symbiosis	EU Circular Economy Action Plan (2015)	Circular Humansphere (2019)	Convivalism (2019)	
Convivality (1973)		Closed-loop Supply Chain				
Steady-state economics (1977)		Biobased Economy/ Bioeconomy				
Permaculture (1978)		The Biosphere Rules (2008)				
Décroissance (1980)						
Deep Ecology (1980)						
Overshoot (1980)						

Table 49.2 Circular economy discourse typology

		Approach to social, economic, environmental, and political considerations	
		Holistic	Segmented
Technological innovation and ecological collapse	Optimist	**Reformist Circular Society** • **Assumptions**: socio-technical innovations can enable eco-economic decoupling to prevent ecological collapse, so a reformed form of capitalism is compatible with sustainability • **Goal**: human prosperity and well-being within the biophysical boundaries of the earth • **Means**: technological breakthroughs and social policies that benefit humanity and natural ecosystems • **Example concepts**: natural capitalism, cradle-to-cradle, performance economy, the natural step, the blue economy, regenerative design, sound material-cycle society • **Proponents**: various international organizations, academics large foundations, and some governments	**Technocentric Circular Economy** • **Assumptions**: technological innovation can enable ecoeconomic decoupling to prevent ecological collapse, so capitalism is compatible with sustainability • **Goal**: economic prosperity and development without negative environmental externalities • **Means**: economic innovations, new business models, and unprecedented breakthroughs in CE technologies • **Example concepts**: industrial ecology, reverse logistics, biomimicry, industrial symbiosis, extended producer responsibility, cleaner production, bioeconomy • **Proponents**: some academics, many corporations, various national and city governments, and international organizations
	Skeptical	**Transformational Circular Society** • **Assumptions**: socio-technical innovations cannot bring absolute ecoeconomic decoupling to prevent ecological collapse, so capitalism is incompatible with sustainability • **Goals**: a world of conviviality and frugal abundance for all while fairly distributing the biophysical resources of the earth • **Means:** the complete reconfiguration of the current socio-political system and a shift away from productivist and anthropocentric worldviews • **Example concepts**: conviviality, steady-state economics, perma-circular economy, degrowth, ecoanarchism, Buddhist economics, buen vivir, ubuntu • **Proponents**: many academics, social movements, bottom-up circular initiatives, and indigenous peoples	**Fortress Circular Economy** • **Assumptions**: socio-technical innovation cannot bring absolute decoupling of economic growth and environmental degradation to prevent ecological collapse, but there is no alternative to capitalism • **Goal**: maintain geostrategic resource security in global conditions where widespread resource scarcity and human overpopulation cannot provide for all • **Means**: innovative technologies and business models combined with rationalized resource use and migration and population controls • **Example concepts**: the tragedy of the commons, the population bomb, overshoot, disaster capitalism, capitalist catastrophism. • **Proponents**: survivalists, a few academics, some geostrategic think tanks, and state policies

Source: Adapted from: Calisto Friant, M., Vermeulen, W. J. V., & Salomone, R. (2020). A typology of circular economy discourses: Navigating the diverse visions of a contested paradigm. *Resources, Conservation and Recycling*, 161, 104917. https://doi.org/10.1016/j.resconrec.2020.104917

In response to the dominance of technocentric propositions, there is a rising movement promoting a holistic CS vision, especially in European civil society. CS visions are gaining greater support, especially from those who have criticized mainstream CE propositions for focusing too much on economic growth and competitiveness and too little on social and environmental justice. Indeed, many see hegemonic CE propositions as forms of **greenwashing**, which create the illusion that "green technologies" will allow us to overcome the biophysical limits of the Earth and continue growing our economies forever (i.e., weak sustainability). Yet, over 50 years of academic evidence show that decoupling economic growth from environmental degradation is impossible on a relevant scale to prevent biodiversity collapse and climate breakdown.

The current domination of technocentric approaches to circularity has curtailed the possibility of a more plural and democratic debate regarding what circular future we want and how we want to get there. Indeed, many governments and businesses have preferred to adopt a depoliticized and uncontroversial CE approach that does not address fundamental issues regarding social equity, political empowerment, and the biophysical limits to economic growth. This prevents tackling key questions, such as who owns CE technologies and innovations, who should reduce their production and consumption levels in line with planetary boundaries, and who should pay and govern the transition. Thus, the dominance of Technocentric CE discourses suppresses alternative discourses and concepts such as degrowth, post-growth, care economics, and post-humanist ideas that highlight the rights of nature and the need for systemic socio-ecological transformations (see Box 49.1).

Moreover, research has shown that grassroots and civil society organizations tend to have a more holistic and socially inclusive vision of circularity than the Technocentric paradigm that governments and companies are implementing (see **Grassroots Innovation** and **Alternative Consumer Cooperatives**). Indeed, surveys have found that people in OECD countries prefer a sufficiency-oriented ecological transition that prioritizes human and ecological well-being rather than profits.

Box 49.1 Status and outlook of CE – moving from a "validity challenge" period to more holistic CE discourses

The CE is in what Blomsma and Brennan call a "validity challenge" period; this means that it must confront its key challenges and limitations to remain relevant. If the CE debate remains stuck in "fairy tales" of "green growth" and doesn't embrace a strong socio-ecological justice agenda, it will lose its social appeal and systemic validity, especially considering the rising inequalities and injustices brought by over 30 years of neoliberal globalization. As we continue to overshoot the ecological limits of the biosphere and the impacts of climate change rise year after year, it will become harder and harder to argue for failed technocentric solutions.

Systemic socio-political change to post-growth circular societies will be necessary, whether we like it or not. Yet, faced with an impending socio-ecological collapse, visions of a Fortress CE will also become more and more appealing. Indeed, as we confront stronger natural disasters and shortages of crucial natural resources, many conservative voices could start arguing for greater nationalism and top-down control over resources and populations. This is already happening today with the rise of far-right movements, and it may become more prevalent in the future. To prevent this, it is key to expand the discussion and shift our debate from narrow visions of a CE to more holistic CS discourses. Improving transformative learning to build "circular literacy" and democratic co-design processes may help develop a greater understanding of the diversity of circularity approaches and break the current dominance of technocentric visions (see **Education for Sustainable Consumption**).

Table 49.3 Value retention and preservation options in a circular economy and society

Transforming cultural practices

Refuse	Re-evaluate	Rethink	Re-conceptualise
Buying, producing or using less and avoiding over-consumption and over-production.	Reconsidering values and worldviews, e.g. from consumerism to sufficiency, from egoism to solidarity	Rethinking how value chains are organized and phasing out linear structures, incentives and mindsets	Redefining the meaning of a good life and shifting away from endless and senseless materialist and individualist aspirations

Transforming economic practices

Redesign	Reduce	Restructure	Redistribute	Re-localise	Restore
Redesigning products and services to improve socio-ecological sustainability	Using less material per production unit (e.g. smaller/lighter product)	Transforming social, political and economic systems to align with the values of a circular society	Redistributing power, technology, knowledge and access to resources, both between and within countries	Producing near consumption centres, a re-localising politics, culture and the meaning of life	Restoring natural ecosystems or cultural heritage to their original or improved condition (also called "regenerate")

Transforming material practices

Replace	Repurpose	Repair	Refurbish	Re Manufacture	Recycle	Re-mine	Recover (energy)
Substituting polluting and toxic materials or chemicals for safe and sustainable ones	Adapting discarded things for another function, such as rubber tyres as fences	Restoring functionality of products, fixing defects	Improving components of a product (also called "reconditioning" or "retrofitting")	Disassembling, checking, cleaning, and repairing in an industrial process (also called "reprocessing" or "re-assembling")	Processing waste products to obtain "secondary raw materials"	Recovering natural resources from landfills	Capturing energy embodied in waste through incineration or bio-digestion

Application

As seen above, there are multiple diverging approaches to a circular economy and society, depending on the interests and objectives of different actors that implement them. To summarize the diverse forms of CE application, Table 49.3 presents a list of the 19 value retention options (also known as Rs), which are the core strategies used in CE implementation. They are divided into three interdependent and interrelated types of socio-ecological transformation: cultural (underlying worldviews and values), economic (social provisioning and distribution systems), and material (production, consumption, and recovery structures).

The above list attempts to provide a comprehensive list of how circularity may be applied at various social, political, and business levels. However, it is in no way exclusive nor exhaustive, so different actors may implement a specific selection of strategies in the list or may develop entirely new strategies not listed above.

Further Reading

Calisto Friant, M., Vermeulen, W.J.V., & Salomone, R. (2024). Transition to a sustainable circular society: More than just resource efficiency. *Circular Economy & Sustainability*, 4, 23–42. https://doi.org/10.1007/s43615-023-00272-3.

Genovese, A., & Pansera, M. (2021). The circular economy at a crossroads: Technocratic eco-modernism or convivial technology for social revolution? *Capitalism, Nature, Socialism*, 32(2). https://doi.org/10.1080/10455752.2020.1763414.

Hempel, N., Boch, R., & Jaeger-Erben, M. (2023). Co-designing a circular society. In *Design for a sustainable circular economy: Research and practice consequences*, pp. 205–232. Singapore: Springer Nature Singapore. https://doi.org/10.1007/978-981-99-7532-7_11.

Rask, N. (2022). An intersectional reading of circular economy policies: Towards just and sufficiency-driven sustainabilities. *Local Environment*, 27(10–11), 1287–1303. https://doi.org/10.1080/13549839.2022.2040467.

Suárez-Eiroa, B., Fernández, E., & Méndez, G. (2021). Integration of the circular economy paradigm under the just and safe operating space narrative: Twelve operational principles based on circularity, sustainability and resilience. *Journal of Cleaner Production*, 322. https://doi.org/10.1016/j.jclepro.2021.129071.

Cluster IV
Value Shifts and Social Activism

50 Alternative Hedonism

Kate Soper

Definition

"Alternative hedonism" challenges dominant conceptions of the "good life" in capitalist societies, and seeks to promote the benefits of less growth-driven, materialistic, and individualized ways of living, working and consuming.

It stands out among theories of sustainable consumption in the attention paid to negative aspects of the "consumerist" lifestyle for affluent consumers themselves (e.g., stress, ill-health, time scarcity, noise, congestion; see **Hedonic Treadmill**), and in the emphasis placed on the pleasures of the slower-paced, less work-centered and more collaborative ways of living that could otherwise be enjoyed (see **Work-Life Balance**). Alternative hedonism rejects the pervasive view of consumerist lifestyles as the ideal of the "good life" to which all other less affluent societies will "by nature" aspire. Instead, the theory argues that consumerism now functions primarily as a means of further enhancing the global reach and command of corporate power at the expense of the health and well-being of both the planet and the majority of its inhabitants.

In opening up a post-consumerist approach to human flourishing, alternative hedonism (i) highlights what people are beginning to experience themselves about the anti-consumerist aspects of their own needs and preferences and (ii) implies a broader systemic opposition to the existing economic and social order.

History

Alternative hedonism has been most fully developed by various authors over the last 15 years (Soper et al., 2009; Soper, 2020; cf. Jackson, 2021). Whereas many critiques of consumerism dwell on its ecologically disastrous impact and appeal to the moral obligation we have to change our ways, the alternative hedonist focus is on the pleasures of doing so. This concept is at odds with hedonism, which is essentially self-oriented and about endorsing personal indulgence (see **Hedonic Treadmill**). Alternative hedonism advocates for more collective modes of living, over the egoism and privatized consumption habits promoted by capitalist consumer culture.

The resulting research has critiqued the proliferation of many goods and services that would otherwise not be needed at all – or could in many cases be supplied more collectively – at less cost to the environment, and in a less socially isolating manner. It has also led to critiques of making consumption the marker of social status and, thereby, creating a competitive spiral of acquisition (see **Conspicuous/Positional Consumption**). Conspicuous and invidious consumption of this kind (buying goods to gain the attention or envy of others) has played a major role in the expansion of many markets (notably in clothing, household goods, and cars), and has in that sense served the growth economy extremely well. However, from the point of view of consumers themselves,

DOI: 10.4324/9781003584056-55

the gratifications of conspicuous consumption are jinxed by what has come to be known as the "hedonic treadmill" – the fact that happiness tends to stabilize, whatever the gains in material goods. Furthermore, the desire to keep pace in the competition for status goods is like a treadmill where no one can finally win, and everyone must keep walking simply to maintain their place. Earlier findings on this have been reinforced by empirical studies suggesting that, beyond a certain point, increased income does not bring about any increase in happiness (see **Household Income Versus Happiness**).

Different Perspectives

Alternative hedonism acknowledges that what counts as happiness is contested, as are the ways in which it is assessed (Soper, 2020: 67–68; Boston Review, 2023: 28–30). It also recognizes that the commitment to a fair and sustainable global order depends upon the emergence of some kind of cross-cultural consensus on human needs and well-being (see **Well-being Economy**, **Buen Vivir and Buenos Convivires**, **Doughnut Economy**, **Consumption Corridors**). It raises, in other words, the complex question of the criteria and norms of a universally satisfying and enduringly available provision for consumption (see **Fair Consumption Space**). Therefore, although it is essential that we do not let ideas of happiness and well-being be defined solely in terms of market provision and sale, we also need to be sensitive to the difficulties of agreeing on the qualities of the "good life".

Most critical responses to "alternative hedonism" have been directed to its focus on consumption rather than production. This has been seen as mistakenly targeting wealthier individual consumers rather than the systemic forces of capitalist industry and commerce as responsible for ecological degradation. The concept of alternative hedonism should instead be understood as propelled by a concern with radical economic and social transformation in affluent democracies: by engaging with emergent forms of disaffection with consumer culture, and giving voice to an alternative "politics of prosperity", it seeks to encourage a cultural revolution that could eventually issue in a political mandate for systemic change (Soper, 2020: 69–76; cf. 164–169; see **Social Norms**, **Social Movements**, **Values and Consumption**). It has also been argued, in this context, that "consumerism" is better viewed not as a middle-class preserve, but as a regime of consumption to whose forms of provision, work ethic, and material aesthetic everyone in affluent societies is currently subject, whatever their income.

Alternative hedonism has also been criticized for advocating **degrowth**. Although a degrowth transition will indeed depend (in the short term) on expanding activity in key areas such as renewable energy and demand reduction, education (see **Education for Sustainable Consumption**), and caring services, what matters is how we view that transitional growth. Do we view it as necessary within an economic system that is being redesigned to foster ways of living and ideas of prosperity very different from those of profit-driven, capitalist consumer culture? Or are we viewing "growth" as an essential and permanent dynamic of any effective economic order, and thus as both compatible with environmental conservation and enduringly sustainable? If the latter is the case, growth must be rejected since more efficient technologies have always led to overall expansion in resource use and commodities (see **Energy Overshoot**, **Rebound Effect**).

Application

Alternative hedonism can help to inspire a more diverse and substantial opposition to prevailing economic orthodoxy by supplying a broader cultural dimension to the existing arguments and outlook regarding the necessity for systemic change. Persuading voters to support the needed changes in consumption will not be easy. But many who are already uneasy about the impact of continuous growth-driven consumption are likely to respond quite positively to an alternative hedonist discourse on prosperity. A compelling representation of the potential for a fairer and more pleasurable way of living for society as a whole can also help to offset the defiant inaction on climate change of the populist right.

Alternative Hedonism provides a framework for those engaged in Commoning and ecosocialist activism and campaigns (see **Social Movements**, **Degrowth**, **Steady-State Economy**, **Foundational Economy**, **Eco-Communities**). Green policies of questionable popularity when first implemented (e.g., congestion charging, "15 minute cities", rewilding) illustrate an "alternative hedonist" dialectic whereby support for them is enhanced through the public experience of the benefits they provide (see **Urban Planning and Spatial Allocation**). In challenging mainstream imagery of the "good life", "Badvertising", Yellow Dot Studios, and similar campaigns are now helping to create a much-needed alternative aesthetic of material culture (see **Subvertising**).

Alternatives to growth-driven capitalism are already being realized in the interstices of the mainstream market through the expanding culture of what has been termed "collaborative" or "connected" consumption: networks of sharing, recycling, exchange of goods and service provision (including banking and other financial services) that by-pass conventional commerce (see **Prosumerism, Sharing Economy, Repair, Alternative Consumer Cooperatives**). These have helped to reduce carbon emissions and waste while at the same time creating more ecosensitive communities and cooperative ways of living. In a transition period, such initiatives act as a check on the individualization of consumption and provide ways of circumventing the obstacles it places in the way of shared and more collective use of goods and forms of transport. They also help to subvert the reach and intrusion of the increasingly personalized address of internet advertising (see **Information and Communication Technology**). More generally, they check the dominant consumerist aesthetics of "newness" by shunning high street-led fashions and mass production in favor of clothes swapping, remakes, and homemade goods. They might also in the process prove to be hubs for exerting pressure on corporations to end reliance on sweat-shop labor and ever-faster turn-over times and to render them accountable for the pollution incurred in production (see **Fast Fashion**).

Further Reading

Boston Review. (2023). *The politics of pleasure: Debating the good life*. Cambridge, MA: Forum 23.

Hickel, J. (2020). *Less is more*. London: Penguin.

Jackson, T. (2021). *Post growth: Life after capitalism*. London: John Wiley and Sons.

Soper, K. (2020). *Post-growth living: For an alternative hedonism*. London: Verso.

Soper, K., Ryle, M., & Thomas, L. (Hrsg.) (2009). *The politics and pleasures of consuming differently*. Basingstoke: Palgrave Macmillan.

51   Well-being and Life Satisfaction Versus Income

Callie Dance, Donna Lybecker, and Nina Szczygiel

Definition

Life satisfaction refers to the overall assessment of one's quality of life, according to their chosen criteria, reflecting long-term fulfillment. More specifically, it is how people evaluate their life as a whole, rather than their current feelings. Happiness, on the other hand, often denotes a more immediate, emotional state characterized by feelings of pleasure and contentment. Life satisfaction and happiness are often referred to as two forms of well-being. This view underscores that well-being is influenced by a complex interplay of personal, social, and environmental factors and clarifies why the relationship between these aspects and income is routinely explored by scholars. As scholars show, the relationship between well-being and income is also connected to consumption. Scholars note that higher income often leads to an increased carbon footprint (see **Carbon Footprint Versus Household Income**). This may lead to cognitive dissonance and less well-being for people with strong environmental values, as they navigate the balancing act of allowing every individual to have their basic needs fulfilled and live a good life while not exceeding the ecological limits of the planet.

History

Income's relationship to life satisfaction, happiness, and/or well-being has long been a point of both fundamental and applied research. Income provides the financial means to meet basic needs, access healthcare, enjoy leisure activities, and achieve financial security, all of which contribute to happiness and life satisfaction. Diener and Tov (2009), among others, have consistently shown a positive correlation between income and happiness, particularly when comparing different countries or individuals within a country. Easterlin (1974) highlighted that richer individuals tend to report higher levels of happiness and well-being, although this correlation has diminished over time and was less pronounced when comparing richer and poorer countries. Overall, higher income is also related to increased consumption, impacting many today and leaving an ecological debt for future generations. Finding a balance within sustainable consumption, reaching satisfaction out remaining mindful of choices, is critical.

In 2010, Kahneman and Deaton challenged the belief that higher income universally leads to greater well-being, by distinguishing between life evaluation (life satisfaction) and emotional well-being (happiness). In their work, higher income was associated with higher evaluative life satisfaction, indicating that individuals with more money tend to perceive their lives more positively. However, above US$75,000 (in 2010), a further increase in income does not increase happiness In contrast, low income gives rise to both low life satisfaction and low emotional well-being. Kahneman, together with Killingsworth and Mellers (Killingsworth, 2021; Killingsworth et al., 2023),

DOI: 10.4324/9781003584056-56

however, corrected these findings as rooted in a methodological error and demonstrated that there is no upper threshold for increasing both life satisfaction and happiness with greater income. They attribute that phenomenon to two factors, that is, social comparisons and a sense of greater stability conferred by higher income (see Box 51.1). They also showed that for the unhappiest, an increase in income staves off unhappiness only to a point.

Current research indicates that while higher income can enhance life satisfaction, the incremental gains (in absolute values) diminish as income rises beyond a certain level. Literature focuses on the *diminishing marginal utility of income.* For example, an increase in income for a person living in poverty can substantially enhance their well-being by providing access to basic necessities. However, the same extra income will have a smaller impact for someone already enjoying a comfortable lifestyle. Additional income for the latter may equate to increased consumption, but of non-necessities, having a smaller impact on overall happiness Finally, factors such as relative income comparisons, social relationships, and cultural context play substantial roles in mediating how income impacts life satisfaction and happiness, and whether individuals choose sustainable consumption.

Different Perspectives

Literature on the relationship between income and life satisfaction relates to a variety of debates. Among the most robust are those about relative income and social comparison, social inequality, spending habits and financial security, and psychological and social factors (see Box 51.1).

Box 51.1   Factors that mediate the income-happiness relationship

Relative income and social comparison

Happiness is not solely dependent on absolute income but is also influenced by relative income – how one's income compares to others. Social comparison theory posits that individuals assess their well-being relative to those around them. Hence, if a person's income is significantly lower than their peers, their life satisfaction may decrease, even if they have ample access to all of life's necessities. Conversely, having a higher income compared to peers can boost self-esteem and happiness, even if in absolute terms that income is not fully sufficient for basic needs.

Societal inequality

Related to relative income, societal inequality is another driver of human unhappiness. Generally, in their drive to establish and maintain their social position, people tend to aspire to be like those "above" them and to distance themselves from those "below". In a highly unequal society, that drive is exacerbated, leading to a perpetual **hedonic treadmill** and general anxiety. Epidemiologists Wilkinson and Pickett have shown in their 2010 and 2024 studies that high income inequality in a society leads to general unhappiness and many social ills, such as poorer health of the population, lower educational achievements, and weaker social cohesion. Thus, while income is important, it is the income disparity that more profoundly affects overall societal well-being, exacerbating social tensions and driving up consumption.

Spending Habits and Financial Security

How individuals spend their money and what they consume also plays a role in happiness. Research indicates that spending on experiences, such as travel, brings more lasting happiness than material purchases. Experiences provide lasting memories and enhanced social bonds, both of which are important for well-being. Moreover, financial security and investing for the future contribute to long-term life satisfaction. Conversely, financial stress can undermine happiness, regardless of income level.

Psychological and Social Factors

Psychological and social factors also mediate the income-happiness relationship. Personality traits, such as optimism and resilience, can influence how an individual perceives their financial situation and overall happiness. Additionally, strong social networks are consistently linked to higher life satisfaction, often outweighing the impact of income alone. People who report strong social connections and support are generally happier, regardless of their income level.

Understanding human life satisfaction and happiness through the income lens has various limitations. First, cultural values and societal norms play a role in shaping the relationship between income and happiness. In collectivist cultures, where social relations are prioritized, the impact of individual income on happiness may be less pronounced, compared to more individualistic societies where personal achievement, consumption, and wealth are highly valued and well-rewarded. Second, during times of societal prosperity, higher incomes for everybody are likely to maintain the same relative happiness as during the periods of economic stagnation. Likewise, shifts in **social norms** regarding wealth and success can alter how individuals evaluate their income over time.

Third, it overlooks other critical aspects, such as mental health, job satisfaction, work-life balance, and engagement with such worldviews as **Ubuntu** or **Buen Vivir and Buenos Conviveres**. The characteristics of high-income jobs, which often bring about increased stress, long hours, and less time for personal relationships, can negate the benefits of higher income, again suggesting the benefits of sustainable consumption rather than extensive consumption. Finally, studies of the hedonic treadmill suggest that people quickly return to a baseline level of happiness following positive or negative changes in their circumstances, including changes in income. This phenomenon implies that while a pay raise may temporarily boost happiness, individuals often adjust their expectations, returning to their prior level of well-being over time.

Application

Applying the concept of *Wellbeing versus Income* involves designing measures that ensure financial stability and reduce inequality (see Box 51.2). This can be achieved through initiatives like progressive taxation, universal basic income, **universal basic services** and accessible healthcare, high-quality education, and other life amenities that enhance opportunities for success in life. These programs increase financial security and may lead to higher levels of happiness and well-being. Financial security allows individuals to meet basic needs without anxiety about financial instability and reduces the anxiety of becoming deemed socially inferior (see **Money**).

Establishing appropriate economic policies is crucial for life satisfaction and a stable and motivated workforce in a society. Implementing progressive taxation to fund social programs, such as unemployment benefits, helps reduce economic inequality, ensuring those with higher incomes contribute a fairer share of earnings (see **Foundational Economics**, **Well-being Economy**, **Ecosocial Contract**). This approach helps create a more equitable society where everyone can access the resources they need to thrive. Similarly, economic policies that ensure fair wages and promote job security increase job satisfaction and financial security. This fosters a more productive and committed workforce, contributing to overall societal well-being.

Implementing universal healthcare systems ensures all citizens have access to medical care without the fear of financial hardship or catastrophe. Good physical and mental health is foundational for life satisfaction, and universal healthcare plays a significant role in reducing health disparities and promoting overall well-being and resilience. Ensuring that medical care is accessible to everyone, regardless of financial situation, contributes to a healthier, more equitable society.

Providing opportunities for adult education and vocational training supports continuous personal and professional development. Lifelong-learning programs can improve job satisfaction, and cognitive functioning, and provide a sense of purpose, all of which enhance well-being. By encouraging continuous learning and skill development, individuals can adapt to changing job markets and pursue fulfilling careers, contributing to a more knowledgeable and resilient society.

Box 51.2   Life satisfaction/happiness – income nexus applications

Economic policies

Sweden and Austria's progressive tax systems require higher-income earners to pay a larger percentage of their income in taxes. This allows tax revenues to fund social programs, including universal healthcare, heavily subsidized education, and unemployment benefits. As a result, citizens enjoy low levels of economic inequality and high levels of life satisfaction. Furthermore, access to essential services provides a safety net and ensures individuals from all economic backgrounds have opportunities to thrive

Social policies

Between 2017 and 2018, Finland conducted a "basic income experiment", providing unemployed citizens with a monthly income of 560 euros. Results showed improvements in mental health and stress levels; however, no significance regarding employment was registered. A similar study in the United States involving unconditional cash transfers failed to show improvement in financial, physical, or mental health. However, results showed increased entrepreneurialism and schooling, and additional effects – such as those on the children of the recipients – may yet be seen.

Healthcare policies

The Norwegian healthcare system exemplifies how universal healthcare can lead to a healthier, more equitable society by prioritizing accessibility, prevention, and early treatment. This system, funded primarily through taxation, ensures access to comprehensive medical care without financial hardship. Government investments in healthcare infrastructure and technology ensure facilities are equipped with the latest medical advancements and that urban and rural communities have access. This approach reduces health disparities and promotes overall well-being, as evidenced by Norway's high life expectancy and low infant mortality rates.

Further Reading

Diener, E., & Tov, W. (2009). Well-being on planet earth. *Psihologijske Teme*, 19, 213–219.

Easterlin, R.A. (1974). Does economic growth improve the human lot? Some empirical evidence. In P.A. David & M.W. Reder (Eds.), *Nations and households in economic growth: Essays in honor of Moses Abramovitz*, pp. 89–125. Academic Press. https://doi.org/10.1016/b978-0-12-205050-3.50008-7.

Killingsworth, M.A. (2021). Experienced well-being rises with income, even above $75,000 per year. *Proceedings of the National Academy of Sciences of the United States of America*, 118(4). https://doi.org/10.1073/pnas.2016976118.

Killingsworth, M.A., Kahneman, D., & Mellers, B. (2023). Income and emotional well-being: A conflict resolved. *Proceedings of the National Academy of Sciences of the United States of America*, 120(10). https://doi.org/10.1073/pnas.2208661120.

Wilkinson, R.G., & Pickett, K.E. (2024). Why the world cannot afford the rich. *Nature*. https://doi.org/10.1038/d41586-024-00723-3

52 Spiritual Consumption

Patrick Elf and Amy Isham

Definition

Spiritual consumption describes the behaviors and processes in which people engage when consuming products, services, and places for spiritual reasons/ends. Whereas William James referred to the "varieties of *religious* experiences", more recent work looks at the "varieties of *spiritual* experiences", demonstrating the role and importance of spirituality in and across societies (see **Values and Consumption**). Indeed, with approximately 60% to 80% of people worldwide adhering to some sort of spiritual or religious belief according to IPSOS's 2023 Global Religion survey across 26 countries, spiritual consumption holds considerable importance for consumer research.

Both spirituality and religious practice involve the human relationship to the divine, sacred, or transcendent. However, a growing number of individuals identify as "spiritual but not religious" (SBNR). Distinguishing between the two terms, religion has an institutional connotation in the form of systems for the monitoring, coding, protecting, and transmitting of information and practices that spirituality does not. Regardless of the specific distinctions between the terms and their underpinning practices, the consumption of spiritual and religious products, services, and places (e.g., pilgrimages) has experienced significant attention from profit-seeking actors that aim to use consumers' spiritual and/or religious inclinations to encourage consumption.

History

Being a translation of the Latin word *spiritualis*, which can be traced back to Hebrew and Greek usage in the Bible, spirituality refers to an intrinsic human experience of being with God. Spiritual life was oriented toward cultivating a personal relationship with God, and involved practices such as prayer, fasting, and charity as well as the consumption of certain services (e.g., spiritual rituals) and products (e.g., eating fish on Christmas Eve). Similar themes can be observed across almost all religious contexts. For instance, in Hinduism spirituality is encouraged in the Vedic texts, emphasizing rituals and meditative and moral practices designed to understand the ultimate reality, Brahman. Likewise, in Islam, spirituality is most prominently expressed through Sufism, the Islamic belief in seeking closeness to God.

The philosophical expansion of the concept of spirituality signified its transition from strictly religious contexts to a broader, arguably more inclusive realm that engages with existential, moral, ethical, and personal development themes. This expansion is especially noticeable from the Enlightenment in the 17th and 18th centuries onwards, as thinkers began to explore spirituality in terms of individual experience and personal growth, independent of institutional religious frameworks. Spirituality came to be seen more as an individual personal journey toward understanding and the essence of being, focused on a search for meaning, unity, connectedness, and transcendence.

DOI: 10.4324/9781003584056-57

Spiritual consumption began to gain further traction in the late 20th century, reflecting a wider mainstreaming of spirituality and a value shift of consumers toward more holistic, authentic, and personally meaningful products and experiences.

Different Perspectives

Spirituality has traditionally induced values and virtues of moderation. From this perspective, the nomos of spirituality, meaning the custom governing human conduct, may reduce the consumption of resources. Religious scriptures from both the East and the West provide commandments to protect the earth as God's creation, while spiritual practices such as mindfulness and certain meditation practices can advance care for other people and the environment (Isham et al., 2022; see **Mindfulness**). Indeed, spiritual people feel more connected to others and experience a greater sense of well-being according to Gallup research in 2023. Just as the presence of religious or spiritual beliefs can steer people away from excess consumption, the absence of such beliefs – especially in modern consumption-driven societies – can prompt unsuitable consumption behaviors. With the deterioration of community and shared meaning in modern life, people often seek to fill this void and the growing sense of anomie, meaning, the breakdown of moral values and standards, through acquiring consumer products (see **Conspicuous/Positional Consumption, Hedonic Treadmill**).

Marketers appear well aware of consumers' desire for spiritual experiences. Seminal work by Belk et al. (1989) discusses "the secularization of the sacred and the sacralization of the secular" in consumer behavior. In line with this, a commodification of spiritual experiences has been observed by Elf et al. (2023), and the embedding of spiritual value in mundane consumption practices. The commodification of spiritual experiences is evidenced through market offerings such as spiritual tourism – from yoga or meditation retreats, pilgrimages to sacred sites, or shamanic tourism whereby people travel to consume substances such as ayahuasca (a spiritual beverage) in structured ritual settings. These market offerings capitalize on the shift toward individualization and bypass the traditional religious gatekeepers to spirituality, allowing the market to create new forms and types of spiritual experience based upon the demands of the spiritual marketplace.

The *commodification* of spirituality raises questions about the (in)compatibility of spiritual and material realms. Historically, spiritual consumption has promoted values of moderation, but modern consumer culture is finding new ways of commodifying spiritual experiences that may lack the underlying spiritual values and act merely as a means of boosting wider consumption. While some argue that commodified spirituality lacks authenticity, moral depth, and access to the interconnected community commonly found within institutionalized religion (Carrette & King, 2004), others suggest that it can help revitalize religion and spiritual ideas in different ways.

Application

Our enormously productive economy . . . demands that we make consumption our way of life, that we convert the buying and use of goods into rituals, that we seek our spiritual satisfaction, our ego satisfaction, in consumption.

(Victor Lebow writing in the *Journal of Retailing* in 1955)

In societies increasingly faced with anomie, the pursuit of meaning becomes a key priority for consumers. Historically, religion has provided a "meaning function" that supposedly helps us to cope with potential discrepancies we encounter daily. It also kept **consumerism** in check through its virtue components stressing the importance of frugality. But, as Jackson and Pepper (2011: 4) ask, "Who provides the meaning function when God departs (or is excluded from) the fray?" The spiritual marketplace is now playing a vital role in our everyday lives, with material goods occupying a key part in processes of sacralization and world maintenance. In this way, consumerist desires themselves have become the object of human striving.

Indeed, as well as offering commodified versions of traditionally spiritual practices, more mundane practices or products can also be transformed to hold spiritual meaning. New Age, marketable versions of centuries-old spiritual practices such as meditation, pilgrimages, and yoga are sold and consumed in the form of apps and often pricey weekend retreats. Other examples include the ritualization of everyday activities to impart a sense of significance and reverence. For instance, Harley-Davidson fosters a sense of community among its customers to create a collective sense of sacredness. The customization of products (e.g., branded leather jackets) can make consumers feel unique and significant while creating a collective sense of sacredness where ownership and use of their products are part of a larger – arguably, at least in part spiritual – lifestyle.

The transition to a sustainable society cannot hope to proceed, therefore, without the emergence or re-emergence of meaning structures that lie outside the consumer realm. If consumers' calls for spirituality are met with material products as a meaning-making go-to solution, then this limits traditional spirituality's potential to facilitate values of moderation and frugality.

So how do we support spiritual striving in the absence of great environmental costs? Numerous practices that can be situated within sustainable lifestyles can support meaning-making and spiritual well-being. Spending time in nature, practicing mindfulness and meditation, decluttering, volunteering, and community engagements are just a few practices that can support sustainable consumption and potentially wider **behavior change** (see also **Voluntary Simplicity**, **Alternative Hedonism**, **Eco-Communities**). Governments and institutions must find ways of fostering opportunities for such practices across society. This may divert attempts to find meaning and connections in acts of consumption.

Further Reading

Belk, R.W., Wallendorf, M., & Sherry Jr, J.F. (1989). The sacred and the profane in consumer behavior: Theodicy on the Odyssey. *Journal of Consumer Research*, 16(1), 1–38. https://doi.org/10.1086/209191.

Carrette, J., & King, R. (2004). *Selling spirituality: The silent takeover of religion*. London: Routledge.

Elf, P., Isham, A., & Leoni, D. (2023). Moving forward by looking back: Critiques of commercialized mindfulness and the future of (commercialized) psychedelics. *History of Pharmacy and Pharmaceuticals*, 5(1), 33–62. https://doi.org/10.3368/hopp.65.1.33.

Isham, A., Elf, P., & Jackson, T. (2022). Self-transcendent experiences as promoters of ecological wellbeing? Exploration of the evidence and hypotheses to be tested. *Frontiers in Psychology*. https://doi.org/10.3389/fpsyg.2022.1051478.

Jackson, T., & Pepper, M. (2011). Consumerism as theodicy: Religious and secular meaning functions in modern society. In L. Thomas (Ed.), *Religion, consumerism and sustainability. consumption and public life*. London: Palgrave Macmillan. https://doi.org/10.1057/9780230306134_2.

53 Values and Consumption

Ian Hamilton

Definition

Values are the deeply held beliefs and principles that guide our decisions and influence our behaviors. They shape how we perceive the world, interact with others, and pursue goals as individuals. They also influence how the protagonists of society – individuals, communities, institutions, and nations – interact. Values are not static. They can evolve, changing as we learn, grow, and are exposed to new experiences. Values can also be a source of conflict in that some strongly held values may conflict with others, forcing the prioritization of certain values over others.

Values are hierarchical in that we hold some values more strongly than others. Similarly, some are considered to be "higher" than others. For instance, values that are considered to contribute to collective human flourishing, such as compassion, generosity, and justice, may be considered higher, whereas values that primarily serve the self, such as greed, avarice, or lust for power, could be considered lower. While addressing the source or origin of values is deeply complex, some key sources include family life, society and social norms, formal education, religion, and spirituality. Values can be considered a cornerstone of society.

Historically, values have served as a moral compass, influenced **social norms**, acted as a basis for developing social cohesion, provided individuals with meaning and purpose, and created a sense of identity. More recently, the wider discourse on values has become predominantly concerned with economic, intrinsic, and utilitarian value – engines of a growth-driven materialistic, economic paradigm.

Considering values through the lens of sustainable consumption, it is important to both unpack and apply currently often overlooked, but societally important, values – such as justice, trust, **sufficiency**, contentment, simplicity, and moderation (see **Climate Justice** and **Voluntary Simplicity**). These and others could prove fruitful territory for empowering a swifter and more inclusive transition to more sustainable patterns of consumption and development. While not unique to religion or spirituality, the aforementioned values are held dear by many religions and spiritual movements.

History

Throughout the ages, values have been a key force in human existence. Looking across religious and spiritual movements, for example, there are clear and direct links between spiritual teachings that lead to values being put into practice. Buddhism teaches that all beings are interconnected and the suffering of one is the suffering of all. This has placed compassion and non-violence at the core of Buddhist practice, often manifested in actions such as vegetarianism or advocacy to abolish nuclear weapons (see **Protein Transition**).

DOI: 10.4324/9781003584056-58

Drawing from Christ's emphasis on love for one's neighbor and selfless service, many Christians express their faith through charitable actions or social justice work, evidenced by the work of organizations such as Catholic Relief Services or Church World Service. The Bahá'í Faith adopts unity as both a goal and a mode of operation, such that many Bahá'í communities around the world, through spiritual capacity-building initiatives for children, youth, and adults, work to create communities in which no form of prejudice is acceptable. Indigenous cultures and beliefs often see a deep connection between the spiritual and natural worlds, which helps foster values such as harmony, moderation, and respect for nature (see **Buen Vivir and Buenos Convivires**).

These are just some brief examples across the breadth of spiritual experience that illustrate how deeply connected spirituality and values are. Even in recent years, as the number of adherents to mainstream religions has fallen in the United States, 83% of US adults believe we have a soul or spirit in addition to our physical bodies and 81% believe there is something spiritual beyond our world. This trend seems consistent across economically advanced countries (see the Pew-Templeton Global Religious Futures project). This trend further suggests that, whether religious or not, a sizable number of people see humanity's existence as being governed not only by physical and material forces but also by spiritual and moral considerations. Spirituality is something more intangible and less organized than traditional religion, but no less a potent force in the promotion and adoption of values.

Different Perspectives

Any discussion of values is inherently complex, multidimensional, subjective, and relative – influenced by personal, cultural, historical, and other factors – and reflects the complexity of the myriad systems underpinning our world. They are difficult to measure and can evolve. In behavioral sciences, values are included and operationalized in seminal theories and models. For instance, Schwarz proposed a two-dimensional typology with 10 value clusters, ranging from hedonism to achievement power and benevolence (see **Hedonic Treadmill**). Based on this and Stern's Model, Steg and colleagues identified and continuously tested biospheric, altruistic, egoistic, and hedonistic values as the drivers for pro-environmental action.

Recognizing this complexity can support a more nuanced and productive conversation on the role of values in modern society (see Box 53.1).

Box 53.1   Toward a global ethic

Discussing the role of values, the World's preeminent interfaith organization, the Parliament of the World's Religions, released the statement "Toward a Global Ethic" which posited that there are universal, shared religious values. The document suggests that justice, equality, and solidarity are key to the creation of a just economic order. To counter overconsumption, the document calls on people to strive for fairness in consumption and to live moderate lifestyles.

See Parliament of the World's Religions. (1993). *Towards a Global Ethic: An Initial Declaration.*

The complexity of value systems is evident in the conflicting, competing, and overlapping value sets inherent to many social actors. The move of many governments toward policy and legislation that promote sustainable lifestyles can be seen to conflict with the goal of corporations to maximize profitability (see **The Role of Business**). Religious actors promoting detachment from material possessions can be perceived as a threat to consumer society. However, many facets of the religious and spiritual experience, such as pilgrimage or **mindfulness**, are becoming commodified (see **Spiritual Consumption**).

By seeking to understand the conflicts and overlaps between different social actors, we may be able to identify shared values that could serve as departure points for united action toward a sustainable society. These values could include the common good, human nobility, and intergenerational justice.

The development paradigm that has largely prevailed since the end of World War II – an approach rooted in the extraction of natural resources and endlessly increasing standards of living – is no longer fit for purpose. The consumerist, technological, and macroeconomic trends that have defined much of the last century have driven increasing global inequality and brought us to the point of environmental collapse (see **Carbon Inequality**). However, just because a certain paradigm is no longer fit for purpose does not mean that suitable alternatives have been identified.

Application

Multinational organizations are increasingly looking toward values-inspired approaches to advancing global systems. Sustainable Development Goal (SDG) Target 17.19 seeks to spur the development of metrics that look beyond Gross Domestic Product and measure sustainable development. In addition to the UN, the EU has begun to champion the "Beyond GDP" movement, which could see the reimagining of prosperity through the promotion of measures such as equality, fairness, inclusivity, and happiness (see **Well-being Economy, Degrowth**). To advance such views in relevant international forums, such as the UNFCCC Conference of the Parties (COP), spaces could be created to explore what values-based measures and metrics are more fit to purpose for systems that seek to center sustainable consumption and, more broadly, sustainable development.

One potential tool in identifying and promoting alternative development pathways is a values-based framework to enable sustainable consumption. Such a framework, built upon the values mentioned above in the definition, and addressing the agency enjoyed by major protagonists of society – the individual, communities/collectives, institutions, and government – could allow these protagonists to understand the needs of each and provide values-based lines of action, as well as concepts, to build more sustainable patterns of consumption.

Figure 53.1 (a simple conceptual framework) is an attempt at a visual representation of what this chapter discusses. It is based upon the author's professional experience of multiple conversations with like-minded civil society, faith-based, and non-governmental organizations who look to principled and values-based responses to pressing societal challenges. It suggests that certain underlying values could serve as the foundations for a range of practical strategies. By considering the implications of **sufficiency** or contentment in their daily lives, individuals could make decisions that prioritize collective interests over individuals. Communities and collectives, by seeking to build trust, could begin to build more social and economic cooperative structures (see **Community Supported Agriculture, Alternative Consumer Cooperatives**). Societal institutions such as universities and private enterprises, predicated on moderation and sufficiency, may begin to look at other forms of shareholder value or develop a deeper understanding of what is meant by equity. Governments, seeking to build justice, could address a great many ills through restructuring social spending or protecting finite resources (see **Universal Basic Services, Foundational Economy**).

Ultimately, recognizing that the climate crisis and other pressing societal challenges are more than just technical problems, but are also the result of incompatible values, could open up a vista

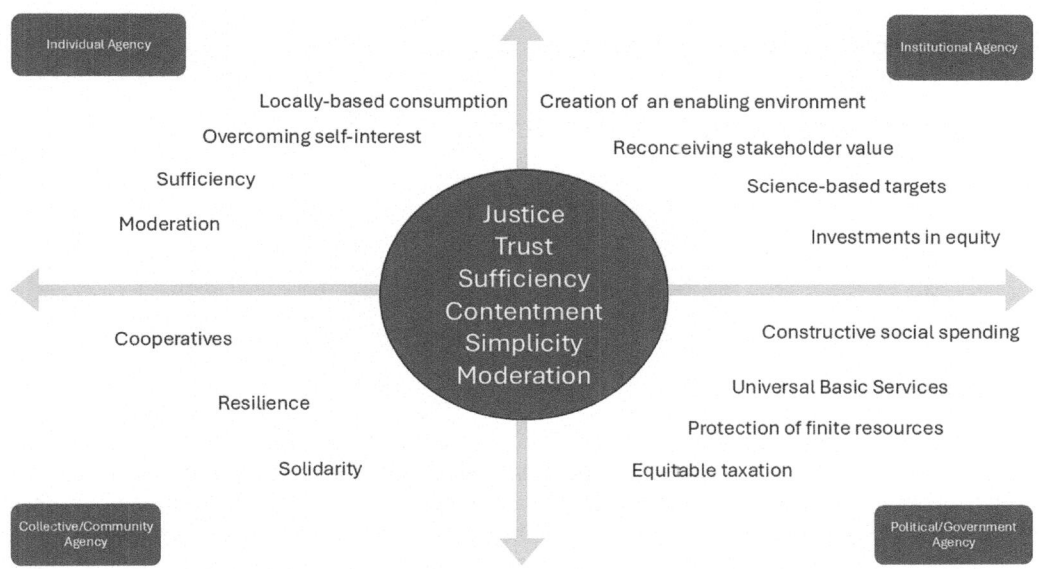

Figure 53.1 A values-based framework for sustainable consumption

Source: By author

for new forms of sustainable action. If people were to consider the climate crisis as a condition borne out of greed, inequality, and injustice, it could provide a blueprint for how they develop their potential to respond to a changing world while also working to refine and enhance the communities and institutions in which they are engaged. The prevalent structures and processes built on a different logic must be reshaped and adapted to facilitate actors to act upon a potentially (newly gained) value-driven agency. To summarize, much remains to be learned about frameworks that center spiritual values and ethical principles that can be both developed and applied to effect change.

Further Reading

Bahá'í International Community (BIC). (2022). *One planet, one habitation: A Bahá'í perspective on recasting humanity's relationship with the natural world*. Available at: bic org (accessed: 13 December 2024).

European Commission. (2007). *Beyond GDP*. Available at: https://sustainable-prosperity.eu. (accessed: 13 December 2024).

Jenkins, W., Tucker, M.E., & Grim, J. (2017). *Routledge handbook of religion and ecology*. London: Routledge.

Pew Research Center. (2022). *Key findings from the global religious futures project*. Available at: https://pewresearch.org (accessed: 13 December 2024).

Tucker, M.E., & Grim, J. (2023). *Yale forum on religion and ecology*. Yale Center for Environmental Justice. Available at: https://fore.yale.edu (accessed: 8 January 2024).

54 Buen Vivir and Buenos Convivires

Susan Paulson and Alberto Acosta

Definition

Buen Vivir is living in equilibrium with the diverse beings and natural cycles that constitute life itself. Such modes of living have long been pursued through social institutions and values that foreground community and ecological well-being; in some Andean and Amazonian communities, they are known by the Quichua term *sumak kawsay* or the Aymara term *suma qamaña*. In recent decades, these and kindred traditions have been interconnected around the Spanish term *buen vivir* and mobilized in dialogue with environmentalist and decolonial movements (Acosta, 2019; Gudynas, 202).

Buenos convivires (multiple ways of living well together) celebrates the coexistence of plural manifestations of a life mode based on principles of relationality, reciprocity, and interspecies care. Practices of buen vivir interact to nourish and mobilize alternative pathways to those promoted by Western development. Buen vivir societies strive to care for and regenerate socio-ecological systems – not to expand economies. They apply knowledge, technology, and organization for the flourishing of life – not the accumulation of capital or material wealth. Individual fulfillment is enjoyed in interdependence with human neighbors and non-human nature – not in exploiting or getting ahead of others. Communities create buenos convivires by (re)generating shared abundance – not by competing over scarce resources.

History

Historians and archaeologists show that populations around the world have developed diverse ways of sustaining community and ecological well-being across generations. In recent centuries, many of these lifeways have been assimilated into and hybridized with different institutions and practices disseminated with (neo-)colonial capitalism and Western development.

During the 19th and 20th centuries, the ascendency of one path (rooted in Western civilization, advancing via industrial transformation of ever more resources) was hailed as "progress". Now it is clear that globalization of that path leads to ecosocial degradation and upheaval on the scale of civilizational crisis.

At the turn of the 21st century, surging Indigenous movements, in dialogue with decolonial, environmental, feminist, and post-development thinking and organization, opened horizons to different futures. In Latin America, that confluence brought perspectives and practices from communities on the peripheries (though never outside) of modern development to the fore. Many come from Indigenous territories where steep mountains and vast rainforests have limited the implementation of colonial-capitalist infrastructures and institutions; others come from communities that have been marginalized by exploitation and adverse conditions at the heart of modernizing processes. In different ways, and to varying degrees, these communities have sustained and

DOI: 10.4324/9781003584056-59

adapted ancestral values, experiences, and practices oriented toward ecosocial well-being. Buen vivir draws on these, not as a route back to romanticized pre-modern worlds, but to forge futures that are pleasurable and sustainable for more people and places.

Different Perspectives

Here we identify some of the perspectives at play in grassroots, academic, and political debates around buen vivir and highlight vital lessons from dialogues among them (see **Grassroots Innovation, Social Movements, Ecosocial Contract**).

Indigenous perspectives advocate modes of living and relating that are aligned with cultural heritages and utopic future visions of Abya-Yala, the territories of Indigenous peoples now called The Americas. They respect culture-specific knowledge, foster dialogue among different ways of knowing and being, and consider the complex dynamics of ancestral traditions in changing worlds. Against common misconceptions that indigenous ways are pre-modern, proponents show they constitute living worlds that continually evolve in tension with colonial modernity, and contribute to debates about global futures.

Environmentalists engage buen vivir in struggles against extractivist political economies and high-consumption lifestyles that degrade ecosystems and earth systems. The challenging co-participation among professional, academic, activist, and local groups can bring Western sciences and technologies (mostly applied to control nature) into a tense dialogue with knowledge and spiritualities (oriented to sustain reciprocity among humans and other nature). In nascent paradigm shifts, ideas of the rights of nature and inter-species kinship are being integrated with conventional scientific management of natural resources.

Anti-colonial, anti-capitalist, anti-racist, and anti-patriarchal perspectives ally with subversive capacities of buen vivir to jointly transform hierarchical institutions and relationships that facilitate and justify unequal access to and control over human and other life. Responses constructed with Indigenous and afro-descendant traditions include moves to re-center care and regeneration of life in ways that subvert hierarchies that prioritize masculine-associated production and subordinate feminine-associated reproduction (see **Gender**).

Statist perspectives, sometimes interpreted as Andean versions of socialism, have led to some high-profile and contentious applications. New constitutions in Bolivia (2009) and Ecuador (2008) garnered global attention with buen vivir frameworks, and these and other governments advanced buen vivir-oriented narratives, policies, and programs.

Pluriversal perspectives do not seek one correct way to define and apply buen vivir but instead support synergy and spiritual alliance among differently positioned, context-dependent paths (see **Values and Consumption**). That calls for dialogue and mutual learning among co-existing traditions in Latin America, identified as *sumak kawsay*, *suma qamaña*, *lekil kuxlejal*, *teko pora*, *ñandareko*, *küme moegen*, *opatssi*, and others. It also calls for constructive dialogue with civic and state expressions, and with resonant pathways elsewhere, such as *eco-swaraj*, *ubuntu*, and *kyosei*.

Application

Experimentation has sparked passionate debates about how to apply, translate, or institutionalize buenos convivires in today's world. Here we mention a few applications operating with different dynamics and scales:

* The Amazonian community of Sarayaku, Ecuador, successfully defended their territory from petroleum exploitation and also rejected models of conservation that exclude humans, by

articulating their worldview *kawsak sacha* (living forest) grounded in symbiotic material practices, and spiritual relationships among human inhabitants, forests, rivers, territory. Bringing together indigenous people from neighboring countries, Sarayaku communities have worked to mobilize alternative responses to the climate crisis, rooted in traditional practices that have long regenerated ecosystems and biodiversity.

- Networks bring together diversely positioned people, organizations, and initiatives. The Argentina-based *Red de Alimentos para el Buen Vivir Sumak Kawsay* works to make food systems healthier for humans and environments by supporting small-scale organic production and facilitating connections among local producers and consumers. The Chile-based *Red Ciudadana Del Buen Vivir* supports collaboration among individuals and community organizations in educational workshops, musical and cultural events, and other initiatives to joyfully co-construct healthy environments.
- National policy can be seen in Ecuador's Constitution, Secretariat of Buen Vivir, and National Plans for Buen Vivir between 2007 and 2015, and in neighboring countries' public commitments to enshrine rights of nature and/or transform relationships with diverse citizens and cultural-linguistic traditions. All have faced challenges implementing visions coincident with buen vivir in current global political and economic environments. Some state strategies to implement these proposals characterized as "neo-extractivist" have been critiqued for pursuing one goal of buen vivir – equitable well-being for all citizens – with funds generated by undermining another goal – living sustainably with nature.

The richness of perspectives and applications outlined here has been obscured by high-profile attention to government efforts to scale up buenos convivires and by critiques of bureaucratization and cooptation of buen vivir to advance political interests. The emerging consensus is that buen vivir cannot be constructed by state-level actors alone; it must be co-constructed in context via interaction among variously positioned actors and processes, interactions through which the state itself is rethought and reshaped (see Box 54.1).

Box 54.1   Example of constructing buen vivir

One promising case involves intercultural co-governance among municipal institutions and a variety of social organizations in Cayambe, Ecuador. Lang (2022) describes territory-focused processes that integrate ancestral knowledge and principles of buen vivir with select dimensions and institutions of modernity to address interconnected issues of gender relations, human-nature relations, and social and epistemological justice.

Taken together, these tangible initiatives and experiences reveal the potential for communities, regions, and governments to collaborate in transforming systems of production and consumption toward worldviews and institutions oriented toward sustaining life itself.

Further Reading

Acosta, A. (2019). Buen Vivir – A proposal with global potential. In H. Rosa & C. Henning (Eds.), *The good life beyond growth*. Routledge Studies in Ecological Economics.

Cortez, D. (2021). *Sumak kawsay y buen vivir, ¿dispositivos del desarrollo? Ética ambiental y gobierno global*. Quito: Editorial FLACSO-Ecuador.

Gudynas, E. (2021). Post-development and other critiques of development. In H. Veltmeyer & P. Bowles (Eds.), *The essential guide to critical development studies*, pp. 49–55. London: Routledge.

Kothari, A., Salleh, A., Escobar, A., Demaria, F., & Acosta, A. (Eds.). (2019). *Pluriverse – A post-development dictionary*. Tulika Books, India.

Lang, M. (2022). Buen Vivir as a territorial practice. Building a more just and sustainable life through interculturality. *Sustain Science*, 17, 1287–1299. https://doi.org/10.1007/s11625-022-01130-1.

55 Ubuntu

Vuyiswa Lamfiti and Joseph Koetsier

Definition

Ubuntu is an African philosophy most closely associated with the Zulu/Xhosa expression "umntu ngumntu ngabantu" – which translates as "a person is a person through other people", or "I am because we are" or "I am because you are". Similar across all nine recognized languages of South Africa, and evident in various forms in other African nations, Ubuntu expresses a fundamental belief that a person's relations with other people are foundational in creating personal identity and significantly contribute to the flourishing of the community. The community precedes the individual, and the self is shaped by relationships with others.

History

Societies across the world have always generated norms and processes that regulate their internal and external interactions and that are embodied in their lived experiences and daily lives (see **Social Norms, Values, and Consumption**). In the African context, this consciousness is often created within the social-cultural context of Ubuntu. Ubuntu was first documented in written form in the 1800s in South Africa, though it is believed to stretch back millennia and is found today in a variety of forms in different Bantu languages of southern Africa (Hailey, 2008).

Ubuntu is often put in historical contrast with the emergence of Western notions of the self, largely derived from René Descartes (1596–1650) who influentially posited, *"cogito ergo sum"* ("I think, therefore I am"). This idea is said to have engendered individuality and promoted the self as the starting point of individual and social life. Ubuntu instead fosters ethical and moral values that are other-regarding and that build common bonds and communitarian ethos, for the benefit of society and all other living things such as plants, rivers, and animals.

Early in the 20th century, Ubuntu was already seen to have implications and threads of connection around the world, also outside Africa. Mahatma Gandhi's (1869–1948) work in India, for instance, emerged from his experiences with community-based practices in South Africa and may have been influenced by it. His Sarvodaya (meaning "upliftment of all") philosophy and practice holds similarities to the philosophy of Ubuntu. After the fall of the South African Apartheid regime, Ubuntu was popularized in the English language through the work of archbishop Desmond Tutu (Hailey, 2008). Tutu was the chairperson of South Africa's post-Apartheid Truth and Reconciliation Commission (TRC), which has been seen as an attempt to implement the empathy, unity, and sense of collectivity inherent in the concept.

Different Perspectives

While seemingly focused on human relations, various scholars and commentators have noted that the philosophy of ubuntu has key sustainability implications. A cornerstone of Ubuntu, for

DOI: 10.4324/9781003584056-60

instance, is the recognition that we are who and what we are because of the interconnections with nature that provide for our sustenance, shelter, and existence.

However, while Ubuntu is held up as a challenge to Western philosophies, social frameworks, and environmental ethics, its broader application has also come under scrutiny. For instance, Ubuntu is often portrayed as a universal framework emerging from Africa, in a way that overlooks the sheer size and ecocultural diversity of the African continent. In relation to sustainability in particular, it is also often discussed in a very broad, philosophical, and impressionistic way, relating to anthropocentrism (a philosophy that centers the human over wider ecology) and environmental ethics, rather than being seen as having direct and concrete implications for sustainable living (see e.g., Etieyibo, 2017).

Concerns have also been raised regarding the manner in which Ubuntu foregrounds the collective over the individual. Depending on the dominant political dispensation in communities, this could reinforce collective tendencies and conformity in a way that may be oppressive of individuality. This, it has been claimed, can result in the quelling of dissent, but also potentially the maintenance and perpetuation of social inequalities and chauvinistic nationalism, with particularly detrimental impacts on women.

Applications

If present lifestyles are all-too-often characterized by competition, materialism, and striving for individual excellence, adopting more of the values central to Ubuntu could mitigate overconsumption and inequality. By providing an alternative and shifting the values underpinning consumption (see **Alternative Hedonism**), Ubuntu has been viewed as a framework and philosophy that might lead to tangible innovations and improvements in various domains. These could extend to the governance and practice of sustainable consumption, oriented around:

- Sharing
- Cooperation
- Compassion
- Empathy
- Collective responsibility
- Respect for others (including the elderly)
- Healing
- Harmony with nature

Given the global nature of pollution and resource depletion, Ubuntu is thought to reorient environmental ethics away from competitive or divisive politics, toward ecological citizenship (see **Citizen-Consumer**). Given its relevance to social support, care for humans and nature, and collective well-being (Letseka, 2012), Ubuntu could also provide the grounds for radical social and environmental policies that reach beyond social or ethnic identities, as seen in innovations such as Universal Basic Income, **Universal Basic Services**, **Foundational Economy**, and **Well-being Economy**.

Another practical instance of Ubuntu would be the arrangement of consumption and lifestyles around commons and the distribution of wealth, rather than unsustainable consumption, privatization, and processes of enclosure (Etieyibo, 2017). Commoning involves the collective stewardship and management of cultural and environmental resources, from shared grazing land or woodlands to water resources or collective infrastructures. In a Commoning system, ecological intelligence and sustainable practice do not reside in one individual but are rather shared and practiced across a community (Shumba, 2011). In everyday life, Ubuntu can be seen operating through the various traditional practices of mutual aid evident in African societies, whereby neighbors take turns helping each other with agricultural planting, harvests, and other labor-intensive activities. By doing so, more work can be done collectively than would be possible individually. For some practical enactments, see Box 55.1.

Box 55.1   Some practical enactments of Ubuntu

Ubuntu and Ukraine

Research scientist Dzvinka Kachur at the Institute for Transition Studies at the University of Stellenbosch and poet Oksana Kutsenko, former cultural secretary of the Ukrainian Embassy, laid the foundation for Ukrainian cultural diplomacy in South Africa and later in other African countries with Ukrainian missions. The policy resulted in an exchange of Zulu poets to Lviv. Ukrainian authors translated indigenous African stories into richly illustrated books in Ukrainian. They are convinced that Ubuntu can contribute to the restoration of humanity in a region now devastated by war. In South Africa, this project resulted in the opening of a special bookshelf in the Main Public Library in Cape Town, which displays the results of this exchange.

Ubuntu and family constellations

In Zulu communities in the province of KwaZulu Natal, Ubuntu and family constellations are closely related. As fellow humanity, we belong to the same field of energy. Problems in families and communities are interrelated. They originate in ancestral bloodline relationships that can go back many generations. These are probed in constellation gatherings in which a whole village participates. In this approach, role players represent the carriers of traumas that have to be probed. This work developed into worldwide practices through the work of Anton "Bert" Hellinger (1925–2019) who stayed in the 1970s for 16 years among the Zulus. He developed a therapeutic method best known as Family Constellations and Systemic Constellations, which aims to heal the wounds of traumas.

Further Reading

Battle, M. (2000). A theology of community: The ubuntu theology of Desmond Tutu. *Journal of Bible and Theology,* 54(2), 173–182. https://doi.org/10.1177/002096430005400206.

Etieyibo, E. (2017). Ubuntu and the environment. In A. Afolayan & T. Falola (Eds.), *The Palgrave handbook of African philosophy,* pp. 633–657. Palgrave MacMillan.

Hailey, J. (2008). *Ubuntu: A literature review.* A paper prepared for the Tutu Foundation, London.

Letseka, M. (2012). In defence of ubuntu. *Studies in Philosophy and Education,* 31, 47–60. https://doi.org/10.1007/s11217-011-9267-2

Shumba, O. (2011). Commons thinking, ecological intelligence and the ethical and moral framework of Ubuntu: An imperative for sustainable development. *Journal of Media and Communication Studies,* 3(3), 84–96. Available at: https://academicjournals.org/journal/JMCS/article-full-text-pdf/23201BA11281 (accessed: 8 January 2025).

56 Education for Sustainable Consumption

Ulf Schrader and Daniel Fischer

Definition

According to the United Nations Environment Programme (UNEP):

> *Education for Sustainable Consumption (ESC) aims to provide knowledge, values, and skills to enable individuals and social groups to become actors of change towards more sustainable consumption behaviors. The objective is to ensure that the basic needs of the global community are met, quality of life for all is improved and inefficient use of resources and environmental degradation are avoided. ESC is therefore about providing citizens with appropriate information and knowledge on the environmental and social impacts of their daily choices, as well as providing workable solutions and alternatives. ESC integrates fundamental rights and freedoms including consumers' rights, and aims at protecting and empowering consumers in order to enable them to participate in the public debate and economy in an informed, confident and ethical way.*
>
> (UNEP 2010: *ABC of SCP.* UNEP-DTIE)

This definition makes clear that ESC goes beyond increasing knowledge, aiming to develop values and skills for action. It refrains from reducing sustainable consumption to green purchasing alone and instead associates it with the adequate use of resources to meet needs and improve the quality of life for all. The definition defines learners as both consumers and citizens, participating in processes that could potentially change the collective contexts in which their individual consumption is embedded (see **Consumer-Citizen**). However, when it comes to specific ESC activities, the UNEP definition still focuses on "information and knowledge".

As a definition, which reflects more recent debates and perspectives (see below), we propose the following:

> *ESC represents a variety of teaching and learning approaches that support learners in developing their own ways to contribute to fair and just needs satisfaction by means of individual and collective, private and civic, economic, cultural and political activities.*

Sustainable consumption and sustainable living cannot be achieved by education alone. However, especially in democracies, the activities necessary to transform consumption and production systems will hardly be possible and accepted without it.

History

ESC has two main origins: Consumer Education and Education for Sustainable Development (ESD).

DOI: 10.4324/9781003584056-61

Consumer education, as a main instrument of consumer policy, has traditionally focused on increasing consumer power by providing knowledge and information to enable consumers to make better decisions for themselves. Over time, this individualistic approach has evolved into a more critical perspective on consumption and self-centered need satisfaction. This eventually led to empowerment to act for mutual and collective interest. While these types of consumer education still exist, the empowerment for mutual and collective interest approach aligns most closely with ESC.

Education for Sustainable Development (ESD) has been strongly promoted by the United Nations (UN). Agenda 21, adopted at the 1992 Earth Summit in Rio de Janeiro brought, for the first time, a dual focus on both Sustainable Consumption and Production and education. The United Nations Educational, Scientific and Cultural Organization (UNESCO) proclaimed the UN Decade of Education for Sustainable Development (2005–2014), seeking to establish a sustainability focus in all areas of education. In this period, UNEP especially underlined the importance of consumption in ESD with the report "HERE and NOW! Education for Sustainable Consumption", written by Victoria W. Thoresen in 2010. The Global Action Programme (2015–2019) continued the UN Decade, before the UN prominently positioned education within its Agenda 2030. Quality Education is thus recognized as a Sustainable Development Goal of its own (SDG 4) and referred to as a "key enabler" of all other SDGs. This is also reflected in the current UN Decade: *Education for Sustainable Development: Towards achieving the SDGs* (2020–2030).

From the 1990s to the mid-2010s, education was seen as a crucial but underrated approach to realizing sustainable development. In recent years, its role in the necessary transformation has been increasingly challenged by scientists and activists, as will be discussed in the following section.

Different Perspectives

Critics have argued that the main motivation for different political actors to eagerly embrace ESC was the prospect of passing responsibility to individuals and reducing pressure to make uncomfortable decisions that would have imposed duties and burdens on businesses and citizens (see **Consumer Scapegoatism**). Accordingly, some scholars doubt the relevance and effectiveness of ESC for sustainable development, suggesting that it concentrates too heavily on individual consumer behavior alone.

While some criticism may apply to earlier or very practical approaches, most ESC scholars and practitioners agree that teaching and learning sustainable consumption must go beyond individual behavior. ESC should involve critically questioning and potentially transgressively confronting consumption (and production) systems, legal frameworks, power structures, lock-in effects, path dependencies, and other drivers of the sustainability crisis. It views learners as consumer citizens and focuses on collective actions while acknowledging the role and limits of individuals' responsibility. It also provides perspectives that allow for a critical engagement in improving quality of life through non-commercialized ways of needs satisfaction, reflecting on both individual and systemic **sufficiency** (see **Foundational Economy**).

To have a realistic understanding of and potential impact on sustainable consumption, ESC must acknowledge the complexity of consumer behavior, its embeddedness in social practices, and its dependency on other determining factors (see **Social Practice Theory**, **Behavior Change**). ESC has been criticized for referring to simplistic, individualistic models of traditional consumer education like the information-deficit model, which fails to capture this complexity. As shown in the definition section above, most ESC learning objectives today promote competencies, which include – in addition to knowledge – motivation and skills. Next to (more individual) problem-solving, prominent competencies in the international ESC discourse are systems thinking, collaboration, and interpersonal skills.

An additional point of contention against ESC is its alleged use of pedagogy to impose predefined "morally correct" behaviors. While this approach exists, most ESC scholars today agree that values, motivation, and skills for transformative sustainability action cannot be taught as such or directly imparted. Rather, ESC can merely offer a conducive environment and favorable conditions for personal development oriented toward sustainability – a process that may yield very diverse and uncertain outcomes. It empowers learners to challenge expert definitions of "sustainable consumption" and encourages reflection on and transformation of their relationship with consumption in the context of the societies and cultures they have been born and socialized into (emancipatory education). This is why transformative learning is widely endorsed in ESC.

Application

Although most ESC studies are situated in schools and universities, ESC can take place in all sectors of lifelong education: formal (pre-school, primary school, secondary school, high school, vocational school, college, university), non-formal (e.g., training on the job, in a youth club, or private language classes), and informal, everyday contexts (at work or in private life).

In formal educational institutions, ESC is not only a "topic" to be taught in the curriculum: it requires the transformation of educational settings so that sustainable ways of living can be experienced and experimented with (Whole Institution Approach). For example, teaching and learning about the importance of regional, seasonal, and low-processed foods in sustainable diets will be contradicted if the school kiosk only sells products of multinational sweets and soft drink companies. Implementing a workgroup or student firm to install and operate a photovoltaic system on the rooftop of the school will probably allow more effective (informal) learning than (formal) teaching in a classroom about challenges and developments in energy supply. Similar effects can be expected by implementing repair cafes or clothes libraries. All of these activities need to include reflections on opportunities and challenges on a system level, to avoid being stuck in individual solutions. For example, teachers might invite reflections on why commercial renting of a dress for one weekend is usually more expensive than new dresses sold in textile discounters or why repairing products is – if possible at all – often more expensive than buying new ones (see **Repair**, **Alternative Consumer Cooperatives**, **Fast Fashion**, **Sharing Economy**).

The guidebook *Teaching and Learning Sustainable Consumption* includes 57 practical ESC examples, many going beyond individual consumption issues and stimulating reflections on public policies, business opportunities, and collective activities. The book also includes more detailed descriptions of the examples in the text boxes, which go beyond the dominant cognitive approach (see Boxes 56.1 and 56.2).

Box 56.1   Education for sustainable consumption through mindfulness meditation

The brochure *Education for Sustainable Consumption through Mindfulness* has been published by PERL (Partnership for Education and Research about Responsible Living) in their Active Methodology Toolkit series (#9). It consists of three modules (personal, social, and ecological dimensions) with 19 consumption-related mindfulness practices and tasks, mainly guided meditations (see **Mindfulness**). They are ready-to-use even for teachers with little meditation experience. The approach follows the assumption that the necessary changes in values, attitudes, and behaviors for a societal transformation require personal transformation beyond increasing knowledge. The brochure was developed within a broader research, practice, and evaluation project which empirically confirmed that mindfulness practices can strengthen non-material values. However, it also made clear that such practices are neither a one-size-fits-all solution nor a silver bullet to instigate transformations in consumption behaviors.

A free download of the brochure is available here:

https://www.oneplanetnetwork.org/sites/default/files/from-crm/toolkit_9_mindfulness.pdf

---

Box 56.2　Real-world experiments for sustainable fashion consumption

In the project "ESC Textile Laboratory", real-world experiments (RWEs) were developed and used as a teaching and learning method with the following stages of intensive learner involvement: (1) co-design to determine the specific questions and tasks the students would like to work on; (2) implementation and active participation; (3) co-evaluation; and possibly (4) re-iteration. Examples of specific textile-related RWEs selected by students in different parts of Germany include repair cafes, upcycling workshops, clothes libraries, or creating a capsule wardrobe (i.e., a collection of a few, essential, versatile clothing items that can be mixed and matched to create a variety of outfits). To open the perspective beyond local activities, the students were also given the opportunity to write a letter to a fictitious but very specific textile worker ("Aadya") to empathize with a person from the Global South, where most textiles are produced. Additional reflections on systemic boundaries of sustainable fashion consumption at schools could be part of the evaluation, even if it bears the risk of reducing the students' self-efficacy due to actions being outside their immediate sphere of influence. The RWEs reflect Johann Heinrich Pestalozzi's triad of learning with head, heart, and hand – which also provides the foundation for other examples of transformative learning (see **Living Labs**).

Further Reading

Fischer, D., Sahakian, M., King, J., Dyer, J., & Seyfang, G. (Eds.). (2023). *Teaching and learning sustainable consumption: A guidebook*. Taylor & Francis.

Gossen, M., & Schrader, U. (2025). Education for sustainable development and sustainable consumption: The role of sufficiency. In L.A. Reisch & C.R. Sunstein (Eds.), *Elgar companion to consumer behaviour and the sustainable development goals*, pp. 56–72. Edward Elgar Publishing.

Grauer, C., Solbakken, I., Fischer, D., & Didham, R. (2025). Education for Sustainable Lifestyles: A powerful tool for what and for whom? In M.J. Cohen, M. Bengtsson, R. Lambino, S. Lorek, & S. McGreevy (Eds.), *Handbook of research on sustainable lifestyles*. Edward Elgar Publishing, forthcoming.

Lotz-Sisitka, H., Wals, A.E.J., Kronlid, D., & McGarry, D. (2015). Transformative, transgressive social learning: Rethinking higher education pedagogy in times of systemic global dysfunction. *Current Opinion in Environmental Sustainability*, 16, 73–80. https://doi.org/10.1016/j.cosust.2015.07.018.

McGregor, S.L.T. (2005). Sustainable consumer empowerment through critical consumer education: A typology of consumer education approaches. *International Journal of Consumer Studies*, 29(5), 437–447. https://doi.org/10.1111/j.1470-6431.2005.00467.x.

57 Social Norms

John Thøgersen

Definition

Social norms play an important role in both preserving the status quo and fostering change to a more sustainable consumption pattern (see Aasen et al., 2024). The term "social norm" generally refers to informal norms as opposed to formal, codified norms such as legal rules (Bicchieri et al., 2023). Social norms are functional in regulating social life and they especially evolve when individual actions cause negative side-effects for others. In these cases, social norms serve the function of restraining egoistic impulses in favor of collective outcomes. Social norms imply that (certain) people should perform or not perform a specific behavior.

A social norm is a behavioral rule for a situation (or type of situation) that lives up to two criteria: a sufficiently large share of the population (1) knows the rule and knows that it applies to this particular type of situation and (2) prefers to conform to the rule in this type of situation (Bicchieri et al., 2023). The second criterion implies that most people acknowledge the need to cooperate for the common good and that they therefore prefer to cooperate. However, a person's preference to cooperate disappears if they believe that (a) an insufficient number of others will conform to the rule in the situation(s) or that (b) an insufficient number of others expect *them* to conform to the rule in the situation(s) – which is an essential mechanism that potentially inhibits the adoption of sustainable lifestyles. Due to the temptation to defect, some individuals may need the belief that others expect them to conform to be backed by the fear of sanctions in case of failure to conform. Others may just conform because they accept the legitimacy of others' expectations and feel an obligation to fulfill them.

History

Social norms have been extensively studied across the social sciences: first, by sociologists such as Durkheim and Parsons focusing on the social functions of social norms and how they motivate people to act; and later, by anthropologists, economists, legal scholars, and philosophers, with each discipline adding their unique perspective to the understanding of social norms (for an elaborate account of historical trends in and the current status of social norms research, see Bicchieri et al., 2023).

It is now common in social norms research, in general and specifically about sustainable consumption, to distinguish between beliefs about what most people do, often termed *descriptive norms,* and beliefs about what others expect one to do, often termed *injunctive norms* (Helferich et al., 2023). Notice that what matters in this connection is not other people's objective behavior or expectations, but the individual's subjective perception of these realities. Reflecting this, empirical research often refers to individual beliefs about others' expectations as *perceived* or *subjective*

DOI: 10.4324/9781003584056-62

social norms. Although they are equally subjective, such qualifiers are usually not added when referring to descriptive norms.

It follows from the two conditions mentioned in connection with the definition of social norms that individuals will *not* prefer to conform to the social norm in a particular situation if either their descriptive or injunctive norm for doing so is weak. Hence, if one of them is weak, the other is of little consequence. Box 57.1 summarizes a field study on the joint impact of communicating injunctive and descriptive norms on water conservation in a shower room.

Box 57.1   Example of a descriptive norm intervention

Save Water: Turn off the water when you soap up! A descriptive norm intervention in a shower room

California struggles with water shortage and has for decades appealed to its citizens to save water. For example, at educational institutions, a common tool is to put up posters in shower rooms, appealing to users to save water by turning off the water while soaping up. Around 1980, social psychologist Elliott Aronson and his team registered compliance with such an (injunctive norm) appeal in a shower room at the University of California Santa Barbara and found it to be only about 5%. Then they conducted a simple experiment to test the effect of being exposed to others complying with the appeal. In one condition, they had one and in a second condition two research assistants standing under a shower with their back to the door, apparently after having finished sport. When somebody entered, they would turn off the water and start soaping up. And then they would register if the person or persons entering the shower room would do the same during their shower. The effect was amazing! When one person demonstrated compliance with the appeal, compliance among those entering increased to 50% and when two persons did so, compliance was nearly 70%. This illustrates, first, that it is often not sufficient that people know a social norm and that it applies to them in a given situation. Second, it illustrates that when knowledge of the injunctive norm is combined with evidence (in this case observations) that others comply with the norm, most people comply.

Source: Aronson, E., & O'Leary, M. (1982). The relative effectiveness of models and prompts on energy conservation: A field experiment in a shower room. *Journal of Environmental Systems, 12,* 219–224

Different Perspectives

In sociology, norm internalization – or the formation of a personal norm – is seen as a social process where society imprints itself upon the individual. A key feature of such socialization is to create coherence between individuals and society. Socialization can therefore be seen as a way to maintain a certain social order, which is relevant to consider when thinking about the consumer society and a transition away from it.

In social psychology, the focus is mostly on how what others do (a descriptive norm) or expectations of negative reactions from others (an injunctive norm) influence people's actions (Bergquist et al., 2019). Here, it is an empirical question of whether a social norm is internalized and becomes a personal norm and how internationalization boosts its behavioral impact, for example, regarding environmentally significant behaviors (Helferich et al., 2023). The strength of norm effects depends on both personal factors, such as the relationship between the norm and core values or social identities, and situational factors, such as the salience of negative consequences of one's actions or inactions and of the norm itself (see Box 57.1).

Application

Most people meet situations where acting to the benefit of society conflicts with their own narrow self-interest. It is more convenient, for instance, to throw litter in the street than to search for a waste bin. It is more convenient to put all one's household waste in the same garbage bin than to source-separate it and bring recyclable fractions to designated collection points. It is usually cheaper and often also more convenient to ignore possible environmental or ethical qualities of consumer products. Hence, although it can be safely assumed that most people prefer a clean environment, that their garbage is recycled, and that the products they buy live up to certain environmental and ethical standards, there is a temptation to litter, to not sort one's garbage, and to ignore possible environmental or ethical problems related to products we buy. However, if everybody behaves in this way, everybody will be worse off than if everybody restrains oneself, for the benefit of the common good. Strong, often internalized, social norms can explain why many – sometimes most – people restrain their egoistic impulses in cases such as these.

In practice, it is difficult to disentangle the influence of different types of norms because they are positively and often strongly correlated. What most people approve of is usually also what most people do. At the individual level, injunctive and descriptive norms may converge because other people's behavior serves as a cue to what is expected of the individual. This mechanism is likely to be strongest for behaviors that are readily observed by others, such as traveling behaviors and to some extent shopping behaviors (and sometimes showering behavior, see Box 57.1). A positive correlation between descriptive and injunctive norms may also be produced by the reverse inference, which is that we expect that most others conform to the injunctive norm. This mechanism may explain the convergence of the two types of norms even for behaviors in the private sphere, such as avoiding food waste or taking shorter showers at home.

Studies using different methods, including surveys, lab experiments, and field experiments have confirmed the impact of both injunctive and descriptive norms on environmentally relevant behavior, such as red meat consumption, littering, recycling, and water conservation. This has also led to "norm nudging" becoming a popular public policy tool (see **Choice Editing**, **Green Nudging**). The few studies testing the proposition that descriptive and injunctive norms interact synergistically to promote cooperation generally support the proposition.

For norms to change, some individuals must start the process as a new type of social pressure cannot establish itself spontaneously (Aasen et al., 2024). Rather, it is an evolving process, which originates among individuals who opt for new norms, and which may lead to societal norm changes to the extent others follow suit. The process may at some stage become self-perpetuating, when a point is reached where the new norm has become "how things are or should be done". This point is increasingly referred to as a "**social tipping point**".

Further Reading

Aasen, M., Thøgersen, J., Vatn, A., & Stern, P.C. (2024). The role of norm dynamics for climate relevant behaviour: A 2019–2021 panel study of red meat consumption. *Ecological Economics*, 218, 108091. ttps://doi.org/10.1016/j.ecolecon.2023.108091.

Bergquist, M., Nilsson, A., & Schultz, W.P. (2019). A meta-analysis of field-experiments using social norms to promote pro-environmental behaviors. *Global Environmental Change*, 59, 101941. https://doi.org/10.1016/j.gloenvcha.2019.101941.

Bicchieri, C. (2023). Norm nudging and twisting preferences. *Behavioural Public Policy, 7,* 914–923. https://doi.org/10.1017/bpp.2023.5.

Bicchieri, C., Muldoon, R., & Sontuoso, A. (2023). Social norms. In E.N. Zalta & U. Nodelman (Eds.), *The Stanford encyclopedia of philosophy*, Winter 2023 ed. Available at: https://plato.stanford.edu/archives/win2023/entries/social-norms (accessed: 8 January 2025).

Helferich, M., Thøgersen, J., & Bergquist, M. (2023). Direct and mediated impacts of social norms on pro-environmental behavior. *Global Environmental Change*, 80, 102680. https://doi.org/10.1016/j.gloenvcha.2023.102680.

58 Consumer-Citizen

Anna Horodecka

Definition

The concept of consumer-citizen mirrors the overlap between private consumption and public citizenship, reshaping the boundaries between consumption and civic engagement. It integrates the roles of consumer and citizen, aiming to bridge the gap between individual and collective interests. Individual interests result from needs, preferences, or personal values that drive consumers' choices at the micro level, for example, size of dwelling, overseas vacations, kind of transport, "green" products, "green" brands. Collective interests are based on shared values of citizens that are addressed by public policies at the macro level, for example, access to affordable public transport, product regulation, luxury taxes, and carbon taxes.

By bringing these perspectives together, the concept seeks to harness the positive aspects of both identities to contribute to the common good, particularly in terms of environmental and social sustainability. Johnston (2008) describes this as "voting with your dollar" to balance individual self-interest with collective social and environmental responsibility (see **Boycott and Buycott**).

Consumer-citizens are consumers who integrate a citizen's perspective into their consumption choices, even if these often remain in conflict and need to be reconciled. Consequently, several challenges emerge, specifically in the context of sustainable consumption when considering collective values and goods. The history of the term reveals further facets and the important role of the wider political context.

History

The idea of combining consumers and citizens in one term, as in "consumer-citizen", has given rise to numerous synonyms and related concepts: citizen-consumer, citizenconsumer (unhyphenated), consumer-citizen, consumercitizen, and consumer-citizenship. Most of these terms focus on how consumers can act as citizens, but the reverse order is also addressed.

The concept has evolved alongside global changes such as the rise of **consumerism**, the global economy, and the influence of corporations: (1) as a process of extending the sphere of citizenship to include consumer activities (e.g., political consumerism) and (2) by giving a civic character to consumer behavior.

The sugar boycott in the 19th century, which contributed to the act of banning the slave trade in Britain, depicted in the film *Amazing Grace*, illustrates the first process: political consumerism. It showed that consumers' collective action could achieve political goals when traditional means were insufficient. It also gave political agency (empowerment, participation) to those who could not vote, especially women, by allowing them to express their civic voice through household purchasing decisions.

DOI: 10.4324/9781003584056-63

Another process leading to the emergence of the citizen-consumer began in the United States in the early 20th century, when the question of consumer rights was raised. To challenge the monopolies that controlled grain milling and food prices, many affected consumers joined together to form cooperatives (see **Alternative Consumer Cooperatives**).

Later, consumer concerns shifted from essential goods (such as food and coal) to the policies and prices of luxury goods (such as electrical appliances). More generally, attention also turned to the role of **advertising**, the legitimacy of marketing, and the free market as an alternative form of governance. Over time, consumer issues shifted toward access to free and fair information ("Naderism") and later in the 1980s, to a process called "alternative consumption" (responses to established forms of consumption) and "ethical consumerism" (a form of political activism based on the premise that purchasers in markets consume not only goods but also, implicitly, the process used to produce them). Sustainable consumption (the use of goods and services that respond to basic needs and improve the quality of life while minimizing the use of natural resources, toxic materials, and emissions of waste and pollutants over the life cycle, so as not to jeopardize the needs of future generations) also came to the fore around this time. The active interest of consumers in sustainability issues emerged and highlighted their role as agents of change and the relevance of environmental citizenship, political consumerism, and lifestyle politics.

Different Perspectives

The different perspectives that emerged around the concept of consumer-citizen originated with the aforementioned events.

The post-war period brought further nuance to the meaning of "consumer-citizen": Consumer action was framed in terms of civic choice, visible in language use like "voting" and "judging" in relation to purchases, and equated with civic engagement (market choice as a form of citizenship). However, the notion of consumers as sovereign voters in a market democracy ("democracy of goods") is criticized as superficial, non-informed, and vulnerable to advertising and marketing, which use the concept to reshape society to benefit corporations more than the public and civil society (see **Greenwashing**).

Neoliberalism introduces new aspects to the meaning of citizen-consumer and citizen as consumer, extending the identity of the consumer to that of a citizen. The terms now refer to individuals who interact with the state and the public sphere from a consumer perspective. This development is also related to the introduction of New Public Management (NPM), which has transformed public service organizations by making them more accountable and legitimate. Citizens became customers who could "buy" certain solutions, services, and goods, while public servants took on the role of public managers. However, this shift institutionalized mistrust, leading to a decrease in trust in government institutions. The role of Members of Parliament (MPs) is replaced by advocacy groups, charities, and NGOs, and shopping has become as important as other civic activities, including democratic participation.

Austerity policies introduced during the 1980s – foremostly in the United States, the United Kingdom, and later in some other countries – further highlight this issue. The concept of the consumer-citizen is used to justify these policies by suggesting a shift of power from the state to individuals. This allows citizens to "vote" with their spending, making important decisions through their purchasing choices. Neoliberalism, as the backdrop for the emergence of this concept, provides a dual rationale: it offers a positive explanation for austerity measures and justifies reducing public spending and taxes by claiming that informed consumers know best what they want.

Most recently, the concept of "data citizens" was introduced, referring to critical and active citizen agency in times of societal datafication and algorithmically driven decision-making. These individuals actively participate in a networked economy, instead of simply purchasing products. They influence social and environmental outcomes through actions such as posting, clicking, and navigating online.

Application

The various perspectives on the consumer-citizen concept shed light on its potential applications, which may inform policy.

- *Questions of identity*: The identities of citizens and consumers often come into conflict. A citizen participates in a system driven by government regulations and policies, rooted in beliefs about the common good and civic responsibility (see **Ecosocial Contract**). In contrast, a consumer exists within a market-driven system, guided by the ideology of individual choice and economic interest (see **Political Economy of Consumerism**). These two identities are irreconcilable because they are based on fundamentally different belief systems. As a result, a person may act as a citizen in some situations and as a consumer in others, but this can create cognitive dissonance. Consumer choices, especially when sustainable options are more costly or less accessible, may not align with civic responsibilities. Therefore, policymakers need to address this conflict, helping individuals navigate the tension between their citizen and consumer identities in a way that promotes sustainability.
- *Erosion of the public sphere and its impact on citizenship, autonomy, and democracy*: Shifting public concerns to the private sphere of consumption choices risks eroding (i) citizens' identity and responsibility and (ii) public space, which may expose societies to authoritarian regimes and uncontrolled market forces. Consumer industries are criticized for infantilizing citizens and threatening their sense of autonomy, which comes from public debate and collective rule-setting. Protecting this autonomy from economic influence is crucial.
- *Justice and inequality concern*: Shifting responsibility to consumers' choices may be considered unfair, especially for consumers with low incomes. Wealthier citizens who have larger carbon footprints can afford green options that appear "cleaner", while low-income citizens – even if they consume less – cannot (**Household Income Versus Carbon Footprint**). Reduced government spending and limited regulations exacerbate this issue, making sustainable options less accessible, and contributing to excessive consumption patterns.
- *Responsibility issue*: Expanding the individual sphere and over-relying on consumer responsibility is short-sighted and can reduce enterprise accountability, and undermine public policy and collective action. It can be used to justify reducing state regulation, public goods provision, and spending (see **Consumer Scapegoatism**).

Further Reading

Clarke, J. (2007). Citizen-consumers and public service reform: At the limits of neoliberalism? *Policy Futures in Education*, 5(2), 39–248. https://doi.org/10.2304/pfie.2007.5.2.239

Fontenelle, I. A., & Pozzebon, M. (2018). A dialectical reflection on the emergence of the citizen as consumer' as neoliberal citizenship: The 2013 Brazilian protests. *Journal of Consumer Culture*, 21(3) (August), 501–518. https://doi.org/10.1177/1469540518806939.

Horodecka, A. (2024). Sustainable consumption as a common good. The citizen-consumer approach challenged. In A. Horodecka & A. Szypulewska-Porczynska (Eds.), *Collective sustainable consumption. The case of Poland*, pp. 51–69. London: Routledge.

Johnston, J. (2008). The citizen-consumer hybrid: Ideological tensions and the case of whole foods market. *Theory and Society*, 37(3), 229–270. https://doi.org/10.1007/s11186-007-9058-5.

Schild, V. (2007). Empowering 'consumer-citizens' or governing poor female subjects?: The institutionalization of 'self-development' in the Chilean social policy field. *Journal of Consumer Culture*, 7(2), 179–203. https://doi.org/10.1177/1469540507077672.

59 Social Movements

*Duncan Crowley, Teresa Marat-Mendes,
and Roberto Falanga*

Definition

A social movement is a force pushing against something, while also pulling its alternative into being. Diani states that "social movements are defined as networks of informal interactions between a plurality of individuals, groups and/or organizations, engaged in political or cultural conflicts, based on shared collective identities". For James and Van Seters, a social movement is also characterized by shared objectives: it is a community that "comes together" around a few minimal conditions,

> as a form of political association between persons who have at least a minimal sense of themselves as connected to others in common purpose and who come together across an extended period of time to effect social change in the name of that purpose.

The essence of a social movement is political organizations pushing for system change in many forms and at a range of scales, from the local (a street, block, or neighborhood), to the regional, national, transnational, and global scale. The relevance of social movements in the context of sustainable consumption and lifestyles is multifold: from communities identifying the complex roots of the ecological and social crises, taking political action to uproot systems that maintain the status quo for short-term power and profit, to creating alternatives that build system change. These can include **Fair Trade**, **Eco-Communities**, **Community Supported Agriculture**, and **Buen Vivir and Buenos Convivires**, to name but a few.

Social movements' organic, non-hierarchical, open structures, often employ high levels of creativity and spontaneity and sometimes enable very intense and strange sparks of hope to emerge (see Box 59.1).

Box 59.1   Exemplary social movement

The 2011 anti-austerity movement in Spain, referred to as 15M or Indignados and by some as the "Spanish Revolution", saw public squares occupied throughout Spain (see Figure 59.1), to tackle economic meltdown. By 2015, these citizen-led movements morphed into a "Rebel Cities" movement that won local municipal elections in many major Spanish cities. In Barcelona, under the city's first woman mayor Ada Colau, a multitude of radical ecourban initiatives were tested and implemented.

DOI: 10.4324/9781003584056-64

Figure 59.1   15M *Indignados* occupy Puerta del Sol, Madrid, during the "Spanish Revolution"

Source: Wikimedia Commons/"Carlos Delgado; CC-BY-SA" (2011)

History

As long as unfair dominant systems of power, abuse, exploitation, injustice, misery, and destruction have needed to be fought against by oppressed communities, social movements have existed. In the wake of social dislocation caused by the Industrial Revolution – which also marked the origin of the climate crisis – rapid urbanization in England, feverish coal mining, and dangerous factory jobs spurred militant efforts to improve worker conditions. These efforts culminated in Marx and Engel's 1848 publication of the *Communist Manifesto* and inspired subsequent political movements advocating socialism, communism, anarchism, nationalism, and democracy.

The West experienced political upheaval in the 1960s: International movements – for example, civil rights, national liberation in the colonized world, radical feminism, and environmental activism – erupted to oppose injustices, such as racism and lack of civil rights in the United States, the war in Vietnam, colonial oppression, and natural degradation. Capitalism's threads connected colonialism, imperialism, racism, exploitation, and war. Its environmental consequences were pollution, nature destruction, and biodiversity loss. Western societies realized their dependence on cheap oil during the 1970s energy crisis. The demise of Soviet Russia, from 1989 onward, led to the global domination of capitalism that fuelled increasingly consumerist lifestyles in the United States, and soon around the world. Driven by the rapid expansion of neoliberal economics, with its perpetual growth addiction and an insatiable hunger for profit, a new, power-hungry, political class carved up the world's finite resources under the banner of free-market economic globalization.

For many, November 30th, 1999 was a day that "permanently changed the political landscape of globalization", when a globally connected social movement, using new forms of communication (e.g., email) and radical peaceful activism techniques, successfully shut down the World Trade Organization (WTO) in what is called the "Battle of Seattle" (see Figure 59.2). A diverse coalition of environmentalists, labor trade unions, women, farmers, community media workers, and students used street theatre and direct action to face off riot cops, violence, and mass arrest. This anti-capitalist social movement, later labeled the "anti-globalization movement", took on corporate globalization and, for a short moment, won.

In August 2018, a 15-year-old Greta Thunberg wearing a distinctive yellow rain jacket held her homemade "*Skolstrejk för klimatet*" (School Strike for Climate) sign outside Sweden's national parliament demanding the government align with the Paris Agreement. She returned every week and sparked the global "Fridays for Future" (FFF) social movement. Their March 2019 global strike gathered more than one million strikers in 125 countries. With the FFF Movement, many young people became encouraged to stop flying, become vegetarians, **repair** things, buy second-hand clothes, and demand quality infrastructure for active mobility and public transport in their communities. While these individual environmental, healthy consumer choices matter, FFF insists that current challenges can only be tackled by collective action. While older European elders turn to the youth for guidance in climate action, FFF highlights the struggles and voices of marginalized, indigenous, and front-line communities threatened by the brutalities of capitalism.

Figure 59.2 Anti-WTO banner drop, by *Rainforest Action Network*, Seattle, 1999 (Citizens Trade Campaign, 1999)

Different Perspectives

Intersectional approaches seek to find common struggles and challenges among a diversity of groups. Asking what is needed most and listening to those at the margins is how privileged social movements can best demonstrate solidarity. Sustainable lifestyle-type social movements often tend to be more white, educated, hetero, male, and affluent. Engagement demands moving beyond comfort zones to reach marginalized, and often invisible, communities found in all modern societies. Going to these local communities and organizing spaces, rather than inviting "them" to come to comfortable, NGO-funded, or university-sponsored spaces, is sometimes challenging and even dangerous. But it builds trust and often is the only way to understand a community and do real work. Relationships can form around playing football, washing dishes, or helping kids with homework.

Instead of quick "sustainable consumption" solutions, it's often better to understand real community challenges, by humbly asking questions, even the hard ones. "Climate solutions" must respond to local challenges (food shortage, gangs and violence, lack of local leadership) first. With culture wars, political polarization, and the resurgence of ethnonationalist movements, great care is needed in defining and defending values and boundaries. White supremacist structures still exist in much of the world and are, despite decades of radical progress, returning to many parts of Europe. Sustainable solutions have to be accessible to everybody. Sustaining human life requires solidarity and action in defending what Indigenous folks have always known and regenerative communities are remembering again: We are nature (see **Buen Vivir and Buenos Convivires**, **Ubuntu**).

Application

Many of today's seemingly basic rights were won by social movements, often after violent repression, imprisonment, and death. This includes votes for women, the eight-hour working day, and the end of apartheid in South Africa. Ecological social movements too have had great victories, here are two we love:

Rights of Nature: The indigenous Māori people's legal victory for their Whanganui River basin in New Zealand came in 2017. They ensured their river was given its own legal identity, with the rights, duties, and liabilities of a legal person, thereby changing legal policy and inspiring more communities, both Indigenous and contemporary, to explore legal pathways to ensure more nature areas are protected by nature guardians.

The Māori say:

From the mountain
to the sea
I am the river
And the river is me

Postgrowth: After decades of being stuck in mostly academia, in May 2023 the Postgrowth community had its "Woodstock moment" when the European Parliament facilitated a three-day "Beyond Growth" conference with a host of diverse speakers from around the world. Building on this, five follow-up conferences are underway in Austria, Denmark, France, Ireland, and Italy, many happening within national Parliaments.

Further Reading

Aufheben. (2001). *Anti-capitalism as an ideology . . . and as a movement? (A critical analysis of the "anticapitalist movement" of the early 2000s)*. Published September 2001 or Aufheben, Added to Libcom, July 24, 2005. https://libcom.org/article/anti-capitalism-ideology-and-movement (accessed: 8 January 2025).

Davison, Isaac. (2017). Whanganui River given legal status of a person under unique Treaty of Waitangi settlement. *New Zealand Herald*, March 15. Available at: https://www.nzherald.co.nz/kahu/whanganui-river-given-legal-status-of-a-person-under-unique-treaty-of-waitangi-settlement/JL3QK-SWVZPA7XW6EN33GKU4JJ4/ (accessed: 8 January 2025).

Shea Baird, K., & Roth, L. (2017). Municipalism and the feminization of politics. *The City Rises, ROAR (Reflections on a Revolution) Magazine*, 6. Available at: https://roarmag.org/magazine/municipalism-feminization-urban-politics/ (accessed: 8 January 2025).

Shiva, V. (1988). *Staying alive, women, ecology and survival in India*. New Delhi and London: Zed Books Ltd. Available at: https://www.arvindguptatoys.com/arvindgupta/stayingalive.pdf (accessed: 8 January 2025).

Thunberg, Greta. (2019). Our house is on fire. World Economic Forum, Davos, January. *Guardian*. Available at: https://www.theguardian.com/environment/2019/jan/25/our-house-is-on-fire-greta-thunberg16-urges-leaders-to-act-on-climate (accessed: 8 January 2025).

60 Subvertising

Eleftheria Lekakis

Definition

Subvertising refers to practices of engaging with advertising critically, both symbolically and materially, and can aim for futures beyond growth-driven economies and consumer-driven cultures. For example, it engages with advertising symbolically through challenging the language of **advertising**, and materially through illustrating its connections with carbon-intensive industries. Subvertising can be understood as both an activist practice and a **social movement**. It is often used interchangeably or in connection with terms such as culture jamming, anti-advertising, adbusting, and ad takeovers, while in Spanish it is known as *contrapublicidad*, and in French *antipublicitaire*.

History

The practice of *détournement* (the mixing of artistic elements to make a political intervention), which was popularized by French radical movements in the 1950s (Letterist International, Situationist International), can be regarded as a predecessor of subvertising. From the late 1970s onwards, subvertising groups and organizations appeared. A non-exhaustive list of subvertising movements includes:

- Adbusters (Canada)
- Berlin Busters Social Club (Germany)
- Billboard Liberation Front (United States of America)
- Billboard Utilising Graffitists against Unhealthy Promotions or B.U.G.A. U.P. (Australia)
- Brandalism (United Kingdom)
- Democratic Media Please (Australia)
- Front de Libération de l'Invasion Publicitaire (FLIP) Switzerland)
- Le Collectif des Déboulonneurs (France)
- Consume Hasta Morir (Spain)
- Proyecto Squatters (Argentina)
- Résistance à l'agression publicitaire (France)
- Special Patrol Group (United Kingdom)

At the same time, subvertising practice has been used by movements such as the Anti-Apartheid Movement calling for the boycott of Barclays Bank in connection to Apartheid and South African goods (see **Boycott and Buycott**). In 2017, Subvertisers International was launched as a network of individuals and organizations taking action to address the effects of advertising, such as overconsumption and natural degradation to name a few. The launch included an international

DOI: 10.4324/9781003584056-65

advertising takeover campaign, public meetings, screenings, lectures, and workshops. Since then, transnational coordinated actions of members (mainly in the Global North), such as the Subvert The City and ZAP (Zone Anti-Publicité) Games have been inviting participatory actions against advertising and **consumerism**. Individual citizens, artists, and collectives come together in these activities to express alternative views and practices and engage in movement-building activities.

Different Perspectives

It is possible to unpack different approaches to subvertising by theorizing it as anti-consumerism. Following Kim Humphery, anti-consumerism is

> a field of alternative social and economic practices, a political stance and current, that traverses movements of various kinds – from the strategically oriented to the experientially based – and invokes an array of political perspectives and strategies – from the liberal to the libertarian, and from the planned to the impetuous.
>
> (2009: 109)

Subvertising tends to be understood as a series of fleeting individual or collective interventions in advertising spaces. Also, research on subvertising has predominantly focused on subvertising as a merely countercultural practice. Subvertising is "*focused on challenging the ideology of consumerism purveyed through advertising*" (Humphery, 2009: 50). Yet, beyond momentary activist intervention, subvertising is also considered a social movement, proposing systems change in response to social and economic inequality and the climate crisis (see **Carbon Inequality**, **Climate Justice**).

Historically, the subversion of advertising texts (a.k.a. culture jamming) has been critiqued for considering citizens to be merely dozing, and activists as attempting to awaken them. The key issue here is the focus of anti-consumerism on what it is "against". Within the context of cities, for example, order is produced by several actors (institutional and non-institutional, legal and social, material and discursive, corporate and non-corporate) to rule out or eliminate disorder. Identifying the dominance of advertising within public (urban) spaces, Thomas Dekeyser writes that "subvertising performs an embodied experiment in cracking open the regime of order" (314). In doing so, the practice of subvertising manifests itself as non-violent civil disobedience, advocating for distance from consumerism and branded culture or against corporate wrongdoings and environmental injustice. Moreover, the range of activities that groups involved in subvertising undertake promote collective dialogue and community by bringing people together to discuss issues related to advertising. These activities range from producing materials for community organizing and advertising literacy material for schools to organizing workshops for making subvertisements.

Application

In the following, two examples of subvertising relating to the systemic politics of consumption are presented. First, the UK-based group Brandalism undertook coordinated actions during COP21, at which the Paris Agreement was adopted, targeting the corporate sponsorship of the UN climate talks. An analysis of visuals that replaced advertisements in bus stops across Paris (see Figure 60.1) demonstrates a range of narratives from (i) corporate greed (highlighting the irony and incongruity of airlines and major fossil fuel polluters sponsoring climate talks) to (ii) inadequate political will (government inability to act in the public interest), (iii) consumer saturation (the role of smartphones in occupying our attention), (iv) Earth in mourning (an unwell planet), and (v) public commitment to the environment (poetic declarations of love and commitment to the environment,

Figure 60.1 Guerrilla bus stop installation in Paris by Brandalism (2015)

Source: Photo by Author

calls for alternative allocation of time, DIY and self-provision and restored value in interpersonal relations and community exchanges) (Lekakis, 2017).

Brandalism's subvertising actions during COP21 used the logic and space of dominant commercial discourses to disrupt and remix them for present and prefigurative aims. In doing so, they went beyond critiques of culture jamming by aligning with other social movements and articulating a systemic critique of economic growth (see **Degrowth**). Importantly, this critique addresses advertising as both a symptom and a cause (intertwining politicians, corporations, and consumer culture) of systemic issues like the climate crisis.

Second, a couple of years later, Brandalism joined the previously mentioned Subvertisers International. While the members of Subvertisers International range in terms of agency (individuals and collectives), tactics (including the use of fire extinguishers to cover advertising screens), and targets (from co-creating and installing subvertisements to lobbying for advertising policy), they share a concern about the influence of advertising, its urgency, and mobilization through non-violent action. In 2022, Subvertising International published a report (*Advertising and its Discontents*) putting forward recommendations for regulations that redefine the definition of acceptable advertising activities, protecting public spaces from the aggressive presence of advertising, and introducing economic measures such as taxation on advertising practice.

Subvertising activism, whether individual, ephemeral, collective, or sustained, is important for transforming consumption and production systems because it brings advertising as a catalyst of overconsumption to the center stage. Subvertising is more than being "against" advertising. It is more than telling people not to buy. It is more than disrupting the presence of advertising (despite that being a central tenet). It is about creating opportunities for thinking beyond what advertising puts in front of us and about creating change through advertising regulation that protects against socially and environmentally harmful advertising. Subvertising has centered on the ethics of advertising in response to the corporatization of culture, but also on socio-economic inequality and the climate crisis. Policy recommendations stemming from social movements such as Subvertisers International pose opportunities for systems change that holds advertising accountable to people and the planet.

Further Reading

Dekeyser, T. (2021). Dismantling the advertising city: Subvertising and the urban commons to come. *Environment and Planning D: Society and Space*, 39(2), 309–327. https://doi.org/10.1177/0263775820946755.

Humphery, K. (2009). *Excess: Anti-consumerism in the west*. Cambridge: Polity Press.

Kozinets, R.V., & Handelman, J.M. (2004). Adversaries of consumption: Consumer movements, activism, and ideology. *Journal of Consumer Research*, 31(3), 691–704. https://doi.org/10.1086/425104.

Leal-Rico, I., Papí Gálvez, N., & Sánchez-Olmos, C. (2024). Evolution of studies on subvertising: A scoping review. *Humanities and Social Sciences Communications*, 11(1), 1–13. https://doi.org/10.1057/s41599-024-03972-9.

Lekakis, E. (2017). Culture jamming and brandalism for the environment: The logic of appropriation. *Popular Communication*, 15(4), 311–327. https://doi.org/10.1080/15405702 2017.1313978.

61 Boycotts and Buycotts

Fabián Echegaray

Definition

Boycotts and buycotts are collective actions by **citizen-consumers** aimed at influencing the behavior of firms or, less frequently, governments to act in favor of public good. These actions involve the motivation of private consumers to choose or avoid one product, brand, or manufacturer based on their ethical, social, or environmental record. Both actions are considered political **consumerism** as they introduce political considerations into private decisions, with consumers aspiring to create a more socially and environmentally just social order. They rely heavily on word-of-mouth though more organized campaigns are also used.

Boycotts can take different forms. The most common is a rejection of products, brands, supplies, or manufacturers who harm the wider interests of society. Other forms include engaging in culture jamming directed at certain brands (see **Subvertising**); activism online or offline negatively affects the reputation or market share of specific brands, producers, sources of production, or products/services. These practices largely take place as private, individual, silent behaviors, but there are also boycotts centrally organized and planned, such as factory sit-ins or blocking access to production or distribution (see Box 61.1).

Box 61.1   Examples of boycotting

Companies that pollute, disrespect social rights, or engage in corruption typically drive individual consumers to engage in boycotts. Avoidance of brands involved in labor or human rights scandals, sitting in front of stores to raise awareness about the company's social or environmental misbehavior, and denouncing products that harm animals in their production processes are examples of boycotting.

Buycotting involves supporting companies or products for ethical, social, or environmental reasons. Usually, this is associated with rallying around brands that align their **advertising**, portfolio, or internal policy with valuable goals, or companies producing environmentally benign goods, engaging in philanthropy, or having active and cogent Corporate Social Responsibility and social inclusiveness programs (see **The Role of Business**). While such support can be criticized for indirectly supporting corporate opportunism or **greenwashing**, buycotting can also extend beyond individual brands, by supporting **social movements** that challenge business-as-usual. The case in point is creating consumer preference for foodstuffs that are certified as organic, **ecolabeled**,

DOI: 10.4324/9781003584056-66

or **Fair Trade** as well as those produced through **community supported agriculture**. It is also about buying products that are locally sourced or handmade, produced by a minority, or through family-based manufacturing. The goals of these actions include ecological sustainability, redress of market imbalances, empowerment of small-scale producers, and strengthening local economies.

Ultimately, both boycotts and buycotts reflect consumers' allocation of incentives in response to agents' perceived role in facilitating or thwarting public benefit through their institutional behavior or their product portfolio. By doing so, boycotts and buycotts influence the distribution of power among companies in the form of reputation, market share, and social license to operate.

History

Interest in individual boycotts and buycotts emerged in the political action literature as scholars identified a shift in civic engagement from formal to unconventional forms of participation. Public opinion studies, such as the World Value Surveys conducted since the mid-1970s, consistently show that, after voting, these forms of political consumerism are among the leading modes of individual participation in industrialized democracies. Initial concerns that boycotting and buycotting would undermine traditional democratic involvement, such as petitioning or street protests, proved unfounded.

Citizens historically targeted states or governments rather than corporations in boycotting behaviors. Early forms of boycotting in Europe targeted goods produced by enslaved people, such as sugar, to protest against violations of human rights and thereby make a political statement. Colonies used the same tactics to gain autonomy or independence. For example, Gandhi's campaign against British textiles persuaded Indians to avoid buying British products and rely on manufacturing their own clothing. Consumers in Nazi Germany, meanwhile, were mobilized by their government to sabotage Jewish stores based on political and moral claims that equated purchasing decisions with an alleged national benefit.

A few examples stand out as acts of consumer mobilization that have successfully raised awareness and pressured corporations and governments to change their practices. This includes high-profile global boycotts of South African products aimed to put pressure on the Apartheid government in the 1970s and 1980s. The highly publicized boycott of Nestlé products, which started in 1977, was a response to the company's unethical advertising of their powdered milk formula to new mothers in the Global South. Other high-profile cases include boycotts over Nike's tolerance for sweatshops and violation of labor rights in their supply chain; the 1988–1990 boycott of yellowfin tuna, driven by concerns over fishing technologies that also killed dolphins; and the 2010 boycott of bluefin tuna, aimed at preventing overfishing and the species' potential extinction.

Different Perspectives

Major factors currently shaping political consumerism and thus boycott and buycott actions include (1) the major shift in values and priorities among younger generations; (2) the growing influence of consumption practices on social identity and citizenship; and (3) the increasing power of corporations as key agents in shaping public and economic life, making them political targets; see Copeland & Boulianne, 2022. These factors reinforce one another.

Regarding the value shift, a substantial transformation is taking place in social and political values: from materialism, utilitarianism, and hedonism to postmaterialism, self-expression, and transcendence (see **Alternative Hedonism**). The value-change theory attributes this shift to the younger generations, who, by living with lower future material and physical security than the elderly, prioritize expressive self-actualization over material abundance. Their growing concern about ecological sustainability also plays a role. As individuals feel less dependent on hierarchical authorities and more empowered to act independently, the authority vested in traditional institutions corrodes. They express growing mistrust of conventional institutions and modes of expression of political representation and action.

The above value shift among younger generations finds an expression in the individualization of political activities and personal engagement through the so-called lifestyle politics, that is, the attribution of fundamental political meaning to day-to-day activities such as shopping, eating, dressing, and spending leisure time. Consumer activities increasingly overlap with civic engagement. In addition to boycotts and buycotts, other forms of participation and bringing about social change have become prominent, such as volunteering and community involvement (see **Alternative Consumer Cooperatives**).

Many scholars argue that younger generations see citizenship more through the consumption of goods and media than through traditional political organizations. The prevailing political discourse equates citizenship to individual access to material goods, improved living standards for the middle classes, and market inclusion of the poor. It creates a convincing narrative that promotes participation in a consumer society as an indication of civic involvement. Individuals exposed to situations where their interests, rights, and values are perceived as better served by endorsing certain products or brands (or by opposing organizations causing negative externalities) easily connect the consumption and the citizenship spheres.

Confronting corporate power is another trend in political consumption. In the processes of economic liberalization and privatization, corporations became more politicized as both actors and targets, mostly because of their visibility as providers of public utility services. Furthermore, the growing corporate disregard for public well-being ignited consumer activism. This led to the institutionalization of legal protections for consumer rights through government policies and to programs in corporate self-regulation, such as the concept of corporate social responsibility (CSR), codes of conduct (for example, CERES), standardized reporting (e.g., Global Reporting Initiative, GRI) and Socially Responsible Investing (see also **The Role of Business**). These processes educated younger generations of citizens to bridge seemingly private consumer issues to wider public affairs and to vote with their wallets to influence corporate policies.

Research suggests that companies deploy symbolic and substantive sustainability performances, such as advertising their CSR programs and adopting rhetoric promoting shared responsibility among consumers and businesses, to gain consumers' favor. This approach uses the institutional ecosystem of market-oriented NGOs, corporate-backed grassroots networks, pro-business think tanks, and business media platforms. Together, these entities disseminate new criteria for market valuation and purchasing decisions. One of the results is that consumers confront their relationship with brands as an exercise of citizenship. For some, it leads to increased political consumerism.

Application

Boycotts and buycotts have created a significant legacy for paving the road toward a sustainable society and disciplining market forces toward a progressive direction. They have played major roles in ending injustices like apartheid and labor rights violations, improvements in environmentally responsible manufacturing and waste management, and driving corporations to align themselves with principles of social responsibility and public accountability for their actions.

Skepticism may abound regarding the durability of this effect. Even when commitments fall short and greenwashing or social-washing persist, the codes of conduct, sustainability pledges, and public scrutiny continue to drive companies to improve their performance or avoid misconduct. This is largely motivated by the need to protect brand value, maintain a good reputation, and safeguard their market share. Similarly, the continued growth of organic, ecolabeled, traceability-based products and brands reflects the long-term positive impact of leveraging consumer power to reward pro-sustainable options in the market. While boycotts and buycotts may often be limited to individualized, focused, and temporary actions, they nonetheless may play a significant role in a gradual transition to a more equal, inclusive, ecofriendly, and economically democratic future.

Further Reading

Bennett, W.L. (1998). The uncivic culture: Communication, identity, and the rise of lifestyle politics. *PS: Political Science & Politics*, 31(4), 741–761. https://doi.org/10.1017/S1049096500053270.

Copeland, L., & Boulianne, S. (2022). Political consumerism: A meta-analysis. *International Political Science Review*, 43(1), 3–18. https://doi.org/10.1177/0192512120905048.

Echegaray, F. (2016). Corporate mobilization of political consumerism in developing societies. *Journal of Cleaner Production*, 134, 124–136. https://doi.org/10.1016/j.jclepro.2015.07.006.

Inglehart, R. (1997). *Modernization and postmodernization: Cultural, economic, and political change in 43 societies*. Princeton: Princeton University Press.

Stolle, D., & Micheletti, M. (2013). *Political consumerism: Global responsibility in action*. Cambridge: Cambridge University Press. https://doi.org/10.1017/CBO9780511844553.

62 Green Parenting

Bianca Stumbitz and Robert Orzanna

Definition

Green parenting refers to a style of parenting that prioritizes environmentally friendly practices in raising children. This approach encompasses various aspects of a child's life, including food choices and other consumer habits, transportation, energy use, waste management, and overall lifestyle. Green parents strive to minimize their family's ecological footprint by making environmentally conscious decisions and teaching their children to respect and protect the environment. Ultimately, green parenting aims to foster a deeper connection between children and the natural environment while instilling values of environmental stewardship and responsibility (see **Social Norms**, **Education for Sustainable Consumption**).

History

Green parenting behavior has long roots in many traditional cultures where parenting naturally had a much smaller ecological footprint than is the case in industrialized economies today. However, the concept of green parenting, that is, the *conscious decision* to adopt an ecofriendly approach to bringing up children, developed with the green political movement in the 1970s and accelerated in the 1980s, alongside the increasing topicality of debates around sustainable development. The term "green parenting" has been in use since the early 1990s and has been growing in scale and relevance in the parenting world, research, and the broader sustainability agenda. For instance, UNICEF encourages the "global community" to support families in green parenting as a measure to reduce the impact of climate change on children and to prepare them for a more resilient future. Although green parenting continues to be predominant in the Global North, the social movement has started to grow in prominence in Global South contexts too. For example, organizations like "Our Kids' Climate" (see Further Reading) bring together a diverse network of parent-led, family-focused climate groups and leaders from across the globe. There are particularly interesting approaches to green parenting that seek to reconcile ancient cultural traditions and ecological knowledge, including from Indigenous communities, with contemporary sustainable consumption practices, with examples coming from Africa, Asia, South America, as well as Canada and Australia (see Further Reading).

Different Perspectives

The nature and levels of involvement in the movement vary widely, and there is no coherent view of what green parenting entails, which can result in misconceptions. Some people assume that green parenting requires a drastic lifestyle overhaul, leading to a perception that it is only accessible to a niche group of environmentally conscious families. This can create a sense of exclusivity

DOI: 10.4324/9781003584056-67

and deter parents who might otherwise be interested in adopting some ecofriendly practices but are uncertain where to begin.

One key debate involves the balance between environmentally friendly practices and convenience. Many parents feel the pressure to use cloth nappies or homemade baby food, but they also face time and resource constraints or emotional turmoil caused by strong self-judgment. Critics of green parenting argue that these practices can be unrealistic for families who are already confronted with a high mental load or those with limited financial means. This raises questions about accessibility and equity in the green parenting movement, where those with more resources may have a greater ability to adopt ecofriendly practices.

Moreover, controversies can arise from perceived judgment within the green parenting community. Parents who do not or cannot fully commit to a comprehensive green lifestyle or engage in a climate activist group might feel ostracized or judged by those who do. At the same time, green parents may feel the desire to continue to belong to the rest of society. Associating oneself with related movements such as Waldorf, Montessori, or forest kindergartens and schools, especially when children grow in age and parental influence decreases, can foster the feeling of belonging and reaffirm the chosen parenting style.

Another area of contention is the impact of green parenting on children's well-being. Some argue that strict adherence to ecofriendly practices or "radical" green parenting can lead to undue stress on both parents and children, potentially creating an environment where children feel pressured to live up to certain ideals. Therefore, parents may have to better balance between teaching their children environmental responsibility while allowing them to experience their childhood without excessive pressure or guilt.

Lastly, the influence of **greenwashing** and marketing on green parenting has sparked criticism (see **Advertising**). Companies may label products as "green" or "ecofriendly" without substantial evidence to support these claims (see **Ecolabeling**). This can mislead parents who are trying to make environmentally responsible choices, contributing to skepticism about the authenticity of green products and practices.

These ongoing debates, misconceptions, and controversies highlight the complexity of green parenting and the need for a nuanced approach that considers the interplay of various family dynamics, economic constraints, and sociocultural contexts and histories.

Application

Entering the world of parenting is one of the most impactful processes in a person's life course. It involves the need to make a range of behavioral and consumer decisions, which can be a challenging, and at times overwhelming, experience (see **Moments of Change**). Green parents can find lifestyle inspiration and advice in a range of books, magazines, blogs, internet forums, and charities set up by green parents for green parents. These sources hold a wealth of tips to help new parents with their choices, often providing a "best practice" list they can adapt to suit their personal circumstances (see Figure 62.1). While some of these tips are age-specific, others relate to broader ecofriendly lifestyle choices regardless of the age of the child.

Some research found that becoming a parent triggered motivations to adopt, continue, and often intensify green lifestyles. The key goals behind this behavior were to reduce the household's ecological and carbon footprint, to protect family health, particularly through diet, and to socialize children to empower the next generation of green citizens. In a consumerist society and an era of digital overload, the overall ethos of green parenting is that "less is more" (see **Sufficiency**, **Voluntary Simplicity**, **Alternative Hedonism**). Nevertheless, green parents face similar challenges as most consumers who lead sustainable lifestyles – costs in terms of time and money. It is also for

Green parenting practices

Younger children
- Use reusable nappies
- Breastfeed
- Make your own baby food
- Adopt and look after a houseplant
- Plant and look after a garden
- Explore the outdoors
- Sort waste together
- Be creative and craft new items
- Switch off the lights and other electrical appliances

Older children & teenagers
- Talk about climate issues and action with your adolescent
- Do not be afraid to let your teen educate you on climate change
- Share news articles with your children about their peers making a difference

Broader ecofriendly lifestyle choices, regardless of the child's age
- Consider ecological aspects in choice of housing size, location, and energy efficiency
- Do laundry the eco-way (line drying, low temperatures)
- Choose eco-friendly cleaning products
- Choose to reuse: e.g., clothes, toys, books, prams, and baby carriers
- Wherever possible, use public transport, cycle or walk
- Most importantly, set an example

Figure 62.1 Selected tips for green parenting

Source: By authors

this reason, and due to the lack of political support, that (intentional) green parenting continues to attract mostly middle-class parents (see **Household Income Versus Carbon Footprint**).

However, while it might be relatively easy to shape the interests and behaviors of younger children, the situation can become more challenging as children grow older and become subject to an increasing range of external factors. The influence of, for example, family (apart from parents) and friends, nursery, school, and the media can introduce children to alternative values, products, and behaviors. In this context, a Chinese study found that parents are most likely to shape green consumption values in their older children if there is a close parent-child relationship (Gong et al., 2022).

Further Reading

Ansari, N.U., Rashidi, M.Z., & Mehmood, K. (2023). "My roots are green": A phenomenological discourse on intergenerational green motherhood in non-Western consumption contexts. *Qualitative Market Research: An International Journal*, 26(1), 19–36. https://doi.org/10.1108/QMR-02-2021-0017.

Auriffeille, D.M. (2021). "Before she was born, I ate Cheerios and beer for dinner": A qualitative examination of green parenting in Lowcountry South Carolina. *Humanity & Society*, 45(4), 439–470. https://doi.org/10.1177/0160597620943195.

Gong, Y., Li, J., Xie, J., Zhang, L., & Lou, Q. (2022). Will "green" parents have "green" children? The relationship between parents' and early adolescents' green consumption values. *Journal of Business Ethics*, 179(2), 369–385. https://doi.org/10.1007/s10551-021-04835-y.

Our Kids' Climate Network. Available at: https://ourkidsclimate.org/about-us/ (accessed: 12 June 2025).

UNICEF. (2023). *Over the tipping point*. Bangkok: UNICEF East Asia and Pacific Regional Office.

63 Grassroots Innovation

Adrian Smith

Definition

Grassroots innovation happens when creative and collaborative people pursue autonomous local development. A widely used definition is "networks of activists and organizations generating novel bottom-up solutions for sustainable development; solutions that respond to the local situation and the interests and values of the communities involved". This takes varied technological, social, and organizational forms in diverse community contexts across Asia, Europe, Africa, the Americas, and Oceania.

History

Grassroots innovation is rooted in both environmentalism in the Global North and development alternatives in the Global South.

Think global, act local emerged as a slogan in the 1970s among environmentalists, mostly in the Global North. It helped animate pioneering activities in community energy, agroecology, ecohousing, **repair** and recycling, and more. Grassroots innovation emphasizes two subsequent aspects of this activity: first, it recognizes the importance of local experimentation, knowledge-production, and skills provision in building more sustainable societies (see **Living Labs**); and second, it counters the more recent monopolization of innovation narratives and resources by technocratic approaches to business greening (see **Greenwashing**). Grassroots innovation stands for local participation and democratic control in sustainable transformations.

The concept has also been vital in long-standing struggles for autonomous alternatives to international development blueprints imposed upon the Global South. Across a wide diversity of so-called developing country contexts, local communities have for generations created appropriate technologies, informal innovations, traditional remedies, and practices. They have culturally rooted community inventiveness in their struggle to resist top-down, technocratic development models – including resisting those models on environmental grounds. In India since the 1980s, for example, the Honey Bee Network has promoted "grassroots innovation" explicitly to heighten awareness and raise support for bottom-up sustainable developments (see Box 63.1). The Network's success has led to national policies increasingly prioritizing grassroots innovation. Whenever policymakers in other countries and multilateral development agencies face social pressure to make their programs more "inclusive", they may become more open to supporting grassroots innovations. However, activists and communities often remain cautious and conflicted about such top-down attention, especially when it misinterprets or undermines the autonomous goals of grassroots creativity.

DOI: 10.4324/9781003584056-68

Different Perspectives

Grassroots innovation contrasts with those business and policy approaches that monopolize ideas about what counts as innovation in society. Innovation policy typically deals with systems of researchers, firms, investors, and entrepreneurs dedicated to commercializing novel products, processes, and services. Sustainable innovation as conventionally promoted by policymakers means working within these systems to develop ecoefficient businesses – without disrupting capitalism's underlying need for increased productivity, managerial control, and competitive accumulation of resources.

Meanwhile, grassroots innovation happens amidst community-based activities. While economic viability is important, its basis is rooted in moral, social, and cultural values operating outside the logic of commodification. Table 63.1 below elaborates on these perspectives (see also **Prosumerism**, **Alternative Consumer Cooperatives**, **Convivial Technology**).

Application

While the sheer diversity of grassroots innovation always requires careful contextualization, it holds in common an emphasis in community well-being, ecological sustainability, and collaborative

Table 63.1 Grassroots innovation movements and conventional innovation institutions

	Grassroots innovation movements	Conventional innovation institutions
Protagonists	Local communities, grassroots activists, civil society organizations, social entrepreneurs, worker co-operatives, NGOs, social movements	Universities, research centers, venture capitalists, firms, science ministries, entrepreneurs
Priorities	Social values, convivial communities, livelihoods, sustainable developments	Codified knowledge, economic growth, competitiveness
Incentives and drivers	Social need, voluntarism, co-operation	Expert authority, reputation, market demand
Resources	Development assistance, social capital, public finance, grassroots ingenuity, local knowledge, activist organization	Public finance, corporate investment, venture capital, scientific expertise and training
Locations of activity	Villages, factories, neighborhoods, community projects, social movements	Laboratories, R&D centers, boardrooms, ministries, markets
Typical knowledge forms	Situated knowledge, tacit knowledge	Scientific and technical knowledge
Appropriation	Knowledge commons, freely shared practices, activist guidebooks and media	Intellectual property, scientific journals, licensed technologies
Emblematic fields of activity	Agro-ecology, community health, small-scale renewable energy, housing	Biotechnology, medicine, nanotechnology, geo-engineering

Source: Smith et al. (2017)

endeavor. It emerges in areas such as agroecology, community energy, hackerspaces, urban food systems, water and sanitation systems, repair cafés, ecohousing, citizen science, or community media (e.g., radio and internet). These initiatives often blend advanced technologies, traditional techniques, and social innovations, prioritizing solutions that align with local needs and empower sustainability transformation in ways that enhance community autonomy.

Intermediary organizations and supportive transnational networks help circulate grassroots innovations by sharing knowledge, experience, resources, and advocacy, helping innovations adapt and be embedded in new contexts. Wider **social movements** can also be of indirect support whenever their goals imply greater attention and support for grassroots innovation. Right to Repair movements in North America and Europe, for example, demand changes that make it easier for consumers to get their products repaired. Some manufacturers are exploring the commercial possibilities in this, but in potentially disempowering ways through extended product-service relationships globally (see **Product-Service Systems**). In response, some in the repair movement press for more radical transformation by aligning with grassroots innovations like repair cafés and community-based remanufacturing (e.g., in maker spaces) that are emblematic of a deeper kind of social repair anticipated in commons-based and postgrowth economies (see **Degrowth**).

Grassroots innovations have diverse material and immaterial impacts over time. Open-source digital technologies in citizen science, for example, enable participants to monitor their environment, build local knowledge, and strengthen community ties. These initiatives foster new community and economic arrangements, social agency, and collaborative skills, and cultivate new participant identities. Indeed, the development of participatory methodologies, alongside organizational skills and collaboration techniques, are important grassroots achievements in themselves. Grassroots innovations embody social values and narratives often ignored by mainstream innovation, which reframe global environmental issues and provide new meanings through local actions. They can thereby offer diverse prototypes for more inclusive and sustainable futures.

Governments and businesses occasionally face pressure to adopt more inclusive and responsible innovation practices, drawing policy attention to grassroots initiatives. The Honey Bee Network demonstrates how policies can support grassroots innovators in becoming more entrepreneurial and scaling up through funding, prototyping facilities, training, incubators, and partnerships (see Box 63.1).

Box 63.1   Grassroots innovation policies in India – The Honey Bee Network

Post-colonial India's focus on industrialization and agricultural modernization often marginalized rural populations. Concerned about a turn away from alternative development models, such as those proposed by Gandhi during the struggle for independence, a movement materialized committed to drawing attention to the inventiveness of informal economic activities in India. As such, the Honey Bee Network, founded by Anil Gupta in 1988, emerged to document and promote grassroots innovations. It acknowledges the inventiveness of individuals and promotes economic support and reward through commercialization and knowledge sharing. Among its activities are regular walks of discovery through India's diverse rural communities, known as *Shodh Yatras* (now on their 48th edition).

In the 1990s, the Honey Bee Network influenced regional and national policies to support grassroots innovation, earning recognition from India's then-Prime Minister. This led to the creation of the National Innovation Foundation (2000), which continues to document inventions, secure intellectual property, offer prototyping facilities, and connect innovators with industry and financial partners. Over 300,000 innovations have been recorded, resulting in 416 patents and enterprises producing off-grid refrigeration, low-cost farming tools, small-scale fabrication technologies, and natural health remedies.

However, policy often overlooks the systems-changing potential of grassroots innovation, instead channeling promising ideas into conventional market-based frameworks for commercialization and scaling. While this can benefit social entrepreneurs, it risks sidelining deeper transformative possibilities (see Box 63.2). These arrangements rarely reinvest in the grassroots conditions and community labor that enabled the original innovation. Policy often overlooks the deeper aspirations of grassroots innovators, who commit to local action within ecological limits and social justice which can confound policy assumptions about scaling-up and requirements for economic growth. This mismatch creates mutual disappointment, as institutions struggle to support the open-ended, place-based autonomy essential for grassroots innovation to drive transformative sustainability.

Box 63.2   Grassroots innovation policies in Brazil – The Social Technologies Network

Launched in 2004 under Lula's first government, Brazil's Social Technologies Network aimed to empower communities through participatory social technologies that gave them greater capacity for subsequent autonomous development. Projects like self-constructed rainwater harvesting systems improved water access while fostering self-organizing capabilities for future community development and resource advocacy. Nationally, the initiative sought to reshape technology policy around these principles.

The Social Technologies Network promoted social technologies through events, awards, and collaborations with over 900 grassroots organizations. including NGOs, universities, government agencies, and financial organizations. Supported by R$400 million from the national government and corporate foundations like *Banco do Brasil*, the Network aimed to empower communities. However, tensions arose as some members favored market-based approaches, promoting pro-poor technologies that could diffuse through markets or be distributed through social programs over its philosophy of liberatory social technologies. These disagreements led to the Network's disbandment in 2012, with its initiatives continuing independently through affiliated organizations.

Radical grassroots innovation calls for autonomous, sustainable transformation. For example, UK defense sector workers in the 1970s prototyped socially useful products like wind turbines and electric vehicles. They were not only demonstrating how jobs could be saved through peaceful redeployment of skills and machinery, but they were also confronting the political establishment with calls for transforming economic institutions based on democratic socialism. Similarly, grassroots innovators in Mexico use community radios, mobile telephony, and appropriate technologies to provide practical benefits while asserting territorial and cultural autonomy. They see in these **convivial technologies** a material expression of territorial and cultural autonomy that resists top-down ideas about innovation and the aims of development programs.

Systems perspectives in sustainable development need to be appreciative of the contradictions, tensions, and synergies highlighted here in relations between the different scales and forms of grassroots innovation, their contexts of operation, and the design of institutional supports. More pluralistic systems perspectives can help explain why grassroots innovation develops and struggles in niches globally and how supportive changes in context can help it to circulate and thrive more widely. Ultimately, grassroots innovation is about acting creatively in communities locally while transforming institutions globally.

Further Reading

Avelino, F., Dumitru, A., Cipolla, C., Kunze, I., & Wittmayer, J. (2020). Translocal empowerment in trans-formative social innovation networks. *European Planning Studies*, 28(5), 955–977. https://doi.org/10.108 0/09654313.2019.1578339.

Kumar, H., & Sharma, G. (2024). *Grassroots innovation: Discourse, policy and practice in the global south.* Abingdon: Routledge.

Roysen, R., Bruehwiler, N., Kos, L., Boyer, R., & Koehrsen, J. (2024). Rethinking the diffusion of grassroots innovations: An embedding framework. *Technological Forecasting & Social Change*, 200, 123156. https://doi.org/10.1016/j.techfore.2023.123156.

Seyfang, G., & Smith, A. (2007). Grassroots innovations for sustainable development: Towards a new research and policy agenda. *Environmental Politics*, 16(4), 584–603. https://doi.org/10.1080/09644010701419121.

Smith, A., Fressoli, M., Abrol, D., Arond, E., & Ely, A. (2017). *Grassroots innovation movements*. London: Routledge.

64 Prosumerism

Thomas S.J. Smith

Definition

Prosumerism refers to a diverse range of processes and practices that involve simultaneous production and consumption. The terms prosumerism and prosumption are often used interchangeably. We could perhaps see the former as an equivalent to an ideological and social system (equivalent to the use of "consumerism"), and the latter as relating more to actual practices (such as how the term "consumption" is used).

Examples of this broad category can include anything from growing your own vegetables and customizing or producing your own clothing to developing and using open-source software and scanning your own produce in a supermarket. Recently, as technologies such as wind turbines and solar panels have become more accessible on a household and community scale, an interest in prosumerism has grown in the field of sustainable energy, whereby people produce and consume (as either households or mediated through community organizations) their own energy (see **Energy Consumption Behavior**, **Sustainable Housing**). By examining such activities and **social practices**, the concept sits in an interesting space at the interface between sustainable consumption and production. It may allow us to avoid the downsides of a focus on individualism, on the one hand, and a focus on technological solutions and green innovation, on the other (see **Consumer Scapegoatism**).

History

Prosumerism was a neologism coined by the futurist Alvin Toffler in 1980, in his work *The New Wave*. While the term did not initially gain widespread use or recognition after publication, it has seen a quiet renaissance in recent years. In part, this revival is thanks to the work of the sociologist George Ritzer, who has examined prosumerism across numerous academic texts since 2010. By tracing the rise of prosumerism, Ritzer sketches out a progression from capitalism's initial focus on production (represented by the predominance of the factory), to its focus on mass consumption in the 20th century (the shopping mall being emblematic here), and finally to what he asserts is today's "prosumer capitalism" (where the prosumer labors on behalf of capitalist enterprise). In the latter case, prosumerism can be seen in increasingly widespread practices such as customers picking up their orders in fast-food restaurants or when they scan their own items at the checkout in a supermarket.

Different Perspectives

Ritzer has been most interested in how prosumerism creates new forms of capitalist exploitation and profit extraction. Developing his ideas around the "McDonaldization of society", Ritzer's

DOI: 10.4324/9781003584056-69

work examined how consumers increasingly take up unpaid labor on behalf of corporations: cleaning up after themselves in fast food restaurants, handling the payment process in supermarkets, or creating user-generated media content on corporate-owned blogs or social media, to take just a few examples. This prosumer labor allows companies to earn greater profits than they otherwise would if they were to employ staff or technology to complete these tasks.

While Ritzer and his co-authors focused on the exploitation that prosumerism entails in a profit- and growth-focused economy, other scholars have begun to also examine the potential importance of prosumerism as part of alternative, sustainable, and collective practices. To take just a few examples, the latter has included a focus on **repair** practices, household and cooperative renewable energy production, and local food prosumerism (see **Community Supported Agriculture**), freecycling, and commons-based manufacturing (see **Convivial Technology, Grassroots Innovation**).

This focus on the possibility and potential of sustainable, grassroots prosumerism accords more closely with the original interests of Toffler. Toffler argued, for instance, that transitioning toward prosumerism would have radical social implications. Reimagining the predominance of self-provision which existed in the pre-industrial world, Toffler pointed to the potential for "do-it-for-yourself" to displace practices of "do-it-for-the-market" (Kosnik, 2018: 126). He further noted that "any significant change in the balance between production for use and production for exchange will set off depth charges under our economic system and our values as well" (Toffler, 1980: 296).

A further difference in perspectives relates to the prevalence (or lack thereof) of prosumerism in society. For some, prosumerism is a marginal topic, of limited application in comparison to more popular topics in sustainable consumption. To be of any relevance, prosumer practices would have to "scale up" enormously. For others, however, prosumerism is a spectrum of activities that is already everywhere; it perhaps even comprises the majority of social practice. Any time someone cooks at home, for example, rather than buying ready-made meals or going to a restaurant, they are engaging in a form of prosumerism. Any time someone grows a herb or other edible plant on their windowsill, they are engaging in a small act of prosumerism. Any time a product is altered or repaired by its user, there exists an opening for prosumerism. As such, autonomous production and consumption have not disappeared with the advent of industrialism but have merely changed form.

Application

Today, facing a need to radically rethink systems of production and consumption, interest is growing in circular and post-growth alternatives to the status quo (see **Degrowth, Circular Economy and Society**). Prosumerism could be of high relevance and interest at this juncture and has been emerging in forms that are grassroots-based and not profit-oriented. Rather than being appropriated by private corporations and shareholders, the proceeds from collective prosumer initiatives can be funneled back into initiatives for social and ecological benefit. As such, prosumerism can provide the ground for a deep form of economic democracy (see **Eco-Communities, Alternative Consumer Cooperatives**).

At their best, prosumer activities and networks can allow greater participation and ownership in the construction of sustainable lives. They allow people to address issues of direct concern to their lives, such as energy poverty, waste production, **food sovereignty,** and much more. As energy cooperatives expand, for instance, this aspect is more visible than ever. **Community supported agriculture** (CSA) could also be seen as relevant in this context. While often structured in a way that maintains a functional distinction between consumers and producers, CSAs also bring consumers into the decisions around what is produced and how and tend to integrate "consumers" into work days and other practical parts of the production process. With this, we can see how prosumerism can lead to a collective co-construction of both supply and demand, meeting needs while avoiding the radical and unsustainable over-production that exists with capitalism.

While a prosumer cooperative or other initiative may start as a way to meet basic needs in energy or food, it can quickly also serve as a crucible of education for democratic governance and a space for engaging with neighbors and other communities. For this reason, some governments and policymakers have begun to incentivize energy coops and other prosumer forms, through local, national, and supranational legislation.

Further Reading

Kosnik, E. (2018). Production for consumption: Prosumer, citizen-consumer, and ethical consumption in a postgrowth context. *Economic Anthropology*, 5(1), 123–134. https://doi.org/10.1002/sea2.12107.

Ritzer, G., & Jurgenson, N. (2010). Production, consumption, prosumption: The nature of capitalism in the age of the digital 'prosumer'. *Journal of Consumer Culture*, 10(1). https://doi.org/10.1177/1469540509354673.

Toffler, A. (1980). *The third wave.* New York: William Morrow.

Veen, E.J., Dagevos, H., & Jansma, J.E. (2021). Pragmatic prosumption: Searching for food prosumers in the Netherlands. *Sociologia Ruralis,* 61(1), 255–277. https://doi.org/10.1111/soru.12323.

Wittmayer, J.M., Campos, I., Avelino, F., Brown, D., Doračić, B., Fraaje, M., Gährs, S., Hinsch, A., Assalini, S., Becker, T., Marín-González, E., Holstenkamp, L., Bedoić, R., Duić, N., Oxenaar, S., & Pukšec, T. (2022). Thinking, doing, organising: Prefiguring just and sustainable energy systems via collective prosumer ecosystems in Europe. *Energy Research and Social Science*, 86. https://doi.org/10.1016/j.erss.2021.102425

65 Alternative Consumer Cooperatives

Beyza Oba, Zeynep Özsoy, and Birgit Teufer

Definition

As defined by Consumer Co-operatives Worldwide (CCW), which represents 26 national organizations of consumer cooperatives around the world, "a consumer cooperative is an autonomous association of consumers united voluntarily to meet their common economic, social and cultural needs and aspirations through a jointly owned and democratically controlled enterprise". In other words, a consumer cooperative is an enterprise that is owned by consumers who unite voluntarily to fulfill their common needs and aspirations. Like in the case of food cooperatives, cooperatives that own retail outlets run by consumers are the most common type of consumer cooperatives. However, there are also consumer cooperatives operating in areas such as childcare, health care, insurance, housing, utilities, and personal finance. The main purpose of a consumer cooperative is not to maximize profit but to provide quality products and services at the lowest price to the customers.

Alternative consumer cooperatives (ACC) can be considered as an extension of Alternative Food Networks (AFNs) that aim to bypass intermediaries in food supply and distribution, support small farmers, empower women, advocate for **food sovereignty**, engage in ecological production, and encourage the preservation of traditional agricultural methods (see **Community Supported Agriculture**). Different from traditional consumer cooperatives, in which "members are the consumers or users of the goods or services made available by or through the cooperative", ACC represents a non-capitalist form of organizing that can be considered an example of "diverse economies", a term coined by J.K. Gibson-Graham. Activists involved in these cooperatives question neoliberal policies, asymmetric power relations, and commodification of labor, and seek alternatives that can lead to transformation to achieve a "fair" distribution of resources beyond capitalist dynamics. Building trust-based ties between producers and consumers (see **Prosumerism**), ACCs are against the existing market logic and experiment with a different governance system that is based on direct democracy and elimination of hierarchies, with practices built on the "ethics of solidarity", in J.K. Gibson-Graham's terms.

History

The cooperative movement began in 19th-century England and France, gaining momentum with the founding of the Rochdale Society of Equitable Pioneers in 1844 by 28 flannel weavers after an unsuccessful strike over soaring prices and poor food quality. Consumer cooperatives that existed before the Rochdale cooperatives were supported by philanthropists and social reformers like Robert Owen and William King. Owen's ideals of using cooperation to create a prosperous and harmonious society influenced the Rochdale Society's formation. Although not the first, Rochdale's principles and practices inspired many subsequent cooperatives, initially in England (see Box 65.1).

DOI: 10.4324/9781003584056-70

Box 65.1   The principles established by the Rochdale Pioneers

The principles established by the Rochdale Pioneers include:

- Democratic Control: Each member has one vote, ensuring equal participation regardless of investment size.
- Open Membership: Cooperatives are open to all who can use their services and accept the responsibilities of membership.
- Economic Participation: Members contribute equitably to the capital of the cooperative and receive limited compensation, if any, on capital subscribed as a condition of membership.
- Autonomy and Independence: Cooperatives are autonomous, self-help organizations controlled by their members.
- Education, Training, and Information: Cooperatives provide education and training for their members, elected representatives, managers, and employees.
- Cooperation among Cooperatives: Cooperatives serve their members most effectively and strengthen the cooperative movement by working together through local, national, regional, and international structures.
- Concern for Community: Cooperatives work for the sustainable development of their communities through policies approved by their members.

Unlike earlier examples, the Rochdale Pioneers were a self-help cooperative aimed at re-organizing the relationships of consumption and production that emerged during the Industrial Revolution (1760–1840). The Rochdale model quickly spread to other European countries and Australia. During World War I, cooperatives grew to counteract war-related commodity shortages, price hikes, and profiteering. This trend continued through the Great Depression of the 1930s, with governments supporting self-help cooperatives. Their principles became foundational for cooperative governance and transactions. In 1937, the International Cooperative Alliance (ICA) drafted seven cooperative principles inspired by Rochdale. These included trading at market prices, selling unadulterated food, using fair weights, cash purchases, and no credit. Governance principles emphasized "one member, one vote" and gender equality, promoting democratic governance and reconfiguring power relations. Profit distribution was done quarterly, based on members' purchases, fostering member loyalty, and providing financial resources, with a portion allocated to education.

However, the 1980s saw a major shift with the rise of multinational supermarkets, forcing cooperatives to compete by focusing on cost-cutting, revenue enhancement, and market share, moving away from their foundational principles of cooperation, sharing, and collective work.

In the early 1990s, a new wave of consumer activism emerged, driven by ethical consumerism and a demand for ecologically safe and ethically produced goods. This movement revived some of the Rochdale Society's ideals, creating alternative consumer cooperatives that emphasized ethical and sustainable practices. Since ACCs can take various organizational forms, including informal ones, it is challenging to provide precise numbers for alternative consumer cooperatives nowadays.

Different Perspectives

Consumer cooperatives play a crucial role in advancing sustainable consumption but are often caught in debates about their effectiveness and motives. These discussions typically focus on the tension between economic survival and a commitment to ecological and social impacts.

As consumer cooperatives increasingly adopt business practices from conventional for-profit enterprises, questions about the erosion of their foundational principles arise. This commercial drift can lead to prioritizing profitability at the expense of social and environmental responsibilities, diluting the cooperative's commitment to its sustainability ethos. Furthermore, internal debates within cooperatives about the true meaning of sustainability, mirroring broader academic discussions that grapple with theories of strong versus weak sustainability – whether to preserve natural capital at all costs or allow for some substitution with human-made capital, allow for in-depth analyses of different perspectives of consumer cooperatives. Stakeholders, including members, managers, and community supporters, frequently hold divergent views on which sustainability aspect – environmental, social, economic – should take precedence. This disparity can lead to internal conflicts about the cooperative's direction and undermine trust and cohesion within the community.

In response to these challenges, there is a growing discussion among the cooperative community and policymakers about reinforcing the authentic elements of cooperatives. Ensuring that cooperatives remain true to their principles of community control, environmental stewardship, and social equity is essential as they navigate modern market complexities.

Application

Consumer cooperatives demonstrate the practical application of sustainability within community-centric commerce, intertwining economic, environmental, and social objectives for production and consumption (see Box 65.2). These cooperatives support local economies by keeping capital within communities and enhancing social structures through job creation and local supplier support. For example, food cooperatives often source from local farmers, supporting regional agriculture and reducing emissions from long-distance transportation (see **Food Miles**, **Food Sovereignty**). This bolsters local economies and promotes environmental stewardship by minimizing the cooperative's carbon footprint.

In addition to economic and environmental impacts, consumer cooperatives influence social dynamics. They challenge traditional consumer behaviors (such as not checking information about producers, labor conditions at the production sites, and fetishization of commodities) and promote a collective ethos toward responsible consumption, shifting from individualistic to communal purchasing practices. This fosters a culture of shared responsibility and sustainability within the community. Cooperatives also serve as educational platforms, informing members about the broader impacts of their consumption choices and promoting sustainable living practices. They encourage members to consider the environmental and social ramifications of their purchases, such as choosing locally produced goods and reducing waste through bulk purchases and reduced packaging.

Box 65.2   Examples of consumer cooperatives

Examples of consumer cooperatives around the world

- In Italy, a significant example of consumer cooperatives is the "Gruppi di Acquisto Solidale" (GAS). These are grassroots groups that collectively purchase goods directly from producers, emphasizing sustainability, ethical production, and local sourcing (see

Grassroots Innovation). Currently, in Italy, it is estimated that there are at least twice as many as the 1,000 registered GAS groups, with typically between 20 and 100 members per group, though some may have several hundred members.

- In Japan, the Japanese Consumers' Co-operative Union (JCCU) is a significant player, with about 30 million members. JCCU is active in food retail and is known for its commitment to food safety, organic farming, and supporting local producers. The cooperative has a strong educational component, offering programs that teach members about the environmental and social impacts of their consumption choices.
- Park Slope Food Coop in New York City, USA, is a smaller yet highly influential example, with around 17,000 members. This cooperative is member-owned and operated. It focuses on offering high-quality, sustainably sourced food at affordable prices, fostering a strong sense of community and shared responsibility.
- In Spain, the Som Energia cooperative is a grassroots initiative with around 80,000 members focused on renewable energy. This cooperative, founded by a small group of individuals in 2010, allows members to collectively purchase and generate green energy.
- In Germany and Austria, the "Solidarische Landwirtschaft" (Solidarity Farming or CSA – **Community Supported Agriculture**) movement consists of small, grassroots cooperatives where members collectively support local farms. Each group typically has around 30 to 50 members who share the costs and harvests of the farm.

These examples illustrate the diversity in size, scope, and geographical presence of consumer cooperatives, highlighting their significant role in promoting sustainable consumption and community development worldwide.

The practical application of consumer cooperatives extends into policy and practice, as demonstrated for example in Italy, where the government supports cooperatives like Coop Italia to promote sustainable consumption. Governments can leverage cooperatives as models for developing sustainable consumption policies, integrating their successes into broader policy frameworks. For instance, Coop Italia has benefited from regional subsidies aimed at promoting organic farming and reducing food waste.

Furthermore, cooperatives can influence local and regional policies, advocating for initiatives that benefit broader community interests. This could include promoting green urban planning or supporting urban gardening projects, enhancing community green spaces, and contributing to urban sustainability (see **Urban Planning and Spatial Allocation**). Through these activities, consumer cooperatives advocate for and shape the landscape of sustainable consumption, demonstrating how integrated community efforts can lead to positive changes in local environments and broader societal structures.

Further Reading

Barbera, F., Dagnes, J., & Di Monaco, R. (2020). Participation for what? Organizational roles, quality conventions and purchasing behaviors in solidarity purchasing groups. *Journal of Rural Studies*, *73*, 243–251. https://doi.org/10.1016/j.jrurstud.2019.10.044.

Belk, R. (2014). You are what you can access: Sharing and collaborative consumption online. *Journal of Business Research*, *67*(8), 1595–1600. https://doi.org/10.1016/j.jbusres.2013.10.001.

Forno, F. (2013). Co-operative movement. In D.A. Snow, D. della Porta, B. Klandermans, & D. McAdam (Eds.), *The Wiley-Blackwell encyclopedia of social and political movements*, pp. 1–3. Blackwell Publishing Ltd.

Hilson, M. (2017). Rochdale and beyond: Consumer co-operation in Britain. In M. Hilson, S. Neunsinger, & G. Patmore (Eds.), *A global history of consumer co-operation since 1850: Movements and businesses*, Vol. 28, pp. 59–78. Leiden and Boston: Brill.

Patmore, G., & Balnave, N. (2018). *A global history of co-operative business*, pp. 26–47. New York: Routledge.

66 Community Supported Agriculture

Bernd Bonfert

Definition

Community-supported agriculture (CSA) is a collaborative **grassroots innovation**, in which farmers and consumers share the costs and outputs of food production. By joining a CSA, consumers agree to cover the costs of a farming operation over a defined period, during which they gain access to a share of the harvest (see Figure 66.1).

The exact business model can differ between CSAs that are (1) run as subscription services by individual producers, (2) based on collective associations of consumers, or (3) run as integrated organizations whose members co-own the entire operation (see **Alternative Consumer Cooperatives**). Similarly, some of the practices in CSAs can differ: Some deliver food to members, while others require them to pick it up. Most CSAs provide uniform food shares, while some offer a choice regarding the purchased products. In some CSAs, members need to spend some time helping out on the farm. Moreover, a few CSAs engage in social activities beyond food provision, such as organizing farm visits, environmental talks, or nutritional education classes for schools (see Table 66.1).

History

CSA originated in Japan in the 1960s under the term *teikei*. It was connected to an organic agriculture movement and involved local consumer associations, mainly run by women, who set up food provision agreements with nearby farmers. The practice was eventually adopted by small farmers in the United States and Europe in the 1970s under the label "CSA". However, it remained relatively unknown until the 2008 food price shock generated a wave of interest in it (Blättel-Mink et al., 2017). Since then, CSA has continuously expanded, developing thousands of local initiatives, as well as national and international CSA networks. CSA has also gained prominence in Latin America and Africa, where it became closely tied to movements for **food sovereignty**, such as *La Via Campesina*. Most recently, CSA received another boost during the COVID-19 pandemic, as disruptions in the retail sector raised people's interest in **prosumer** activities.

Different Perspectives

Farmers benefit from CSA by gaining financial security and in some cases practical support. By matching food production and demand, CSA enables small farmers to avoid volatile market competition, thus raising their resilience. In turn, consumers receive access to locally produced food, and in many cases can become involved in the production process. As a business model, CSA is viewed as fostering the development of small, ecologically sustainable food supply chains and

DOI: 10.4324/9781003584056-71

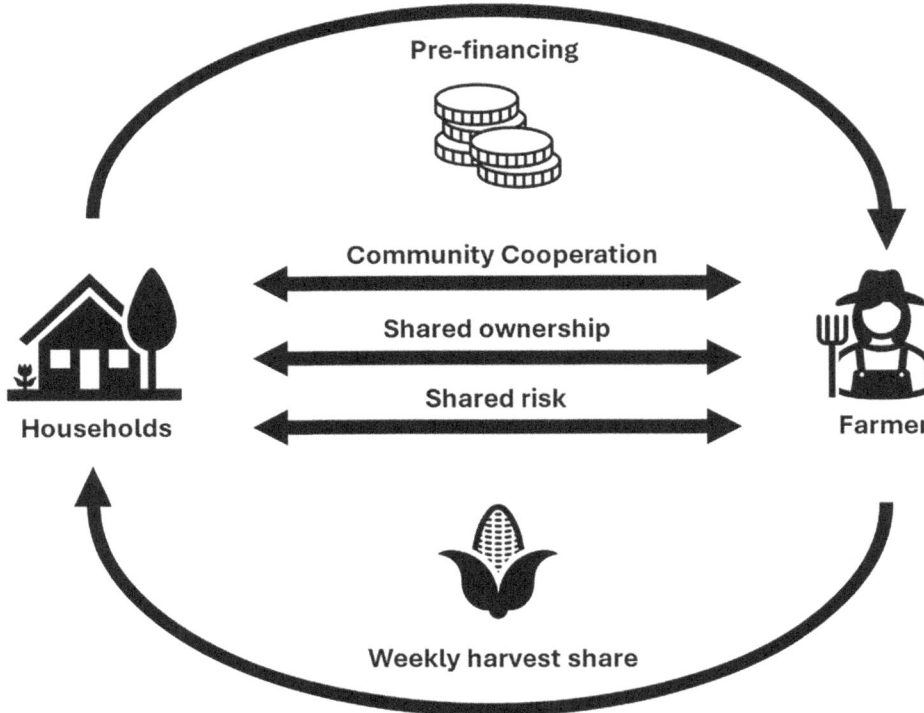

Figure 66.1 CSA model
Source: By authors

prefiguring a non-capitalist alternative to essential needs provision (Hitchman, 2016; see **The Role of Business**, **Foundational Economy**). Moreover, the practice is often praised for developing closer social ties between producers and consumers, aiming to raise people's awareness of the need for local food systems and encouraging more environmental and healthy nutritional behaviors.

On the other hand, critics of CSA highlight various shortcomings. An oft-cited problem is the practice's relative inaccessibility, as CSA membership requires a degree of financial security and cultural capital, including nutritional awareness and cooking skills. Consequently, in most countries, CSA members are overwhelmingly white and from well-educated middle-class backgrounds, with progressive-environmentalist values (Mert-Cakal & Miele, 2020), making up a relatively small "bubble" (see **Social Movements**). Even among CSA members, there is usually only a small core of dedicated activists who strongly share the practice's political principles and engage in voluntary activities, whereas the rest act as passive consumers. This raises concerns about CSA's ability to compete with more conventional organic retailers or box schemes and, by extension, its ability to transform the food system at scale.

Conversely, in regions where multiple CSAs have clustered, they are more capable of offering a significant level of food provision, as well as connecting and diversifying their supply. However, some CSAs in such areas have also begun competing with each other, highlighting their need to attract a broader clientele in general.

Table 66.1 CSA models and activities

CSA models activities	
Governance model	
Producer-run	CSA is run by individual farmer
Consumer-run	CSA is run as association of consumers in collaboration with farmer(s)
Integrated	CSA involves all farmers and consumers, e.g., as farming cooperative
Financing	
Fixed amount	All members contribute the same amount
Variable/solidarity pot	Member contributions can vary (e.g., based on income)
Bidding round system	Members determine their own contribution to meet a common budget
Food distribution	
Self-pickup	Members pick up/harvest their own food share
Collection points	Members pick their share from central collection points
Delivery	CSA delivers food to members
Food shares	
Uniform	All members receive the same share
Selection	Members can choose specific types of products
Work contribution	
None	No work necessary
Mandatory	All members help out on the farm
Optional	Members can help out to receive benefits, e.g., lower funding contribution
Additional activities	
Farm visits	CSA organizes farm visits for members or the public
Education	CSA offers educational activities, e.g., for local schools
Talks	CSA hosts public talks on, e.g., agroecological subjects

Application

CSAs exist around the world, often under different names and with country-specific characteristics (see Mayer & Peréni, 2021). In francophone contexts, the practice is known as AMAP (*Association pour le Maintien d'une Agriculture Paysanne* – "Association for the Maintenance of Peasant Agriculture"). This constitutes the most numerous form of CSA in one country, as at the time of writing there exist over 6,000 AMAPs across France.

The US hosts the second-largest national CSA movement, counting over 2,500 enterprises. This movement has recently developed a strong focus on combating racial inequalities in CSA membership and farming. The country's "CSA Innovation Network" has developed practical tools for integrating antiracist practices into the model, and many CSAs work particularly with farmers and communities of color.

The German CSA model, which is known as Solawi (*Solidarische Landwirtschaft* – "Solidarity-based Agriculture"), presents a different approach to tackling social inequalities. Most Solawis organize "bidding rounds", in which members anonymously pledge how much money they want to contribute. If they fall short of reaching a minimum budget, members are asked to pledge again. At this point, most Solawis succeed at meeting their target. This system accommodates members with different income levels without creating a stigma of charity, thereby making CSA more accessible by practicing solidarity.

In Italy, CSA has only recently taken root, but another food prosumer practice has flourished for many years under the term GAS (*gruppi d'aquisto solidale* – "solidarity purchasing groups"). Such groups collectively buy food and other goods from sustainable sources, but they do not co-finance farming operations. Due to the recent spread of CSA in the country, many GAS are considering "upgrading" to what they perceive as the more ambitious model.

Moreover, the original Japanese *teikei* movement is still active and recently celebrated its 50th anniversary. However, its reach has declined since the 1990s, when the number of *teikei* groups peaked at around 300. This regression reflects a concentration of demand around fewer, more professionalized *teikeis* and a shift toward a more passive consumer base, indicating what risks CSA movements may face if they fail to address their shortcomings.

To help CSA grow beyond its niche, initiatives in many countries have developed country-wide CSA networks (Guerrero Lara et al., 2024). These mainly offer a way to engage in mutual learning and resource sharing, thereby helping new CSA initiatives off the ground. Some CSA networks also collaborate economically to develop common supply chains and food processing facilities, such as bakeries. Most recently, larger CSA networks also started advocating for policy changes, mainly to receive agricultural subsidies and provide CSA access to lower-income households. While most CSAs collaborate around themes of sustainable farming and **food sovereignty**, CSAs in Southern Europe are embedded in networks of the "social and solidarity economy", thus connecting CSA with non-market-based economic approaches across other sectors.

To share knowledge and skills between these different national approaches, CSAs also engage in international exchange, for instance, via the *LSPA MedNet* (Mediterranean Network of Local Solidarity Based Partnerships for Agroecology), which brings together CSAs from Southern Europe and North Africa, as well as the international *Urgenci* network, which connects CSA networks across the globe and provides information about the movement's worldwide development.

CSA is unlikely to replace dominant agricultural practices on its own, but it offers a powerful vision for how to organize more sustainable, resilient, community-based food supply chains. As such, it not only attracts consumers but could also inspire policymakers and green enterprises to develop local partnerships to grow and distribute food outside conventional market channels.

Further Reading

Blättel-Mink, B., Boddenberg, M., Gunkel, L., Schmitz, S., & Vaessen, F. (2017). Beyond the market – New practices of supply in times of crisis. The example community-supported agriculture. *International Journal of Consumer Studies*, 41(4), 415–421. https://doi.org/10.1111/ijcs.12351.

Guerrero Lara, L., Feola, G., & Driessen, P. (2024). Drawing boundaries: Negotiating a collective 'we' in community-supported agriculture networks. *Journal of Rural Studies*, 106, 103197. https://doi.org/10.1016/j.jrurstud.2024.103197.

Hitchman, J. (2016). *How community supported agriculture contributes to the realisation of solidarity economy in the SDGs*. UN Inter-Agency Task Force on Social and Solidarity Economy.

Mayer, B., & Peréni, Z. (2021). *Food & More toolkit. From beginner to advanced*. European Commission. Available at: https://cloud.urgenci.net/index.php/s/MsXw74WZcRRaFGt?dir=undefined&path=%2F&opennfile=5519 (accessed: 8 January 2025).

Mert-Cakal, T., & Miele, M. (2020). Workable utopias' for social change through inclusion and empowerment? Community supported agriculture (CSA) in Wales as social innovation. *Agriculture and Human Values,* 37(4), 1241–1260. https://doi.org/10.1007/s10460-020-10141-6

67 Fair Trade

Lindsay Naylor

Definition

Fair Trade is simultaneously a multi-faceted movement and a market whose proponents are in favor of sustainable economic, social, and environmental well-being for people and the planet. As a solidarity movement, advocates for fair trade support (i) producers being paid a fair wage or price, (ii) the sustainability of production, and (iii) opportunities for dignified livelihoods through reducing the number of actors in the commodity chain in the economic exchange of goods. The fair trade market is populated by consumers, producers (who are also consumers), and third-party certifiers. Consumers, for their part, seek to act out ethical and sustainable purchasing practices by buying third-party certified fair trade labeled goods in the marketplace. Producers, who sell their products under the label, are tasked with upholding a variety of standards for sustainability (e.g., protection of biodiversity; prohibition of use of hazardous substances) in their production practices and community development efforts. Third-party certifiers create, maintain, and surveil the standards for production under fair trade labels and certify producers.

History

The exchange of goods under a solidarity model that fed both the movement and the market that is now called fair trade began in the 1940s. This took place with the direct trade of goods from Latin America to the United States, where a group concerned with poverty sold goods in the United States and paid the producers directly. In 1958, the first fair trade store, *Ten Thousand Villages,* opened in the United States and proliferated, with dozens of stores in operation today. In the 1980s, when the International Coffee Agreement collapsed and prices of coffee dipped, a Dutch solidarity group began a certification system for the exchange of coffee. Coffee became the first product that had a label saying it was fairly traded under the Max Havelaar label in 1988. New labels and production standards for certification emerged in the 1990s with other national fair trade organizations. These groups were then consolidated under the international Fair Trade Labeling Organization (FLO) in 1997. In 2011, FLO reorganized as Fairtrade International with the separation of Fair Trade USA from the consolidated group. There are now dozens of fair trade labels that are overseen by third-party certifiers. In addition to many different certification labels, there are over 35,000 products ranging from primary products – such as coffee, cotton, tea, and bananas – to composite products made up of many products, such as chocolate, ice cream, and cookies. In addition to the many fair trade certified food and drink products, textiles, precious metals, and other consumer goods like sports balls are also now certified.

DOI: 10.4324/9781003584056-72

Different Perspectives

As a sustainability measure, fair trade is an excellent example of mobilizing many actors to under-take sustainable practices, including both producers and consumers. However, as a method of exchange, it is uneven.

Fair trade is often thought about as an economic and social development mechanism that pro-motes sustainable livelihoods for "disadvantaged producers" and sustainable purchasing practices for consumers with greater economic advantages. In part, this is problematic as it means for fair trade to exist there always needs to be a gap in economic wealth between producers and consum-ers. Many studies show that while fair trade does assist with keeping producers from falling into increasing debt and impoverishment, it does not significantly impact their livelihoods in a way that "lifts them out of poverty". However, third-party certifiers often make the case that this is what the fair trade market does by using the justice and solidarity narratives that are promoted by the move-ment. For example, Fair Trade USA encourages consumers to adopt a "fair trade lifestyle" as part of sustainable, ethical consumption for a more just world. The focus on the market suggests that we can buy our way to a better world (see **Boycott and Buycott**), while the movement suggests that we have to dismantle unjust systems to improve living conditions.

Environmental sustainability may be among the more promising components of fair trade certi-fication – but it is also not without its problems. Environmental standards are based on the sustain-ability desires of consumers that can be burdensome but must be adopted by producers. Fair trade products are produced using standards for environmental sustainability that are both stringent and changing as more products are certified. These standards may be applied in conjunction with other certification standards, for example, organic or bird-friendly (see **Ecolabeling**). Fair trade prod-ucts aim for a shorter supply chain, to limit transportation emissions (see **Food Miles**). It thereby aims to provide an alternative purchase motive to consumers in preference to less sustainable options – for example, fair trade cotton clothing instead of **fast fashion**.

Application

The most important aspect of fair trade, when considering sustainable consumption, is in consum-ers learning what is being certified or what standards are being applied in the production of the product. Not all certifications are the same, even though the labels look very similar. As the market continues to expand it is critical to understand that purchasing a fair trade good is not synonymous with having a sustainable lifestyle, as the environmental footprint of each product and where it is produced differs. There are four certified labels that consumers in the global core (the United States and Canada, Western Europe, Japan, and Oceania) most often see in stores: Fairtrade International, Fair Trade USA, Fair for Life, and the Small Producers Symbol (see Figure 67.1).

Fairtrade International remains an umbrella for over 20 member countries found in the global core. These third parties work with over a million producers across 75 countries in the global periphery. They claim to connect producers and consumers through fair trade as a sustainable development model. Certified producer organizations are audited annually to assess compliance with the standards set by the Fairtrade International Board. The board does have producer members who have a vote.

Fair Trade USA, based in the United States, works in 21 countries across the global periphery and the global core. They have over a million certified producers. They suggest that all produc-ers deserve a fair price to send their children to school, improve their communities, and produce sustainably. Their standards apply to large- and small-scale production, including plantations and factories, to meet environmental, social, and governance goals set out by their board, which does not have producer representation or votes.

Figure 67.1 Commonly available certifications

Source: Fairtrade, Fairtrade Certified, fair for life, and Tusimbolo.org, yoursymbol.org

Fair for Life started in Switzerland, is now based in France, and works with hundreds of thousands of producers and over 700 companies in 70 countries globally. They are unique among the larger labels in that they certify each node in the supply chain including producers, manufacturers, traders, and retailers. They focus on human rights, corporate social responsibility, and resilience (see **The Role of Businesses**). The board that sets and assesses its standards has representation from each stakeholder group that can be certified.

The Small Producer's Symbol is based in Latin America, and they work with just under 100,000 small producers across 17 countries in the global periphery. Their label differs from the other three in that it was created by and for small producers. The standards are focused on dignified living and supporting market entry for small producers. Only small producers can be certified, and their board representation is entirely made up of small producer representatives.

Fair trade as a movement brought attention to inequities in economic exchange and the market allows producers to benefit from stabilized income and consumers to contribute to non-conventional economic exchanges.

Further Reading

Archer, M. (2022). How to govern a sustainable supply chain: Standards, standardizers, and the political ecology of (in)advertence. *Environment and Planning E: Nature and Space*, 5, 881–900. https://doi.org/10.1177/251484862110145.

Bennett, E.A. (2020). The global fair trade movement: For whom, by whom, how, and what next. In K. Legun, J. Keller, M. Bell, & M. Carolan (Eds.), *The Cambridge handbook of environmental sociology*, pp. 459–477. Cambridge University Press. https://doi.org/10.1017/9781108554558.029.

Lyon, S. (2021). Anthropological perspectives on fair trade. In *Oxford research encyclopedia of anthropology*. Oxford University Press.

Naylor, L. (2014). "Some are more fair than others": Fair trade certification, development, and North–South subjects. *Agriculture and Human Values*, 31, 273–284. https://doi.org/10.1007/s10460-013-9476-0.

Raynolds, L.T., & Bennett, E. (2015). *Handbook of research on fair trade*. Northampton, MA: Edward Elgar Publishing.

68 Food Sovereignty

Giovanna Micarelli

Definition

Food sovereignty is a plural concept and practice that emerged out of resistance to the dominant model of agribusiness globalization and its harmful effects on people and the planet. Food sovereignty argues that to achieve the realization of the human right to food it is not enough to focus on the availability and access to food without touching the question of how food is produced and by whom. In the perspective adopted by a wide variety of **social movements** and initiatives, both locally and globally, food sovereignty implies prioritizing local agricultural production; access of peasants and landless people to land, water, and seeds; the right of consumers to be able to decide what they consume, and how and by whom it is produced; and the right of countries to protect themselves from "dumping" (low priced agricultural and food imports). Food sovereignty entails using and managing land, territory, seeds, and biodiversity according to autonomous and sustainable agroecological choices; fighting against GMOs (Genetically Modified Organisms); establishing or recovering local circuits of production and consumption; reclaiming local knowledge, practices, and food traditions; and creating cosmopolitan solidarity among social groups struggling for dignity and justice in food production and consumption.

History

Driven by the interests of large transnational corporations and the world's political superpowers, international institutions such as the IMF (International Monetary Fund), the World Bank, and the WTO (World Trade Organization) implemented neoliberal agrifood policies that enabled these corporations to dominate the global food market. As a result, local agricultural economies are being destroyed, and the genetic, cultural, and environmental heritage of our planet, along with our health, are being jeopardized. Land grabbing, large-scale land concentration, an emphasis on raw material exports over local economies, and the spread of monoculture farming reliant on heavy use of fossil fuels, agrochemicals, and GMOs have caused significant and irreversible socio-environmental damage. This has included the destruction of ecosystems, soil and biodiversity depletion, and the exacerbation of global warming. In parallel, there is a disturbing pattern of violence against groups claiming land rights and territorial autonomy. Traditional knowledge and food systems are being systematically eroded, and exploitative practices such as slave labor persist. Hundreds of millions of farmers are forced into debt or driven to migrate from rural areas. Despite claims of modernization, agribusiness essentially replicates the colonial exploitation of land, nature, and people.

It is in response to this situation that food movements worldwide have advocated food sovereignty as a way to radically redesign more just, democratic, and sustainable food systems (see

DOI: 10.4324/9781003584056-73

Community Supported Agriculture, **Alternative Consumer Cooperatives**). The concept of food sovereignty was first introduced to public debate by *La Via Campesina* (LVC), a global network representing approximately 200 million farmers, during the World Food Summit in 1996. Despite growing calls from social movements, it wasn't until 2012 that FAO initiated discussions on food sovereignty as an additional paradigm to food security. In 2014, the UN Special Rapporteur on the Right to Food helped draw attention to food sovereignty as a necessary condition for achieving the full realization of the human right to food.

The 1996 definition of food sovereignty by LVC refers to the right of peoples and countries to define their agricultural and food policies without dumping from third countries. Over the years, food sovereignty has acquired broader meanings of local, self-dependent, biodiverse, sustainable, communal, fair, healthy, and culturally appropriate production and consumption.

First, social movements have placed subsistence networks and peasant knowledge systems at the core of food sovereignty. This entails a biocultural understanding of food systems, in which farmers are not merely food producers, but upholders of invaluable knowledge and expertise in agroecological management.

Second, there is a shift from land to territory. Food sovereignty entails the sustainable management of lands, territories, watersheds, seeds, livestock, and biodiversity. It is based on farming people's right to freely use and protect genetic resources they have developed, and to defend their territories from the actions of transnational corporations. Transcending mere geographical boundaries, the territory refers to the collective wisdom and communal stewardship practices honed over generations by its inhabitants. Intertwining a deep sense of identity with particular landscapes and agroecological traditions, food sovereignty brings forward a place-based understanding of rights.

Third, food sovereignty movements have placed the commons at the core of food rights. While the FAO qualified food security as a global public good in 2009, the commons bypasses the private/public divide, addressing the rising tide of privatization and dispossession of shared resources. Food sovereignty movements join in the transformational pathway undertaken by a range of socio-environmental movements organizing to defend the commons: these include water, seeds, landscapes, and biocultural diversity. In so doing, dominant narratives of nature, the economy, development, and democracy perpetuated by market forces and the State are challenged. This lays the groundwork for a new political discourse centered on the conditions necessary for achieving social justice, sustainability, and dignified life for all (see **Foundational Economy**, **Steady-State Economy**, **Ecosocial Contract**, **Buen Vivir and Buenos Convivires**).

Different Perspectives

A commitment to diversity and intercultural dialogue has greatly contributed to the meaning of food sovereignty. In this regard, Indigenous perspectives are especially relevant for advancing a plural understanding of food rights and sovereignty (see **Buen Vivir and Buenos Convivires**). Indigenous food movements stress that food sovereignty depends on nurturing healthy and interdependent relationships with a multitude of natural communities working together. Even though Indigenous territories maintain the world's greatest biodiversity, the food insecurity of Indigenous populations is twice that of the non-Indigenous population, according to the FAO. Hunger and malnutrition result from the complex interplay of colonial and neo-colonial power structures that are responsible for the dispossession of lands and territories, environmental degradation, cultural erosion, and the breakdown of community economies. This suggests that a plural reimagining of food rights cannot be achieved solely by considering the cultural acceptability of food. Recent discussions have underscored the importance of non-nutritional values associated with food in shaping the overall image and well-being of individuals and communities. What's more, for Indigenous peoples, the right to food is inseparable from their right to the territory and self-determination. Food sovereignty is, by all accounts, an anti-colonial struggle.

Applications

Food sovereignty positively impacts the multiple dimensions of sustainability from the bottom up, bridging the gap between urban consumers and farmers, enhancing community food systems, promoting active citizenship and democratic participation (see **Citizen-Consumer**), and fostering resilience and cooperation while reducing dependency on the market. In both the Global North and the Global South, spanning rural and urban areas alike, initiatives are on the rise. Local organizations, Indigenous and Afro-descendant communities, peasants, urban farmers, and consumers connected across different scales of collective action and solidarity are reshaping food paradigms, nurturing healthier, more equitable foodscapes while promoting socio-environmental justice and sustainability. Through agroecological practices, local supply chains, school gardens, ethical buying groups, urban agriculture, and the recovery of food landscapes and traditions, among other initiatives, they both resist agribusiness food systems and foster social innovation (see **Grassroots Innovation**), where people and local communities are at the forefront as agents of change. These diverse initiatives build a critical mass to drive the transition to fair and sustainable policies that address hunger, poverty, and malnutrition, while also working to reduce biodiversity loss, food waste, land degradation, and global warming.

Further Reading

Edelman, M., Scott, J.C., Weis, T., Baviskar, A., Borras Jr., S.M., Kandiyoti, D., Holt-Giménez, E., & Wolford, W. (2014). Global agrarian transformations, volume 2: Critical perspectives on food sovereignty. *The Journal of Peasant Studies*, 41(6). https://doi.org/10.1080/03066150.2014.963568.

Local Futures. (2024). *Food*. Available at: https://actionguide.localfutures.org/themes/food (accessed: 11 August 2024).

LVC. (2021). *Food sovereignty, a manifesto for the future of our planet.* Available at: https://viacampesina.org/en/food-sovereignty-a-manifesto-for-the-future-of-our-planet-la-via-campesina/ (accessed: 21 April 2024).

Patel, R. (Ed.). (2009). Food sovereignty. *The Journal of Peasant Studies*, 36(3). https://doi.org/10.1080/03066150903143079.

Wittman, H., Desmarais, A., & Wiebe, N. (Eds.). (2010). *Food sovereignty: Reconnecting food, nature and community*. Halifax and Oakland: Fernwood Publishers and Food First Books.

69 Eco-Communities

Jenny Pickerill

Definition

Eco-communities are grassroots movements for social and ecological change. They are self-organized initiatives, independent of government control, that develop practices, infrastructure, and spaces to live more sustainably and in harmony with each other and the environment (see **Grassroots Innovation, Alternative Consumer Cooperatives**). Eco-communities are collectively organized housing, livelihood, and educational spaces purposefully designed to regenerate ecological and social environments from resource depletion and inequality. Eco-communities bring together residents with a collaborative spirit to build a better future. They seek (i) to self-provide, as far as practically possible, their own energy requirements, food, livelihoods from the land or locally, and resources for building their homes, and (ii) to minimize waste generation. The social dimensions of eco-communities are just as vital as their ecological features. Eco-communities practice participatory self-governance, share common resources, have systems of sharing between residents, and have expertise in non-violent communication and conflict resolution (see **Sharing Economy**).

History

Cattaneo (2014) first used the term eco-communities to reference human communities that shared an intent to reconfigure their environmental impact in how they lived and worked in their everyday lives. This coalesced the commonalities between ecovillages, intentional communities, ecological co-housing, and low-impact developments. What began as back-to-the-land experiments and ecological intentional communities in the 1960s and 1970s evolved into ecovillages such as Auroville (India, 1968), The Farm (USA, 1971), Damanhur (Italy, 1975), SEKEM (Egypt, 1977), Findhorn (Scotland, built late 1980's), Huehuecoyotl (Mexico, 1982), Gaia Association (Argentina, 1991), Dyssekilde (Denmark, early 1990s), and Sieben Linden (Germany, 1997). These communities interconnected informally but were not part of an international movement until the formation of the Global Ecovillage Network in 1991.

Historically, eco-communities have thrived in rural settings due to ample space, privacy, freedom, and affordable land. However, there's a growing trend of urban ecovillages. Pioneering examples include Freetown Christiania in Copenhagen (founded in 1971) and UfaFabrik in Berlin (founded in 1979). This shift toward urban locations reflects a desire to influence sustainable urban development and gain greater political influence due to their proximity to city centers. This evolution encouraged the inclusion of residential co-housing, the design and density of which suited smaller urban spaces (see **Urban Planning and Spatial Allocation**). There has also been a move away from manual-labor-intensive land-based income, which some struggled to sustain as they aged. Recent examples of these urban experiments include LILAC in England (see Figure 69.1),

DOI: 10.4324/9781003584056-74

Figure 69.1 LILAC, England
Source: Pickerill et al., 2024

the Kailash ecovillage and Los Angeles Ecovillage in the United States, and Cascade CoHousing and Christie Walk in Australia.

Different Perspectives

Eco-communities are experiments in building a more sustainable and equitable world. They, however, vary significantly in size, social and economic structures, residents' composition, ecological impact, and how they relate to space and their neighbors. Eco-communities share a collaborative spirit, innovative approach, environmental focus, and purpose. They are about building and living overlapping lives, finding ways to exist together, and mutually supporting each other through collaboration and shared living. This means residents work together, share resources, and make decisions collectively. Eco-communities are also spaces of experiments and innovation. Eco-communities act as laboratories for new social and environmental solutions (see **Living Labs**). They explore alternative economic models and ecofriendly practices. Given this experimental approach, it is not surprising that eco-communities are also dynamic, in constant flux, messy spaces of doing, making, and creating. Eco-communities are unfinished projects, adapting to new circumstances and needs. Reducing environmental impact is a core principle. People living in Dancing Rabbit Ecovillage (Missouri, USA), for example, use significantly fewer resources (less than 10%) per person than the average American across most categories (Boyer, 2016). This is achieved through shared resources, local production, and sustainable living.

While many eco-communities might not articulate themselves as explicitly anti-capitalist, the majority start from a quest to remove themselves from capitalist extractive systems *and* seek economic alternatives to traditional capitalism (see **Well-being Economy**, **Degrowth**, **Ecological Economics**, **Foundational Economy**). They promote local livelihoods, minimize consumption, and explore non-monetary systems (see **Sufficiency**). This reflects a desire for a more sustainable and equitable economic model.

Eco-communities, while offering a promising future, face several hurdles. The very idea of community can be divisive. Eco-communities often seek a specific kind of community, geographically close-knit with shared values. Community can be a powerful political project of togetherness, but it can also be exclusionary and limit participation. A significant critique of eco-communities is their apparent homogeneity. They tend to attract a narrow demographic, often white, educated, and well-off. This lack of diversity limits the applicability of their innovations and can make them unwelcoming to outsiders. The focus on shared identity can also lead to a homogenous aesthetic across eco-communities globally. In other words, an eco-community kitchen in Thailand (Panya Project) looks much the same as one in Spain (La Ecoaldea del Minchal). Such homogeneity is not only potentially off-putting to newcomers but also suggests that eco-community ideas are replicated across the globe without necessarily emerging as distinctively place-based projects that adequately reflect local specificities. Equally, eco-community living is demanding. The romanticized view of a simple life overlooks the realities of collective decision-making and manual labor. Reaching consensus takes time, and the work can be physically challenging. Additionally, some eco-communities rely on volunteers, raising questions about long-term sustainability.

Eco-communities also tend to operate on a small scale, leading some to dismiss them as insignificant. However, many actively share their knowledge and resources beyond their borders, acting as hubs for wider environmental and social activism (see **Social Movements**). The challenge lies in "scaling out" their positive impact while maintaining the core values that make them successful. Finally, as eco-communities become more popular, their radical ideals might weaken. Some communities adopt more individualistic and market-oriented practices to survive financially. Additionally, many rely on younger residents for manual labor, raising questions about how they will adapt to an aging population. These challenges highlight the need for eco-communities to find ways to be more inclusive, adaptable, and resilient while staying true to their core principles.

Application

Many of the practices, infrastructures, and designs eco-communities have developed can foster more sustainable places to live where everyone benefits. This is best demonstrated in tackling the problems of intensive resource use and pollution in cities. Current solutions, often focused on green development and technology, are dominated by corporations, and exacerbate social inequalities. An approach is needed that fosters ecologically sustainable and equitable cities for the future, and eco-communities offer five strategies (see Figure 69.2).

First, community-based economies: instead of relying on resource extraction, cities can build local economies based on shared resources and collaboration. Second, cooperative action: decision-making shouldn't be top-down. Democratic processes and cooperation will empower communities to find solutions together. Third, social resource management: cities can prioritize social approaches to managing resources, ensuring everyone's needs are met sustainably. Fourth, participatory governance: collaborative governance allows residents to contribute directly to urban planning and policies, promoting equity and fairness. Finally, embracing diversity: cities thrive on heterogeneity. Recognizing and celebrating social differences is crucial for building a just and sustainable future. By exploring these strategies and their real-world applications, we can equip

Figure 69.2 Five strategies from eco-communities to build more equitable sustainable places to live
Source: Pickerill et al., 2024

cities with the tools they need to become models of sustainability. This will pave the way for more sustainable places where all residents can enjoy benefits.

Further Reading

Boyer, R.H.W. (2016). Achieving one-planet living through transitions in social practice: A case study of Dancing Rabbit Ecovillage. *Sustainability: Science, Practice and Policy*, 12(1), 47–59. https://doi.org/10.1080/15487733.2016.11908153.

Cattaneo, C. (2014). Eco-communities. In G. D'Alisa, F. Demaria, & G. Kallis (Eds.), *Degrowth: A vocabulary for a new era*, pp. 165–168. Routledge.

Christian, D.L. (2003). *Creating a life together: Practical tools to grow ecovillages and intentional communities*. New Society Publishers.

Pickerill, J., Chitewere, T., Cornea, N., Lockyer, J., Macrorie, R., Malý Blažek, J., & Nelson, A. (2024). Urban ecological futures: Five eco-community strategies for more sustainable and equitable cities. *International Journal for Urban and Regional Research*, 48(1), 161–176. https://doi.org/10.1111/1468-2427.13209.

Sanford, A.W. (2019). *Living sustainably: What intentional communities can teach us about democracy, simplicity, and nonviolence*. University Press of Kentucky.

Cluster V

Governance, Policy, and Choice Architecture

70 Product-Service Systems

Arnold Tukker

Definition

The business literature has shown that in many markets, products from competing producers perform equally well. In such cases, the basis of competition has shifted from functional performance to price, which in turn reduces profit margins. Enhancing production volumes then becomes the only way to enhance overall profits. The result is that in many markets we have seen a shift toward "fastness" and "planned obsolescence". **Fast fashion**, fast furniture, and so on, incentivize consumers to buy new stuff that fits the newest fashion before old products are worn out.

Another manifestation of such product-oriented business models is that products are not repairable or reusable by design, even when a minor component fails (see **Repair** and **Ecodesign**). A washing machine with a failing ball bearing will see labor costs for repairs as high as the costs of a new machine. A breakdown of a key on a computer requires replacing the entire keyboard or even the computer.

From a sustainable consumption perspective, product-service systems (PSS) form an interesting alternative to such product-oriented business models. PSS aim to satisfy user needs by providing an integrated service, in the form of a unit of product use or a desired result sought from the product. The user pays for the result or unit of use. Any product used in the PSS is owned, serviced, and maintained by the provider (Box 70.1; also see **Extended Producer Responsibility**).

With the profit center now located in service provision instead of product sales, providers have an incentive to minimize the life-cycle costs of the product-service. It is in their interest to have products that are sturdier, easier to repair, and, if the provider also takes responsibility for consumables in the use phase, that products are energy efficient (see **The Role of Business**). In theory, PSS thus creates a win-win: lowering the overall life cycle costs of achieving the product's functionality, which can be shared by both users and providers, and lowering environmental impacts. PSS can also enhance the competitiveness of providers by meeting clients' needs in an integrated and customized way; building unique relationships with clients and engendering their loyalty; and allowing providers to innovate faster by being able to easily monitor clients' needs.

Box 70.1   Some illustrations of the success and failure of PSS

Industrial electric engines. Around the year 2000, a major European industrial electric engine producer faced competition from Asian companies. The former's engines were higher priced, but much more energy efficient, and so had lower costs per use-hour. They

---

DOI: 10.4324/9781003584056-76

developed a result-oriented business model: install a motor, pay the electricity bill, and price per hour of use. They immediately ran into purchasing mandate issues. Engines were paid for from a central investment budget, and electricity from an operational budget. The provider further took the risk of rising electricity prices. Once the contract was closed, the provider inevitably faced a loss when electricity prices rose.

Copiers. Major copier producers don't sell their copiers anymore, but let users pay per print. The copier is remotely monitored, giving excellent insight into user behavior. When signs of malfunctioning are seen, the copier is replaced. Many old parts are used in the next generation of copiers, saving significant costs.

Car sharing. Car-sharing systems lower transport emissions, since people tend to use public transport more often and use cars with more passengers (see e.g., Amatuni et al., 2020). However, so far, car sharing has remained a niche market in many places. Access to the car is more difficult, for instance, while car sharing lacks the prestige of car ownership.

Car hailing. Car-hailing platforms like Uber and Didi became very successful competitors of regular taxi services. They give access to a car in minutes, offer easy and secure electronic payment, and are relatively cheap (sometimes for questionable reasons, such as taxes or labor being underpaid).

Communal washing services. The above four examples focus on business-oriented PSS since most products in modern society are provided via markets. PSS work well in other contexts, however. In Sweden, common washing machines in multifamily residential buildings are the norm, included in a service package of the housing association.

History

Already in the 1980s, Walter Stahel proposed business models that sell the service or performance of the product instead of the product as such, as a method to lower resource use and emissions related to final consumption. This concept was further elaborated between 2000 and 2010 by authors such as Oksana Mont, Tim Baines, and Arnold Tukker, and more recently by Nancy Bocken. In parallel, various industries started to experiment with PSS business models, with varying levels of success.

These academic and practical experiences led to a now fairly generally accepted typology of PSS in three broad classes, as illustrated in Figure 70.1. *Product-oriented services* are geared toward sales of products, but some extra services are added. In *use-oriented services,* the traditional product plays a central role but it is owned by the provider and made available for use in different forms, and sometimes shared by several users. In *result-oriented services,* the client and provider agree on the result, and there is no predetermined product involved.

The use-oriented and result-oriented services are the most promising ones ecologically because the provider is in general incentivized to reduce the cost of the services by reducing resource use and prolonging the life of the product and its components.

Different Perspectives

The examples in Box 70.1 have shown that use- and result-oriented services indeed can be successful and can lead to lower environmental impacts. But we also see that many PSS stay in niche markets due to various problems inherent to the business model.

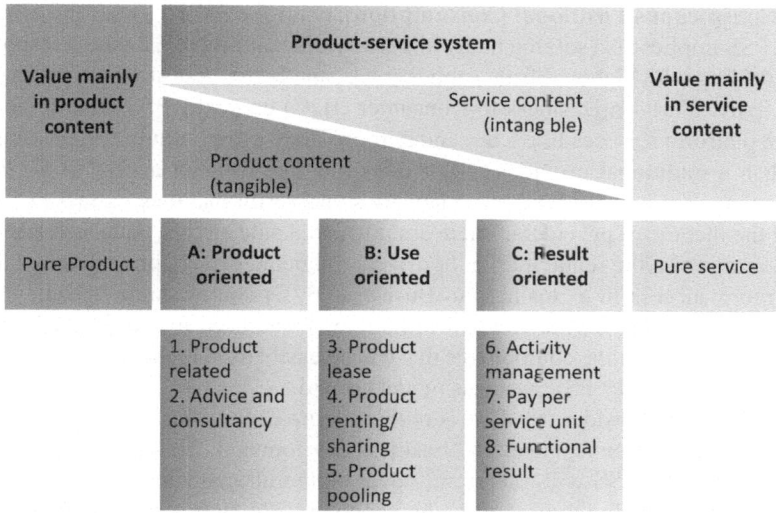

Figure 70.1 Primary and sub-categories of product-service provisioning systems

Source: Tukker, Tischner (eds.), *New business in old Europe*, Greenleaf Publishers (2006)

First, the *market value* of the product-service can be lower than ownership of the product. While the functionality of car sharing and owning a car may be similar, the intangible value of car ownership gives the latter an advantage. Also, access to a car in car sharing is not as instantaneous as when it is owned.

Second, the *life cycle costs* of the product-service may be higher. This is because a provider takes responsibility for the use stage of a product. In the case of car sharing and pooling, the provider must factor in drivers who treat the car with less respect since they do not own it. While some protection against such misbehavior can be arranged via legal agreements, the transaction costs of such enforcement and lawsuits can become very expensive.

Further, in the case of product sales the full costs of the product are recovered directly. In the case of pay-per-use, this only happens over the lifetime of the product, implying a producer has much higher *upfront investment and capital needs*, which it needs to obtain from a bank that may not understand this new business model. Finally, product-service systems need a whole *business model overhaul* of the provider, for example, toward now also offering repair services. It may be that the provider currently has other core competencies and needs to invest heavily to make the new business model work.

Applications

In principle, use- and result-oriented PSS have the potential to create systems of provision that give both producers and consumers incentives to counter the unsustainable practices of planned obsolescence, fast change of fashion, and providing products that are by design not repairable or reusable. By offering users a result of having them pay per unit of use, the source of a service becomes the profit center. What then determines profit is providing a service with minimal life cycle costs – hence minimal use of resources.

As shown in Box 70.1, in the past, various attempts have been made to put PSS on the market. Some were successful, others were not. We see that owning a product has many advantages:

prestige (see **Conspicuous/Positional Consumption**), control over the use of the product, and so on. Offering a PSS implies also solving many hitherto irrelevant problems: taking responsibility for the use stage, dealing with higher upfront capital needs, and higher transaction costs in general. For these reasons, particularly in Business-to-Consumer (B2C) contexts, PSS tend to stay in niches. Exceptions are platform services like Uber and Airbnb. They use existing capital assets and hence are cheaper than a traditional provider and/or offer more convenient access to the service. Yet, these platforms have many disadvantages. They are so powerful that they extract excessive value, marginalizing the income of providers. There are unwanted side effects, such as houses in popular places disappearing from the rental market or drivers not being drivers protected by labor laws.

PSS have more success in a Business-to-Business (B2B) context, although they are still not mainstream. Mature products with standardized components, used in a predictable environment that can be easily repaired or the components that can be easily used in the next product generation, are good candidates to build a PSS business model around.

One has to, however, consider that PSS is nothing more or less than a business model that has to be successful in the current societal and business environment. This is a system that is unstable without GDP growth. PSS will, therefore, never be the ultimate answer to broader questions around how to create a degrowth or well-being economy (see **Degrowth** and **Well-being Economy**) – these imply a radical change to the rules of our economic game.

Further Reading

Amatuni, L., Ottelin, J., Steubing, B., & Mogollón, J.M. (2020). Does car sharing reduce greenhouse gas emissions? Assessing the modal shift and lifetime shift rebound effects from a life cycle perspective. *Journal of Cleaner Production*, 266. https://doi.org/10.1016/j.jclepro.2020.121869.

Bocken, N.M.P., de Pauw, I., Bakker, C., & van der Grinten, B. (2016). Product design and business model strategies for a circular economy. *Journal of Industrial and Production Engineering*, 33(5), 308–320. https://doi.org/10.1080/21681015.2016.1172124.

Mont, O., & Tukker, A. (2006). Product-service systems: Reviewing achievements and refining the research agenda. *Journal of Cleaner Production*, 14(17), 1451–1454. https://doi.org/10.1016/j.jclepro.2006.01.017.

Stahel, W.R. (2016). The circular economy. *Nature*, 531(7595), 435–438. https://doi.org/10.1038/531435a.

Tukker, A. (2015). Product services for a resource-efficient and circular economy – A review. *Journal of Cleaner Production*, 97, 76–91. https://doi.org/10.1016/j.jclepro.2013.11.049.

71 Universal Basic Services

Anna Coote and Ian Gough

Definition

Universal basic services (UBS) describes a set of proposals for achieving universal access to life's essentials within planetary boundaries. The component terms can be defined as follows:

Services: Collectively generated activities that serve the public interest
Basic: Services that are essential and sufficient – rather than minimal – to enable people to meet their needs
Universal: Everyone is entitled to such basic services according to need, not ability to pay.

An inherent normative assumption is that governments should ensure that every individual has access to the core necessities that make life possible and worth living. The concept has emerged in the Global North, though its influence is spreading globally.

There is broad agreement on what "necessities" or "essentials" are: nourishing food, clean air and water, a decent home to live in, domestic energy, education, people to look after us when we cannot do so ourselves, healthcare when we are ill, transport to take us where we need to go, access to the internet, and – underpinning them all – a safe environment.

None of us (not even the rich) can meet all our needs without sharing risks and pooling resources to generate a *virtual income* or *social wage*. This is made up of "in-kind" benefits – collective measures provided or financed by public authorities. They include a range of social and economic services, commonly described as the "welfare state". UBS aims to improve existing services, such as healthcare and schools, and to extend collective measures to areas where basic needs are not met universally or sufficiently (see **Sufficiency** and **Consumption Corridors**). The scope and priorities advocated will differ across time and place according to social, economic, and political contexts (see also **Doughnut Economics** and **Fair Consumption Space**).

Advocates of UBS contend that "sustainable consumption" can best be advanced by further expanding the share of collectively provided services in total consumption. UBS reflects and builds on experiences of high-quality universal services in health, education, and other areas of need that have endured in many countries for 70 years or more. The case for such an expansion combines social and environmental arguments; hence, the centrality of UBS for sustainable consumption.

History

The idea of UBS was first outlined in a 2017 report by Moore et al. entitled *Social prosperity for the future: A proposal for Universal Basic Services*. In 2019, the idea was theorized in terms of human needs by Gough (2019). It received book-length elaboration in 2020 by Coote and Percy. Subsequent work has developed the ecological case for UBS (e.g., Vogel et al., 2024).

DOI: 10.4324/9781003584056-77

The UBS label was initially designed as a critical alternative to the campaign for Universal Basic Income (UBI). The case for UBS recognizes that everyone should be entitled to a sufficient cash income, but that an unconditional grant to all would be insufficient or unaffordable or both. UBS has aimed to reassert the value and efficacy of public services and other *in-kind* benefits at the heart of policy development. Debates continue about whether UBS could be combined with some version of UBI, or (more plausibly) a guaranteed minimum income.

Different Perspectives

Advocates of UBS make the following arguments:

First, it would directly meet needs via collectively provided "needs satisfiers". This contrasts with the "transfer arm" of the welfare state, which augments people's incomes but leaves provisioning to market forces. The latter may work well for providing tomatoes or gyms, but much less well for essential goods and services. There are strong efficiency arguments for the latter based, inter alia, on economies of scale and network benefits.

Second, free or low-cost delivery of life's essentials is inherently equalizing because necessities *by definition* account for a greater share of the expenditures of lower-income households. This is the case in all tax-funded systems, even when the tax burden is not progressive.

Third, it achieves redistribution without moral or consequentialist drawbacks. State transfers are also progressive on balance, but this is mainly achieved by targeting, which typically involves "means testing" with its associated problems of demeaning treatment and conditionality. Public services and their extension via UBS automatically redistribute the social part of consumption according to need, without these disadvantages, while enhancing the total value of households' disposable real income.

Fourth, many of these essential items, such as education, health, and social care, are not simply about consumption. They are investments that yield a stream of social benefits into the future, for example by enhancing health and wellbeing, strengthening social cohesion, and building knowledge and skills. Many of these benefits have been evaluated in **money** terms, and research suggests their rate of return can exceed that of some "productive" investments in manufacturing. Such returns will be critical in moving toward a less commodified economy.

Fifth, social consumption and provisioning can directly contribute to the abatement of greenhouse gas emissions and other environmental goals. UBS is a needs-based approach that identifies and prioritizes satiable needs over some unsustainable wants (see **Conspicuous/Positional Consumption** and **Personal Carbon Allowance**). Comparisons of the carbon footprint of health care show a significant excess in the USA compared with socialized healthcare in Europe, while delivering worse outcomes in terms of need-satisfaction. Social provisioning can embed sustainability goals in day-to-day practice.

Advocates of UBI argue that UBS "pre-selects what people need", rather than offering a choice via adequate incomes and market provisioning (see **Choice Editing**). Against this, the case for UBS singles out everyday necessities from other consumption goods that would continue to be distributed via market mechanisms. It has affinities with the concept of a distinct "**foundational economy**": mundane, taken-for-granted networks and services that people depend on in their daily lives.

Application

UBS is inherently disaggregated and context-specific. Can any general guidelines be agreed upon to achieve a coherent public system of universal services? Determining what constitutes universal basic services in a specific context must involve a collective democratic process, a role that has traditionally been the prerogative of representative governments at national and sub-national levels. Many supporters of UBS now advocate a complementary role for dialogic methods such as deliberative assemblies, which can more effectively tap into the practical realities of day-to-day life, for consideration alongside expert and practitioners' knowledge (see also **Living Labs**).

Coote and Percy (2020) outline three key government functions required for implementing a UBS strategy:

- Guarantee the entitlements of citizens/residents to essential services and ensure equality of access: The codification of rights and entitlements will differ across services and between countries, for example through constitutional law, other justiciable laws, regulatory bodies, professional ethics, and other processes.
- Raise taxes or borrow money and distribute resources: Much of the focus will be on current expenditure, but capital and infrastructure spending will gain new importance, for example, the UK's Infrastructure Strategy Commission calling for Universal Basic Infrastructure (see **Sustainable Finance**).
- Set standards and regulate: This would require meaningful devolution of power to regional and local authorities, combined with strong central direction to ensure equal access and quality through investment, distribution, regulation, and coordination.

While no city or country fully exemplifies UBS in practice, Coote and Percy point to many examples of services delivered in ways that reflect the ambitions of this agenda – from childcare in Norway to housing in Copenhagen, transport in London, social care in Germany, and much more. There are signs of growing interest in policy circles: the report by Enrico Letta (2024) to the European Council urges the EU, for the sake of social cohesion and a viable single market, to prioritize the pursuit of "universal access to essential services – including water, sanitation, energy, transport, financial services and digital communications – to meet basic human needs to live and to participate in society".

Further Reading

Coote, A., & Percy, A. (2020). *The case for universal basic services.* Cambridge: Polity Press.

Gough, I. (2019). Universal basic services: A theoretical and moral framework. *Political Quarterly*, 90(3), 534–542. ISSN: 0032-3179.

Letta, E. (2024). Much more than a market. *Report to the European Council*. Available at: https://www.consilium.europa.eu/media/ny3j24sm/much-more-than-a-market-report-by-enrico-letta.pdf (pp. 90–120) (accessed: 8 January 2025).

Stakelum, R., & Wiese, K. (2024). *Universal basic services: Road to a just transition*. Available at: https://www.socialeurope.eu/universal-basic-services-road-to-a-just-transition (accessed: 8 January 2025).

Vogel, J., Guerin, G., O'Neill, D., & Steinberger, J. (2024). Safeguarding livelihoods against reductions in economic output. *Ecological Economics*, 215, 107977. https://doi.org/10.1016/j.ecolecon.2023.107977.

72 Urban Planning and Spatial Allocation

Valerie Brachya

Definition

Many decisions on lifestyles result from the ways in which space has been allocated in the past and is being allocated in the present. A sustainable urban lifestyle requires, for example, the availability of public transport, which cannot function effectively and efficiently without relatively high dwelling density. It requires spaces for public institutions close to dwellings and corridors for efficient infrastructure alignments. Spatial allocation can enable proximity to convenient and affordable essential facilities, for home, work, shopping, schooling, and health care. It can also combat unsustainable lifestyles by refusing to allocate space for uses, such as plots for single large houses and spaces for parking private vehicles.

The allocation of space for various types of development and activities determines the location, scale, and design of residential, commercial, and industrial land uses. It essentially creates the physical preconditions for choice architecture (see **Choice Editing**). It can enhance beneficial "provisioning factors", such as proximity, or mitigate harmful ones, serving as a mediator in achieving more sustainable lifestyles.

Responsibility for spatial allocation is usually governed by a regulatory system, which varies between countries, regions, and even municipalities. The system is based on a separation between ownership and rights of use. The rights to change land use or to develop a site do not necessarily belong to the landowner but to the public, represented by the state. They are allocated according to approved statutory plans and policies. The regulatory system includes (i) the designation of areas for development and conservation, (ii) where development may be prohibited or restrained, and (iii) the designation of space for sites and alignments for infrastructures. The land use regulatory system enacts building codes, which can incorporate mandatory requirements and compliance with standards for energy efficiency in buildings, solar energy, resilience to seismic risks, green building, and preventing building in areas of high risk, including risk to climate disasters, such as in floodplains.

History

The evolution of spatial allocation can be traced back many centuries, for instance where central city spaces were allocated to governmental, religious, and trade activities. Examples include the Greek cities planned by Hippodamus of Miletus in the 5th century BC. However, the modern land-use regulatory planning system in cities is often attributed to reformers in the mid-19th century, responding to illness, epidemics, and pollution generated by industrial development in crowded towns. They proposed creating decent conditions for workers, such as "garden cities" with good quality living conditions and healthy green open spaces. This separation of residential

DOI: 10.4324/9781003584056-78

neighborhoods from mass production work locations created the need for daily travel to work. While it provided affordable and decent housing, it had profound impacts on the choice architecture for lifestyles.

This was accelerated in the post-World War II era, when the upsurge in demand for single-family houses resulted in urban sprawl in places like the United States, initially along suburban train routes and later along highways, facilitated by mass car ownership. This resulted in the outward dispersal of low-cost, low-density residential development. It was encouraged by governments through low-interest mortgages and the availability of credit and loans and was seen as fulfilling the dream of homeownership for all.

By the turn of this century, many cities continued to spread outwards, with ever-increasing commuter distances and loss of land to buildings. But that was not universal. Others took a more sustainable approach, creating vibrant and lively city centers, providing for walkability and the use of bicycles, and giving preference to efficient, affordable, comfortable, and convenient public transport (see **Sustainable Mobility**). Cities such as Barcelona, Vancouver, Paris, and Copenhagen not only are regarded as cities that are attractive and provide a high quality of life but also enable their residents to live sustainable lifestyles. However, property prices and rentals in these attractive and vibrant cities are increasingly high, and consequently, sustainable lifestyles may not necessarily be affordable.

Different Perspectives

Spatial allocation has always been a controversial and contentious issue, both over its content and over the process of decision-making. Many controversies, for instance, end up in the courts. Who should make decisions on how to allocate space? Should it be determined by national or local government? Whose interests do they represent and how can civil society be engaged and involved in the processes of decision-making? How are the interests of future generations taken into account?

Early regulatory systems allocated the powers of decision-making to appointed or elected representatives, supported by professional architects, engineers, and town planners. Civil society could object to development proposals and hope their views would be taken into account, but were often regarded negatively as NIMBYs ("Not In My Back Yard"). Over the decades, civil society gained influence, initially through demands for transparency but later through involvement in the decision-making process by forming coalitions, obtaining support from non-governmental organizations with expertise, and financial support from philanthropic funds (see **Social Movements** and **Consumer-Citizen**).

Although the economic interests of developers and the revenue from local rates and betterment taxes provide strong incentives for approving projects and proposals, spatial allocation has become far more responsive to community, local, social, and public interests. Perspectives vary over time and across countries and often reflect political differences toward a neoliberal free market for property rights versus state responsibility for public welfare and the provision of public services.

The following are some approaches taken by professional experts involved in urban development:

Demand Versus Need

Land use planning has traditionally been based on forecasts of demand – for housing, industrial and commercial space, public services, and for meeting transport requirements – based on market predictions and real estate values. Environmental and social considerations were integrated into decision-making, initially as externalities but later as valid public considerations. However, when an urban planning committee reviews a development proposal, it typically assumes that market demand is valid and does not consider whether that demand should be restricted, even if public infrastructure is available and potential environmental impacts are insignificant. Even when objectors turn to the courts, they are typically required to justify why a proposal should *not* be approved (e.g., due to harm to the habitat of a rare species). A focus on sustainable lifestyles would shift this burden of proof: a development proposal (whether private or public) would be considered *unnecessary* unless the proponent could demonstrate its necessity for serving the public benefit (such as providing decent housing for all) and prove that it is the best alternative to fulfill that need.

Separation of Land Uses (Zoning) Versus Mixed Use

Spatial allocation through land-use planning, commonly referred to as zoning, designates permitted uses intending to ensure high-quality living conditions and maintain stable real estate values. It provides long-term security for land values and helps the market provide clarity to buyers and sellers about what would be permitted on adjacent plots and in their neighborhoods.

It has, however, in some cases, resulted in social and racial segregation and prevented the mixture of communities, even creating barriers to the interaction between peoples and activities. Urban policies today favor mixed uses, enabling homes, work, and commercial activities to take place in close proximity, strengthening urban street life from morning through evening, and enabling non-motorized circulation.

Urban Renewal Versus New Towns and Outer Suburbs

After favoring outward expansion, many cities found themselves burdened with land and buildings that had deteriorated over time and were even abandoned. Inner urban areas suffered from industrial contamination and dilapidation as inhabitants and activities moved to outer suburban, peri-urban areas or new towns. Governments and cities recognized that they had to bring people back into the inner city by providing high-density and high-quality development, well served by public transport. However, this is often no longer affordable. A shift has occurred whereby those with high incomes who can afford to live in city centers have access to public amenities funded by public money (e.g., public transit, parks, inexpensive commutes, libraries, etc.), while those with lower incomes, who live further and further away, have little access to these public amenities.

Underground Versus Overground

Many cities grew upwards, with ever-increasing building heights as real estate values rose. Foundations and parking spaces were required by building codes to be below ground, but few cities developed their underground space beyond the provision of public transport systems. Singapore is an exception, with such a shortage of space that underground space was seen as essential to its development. Montreal developed its underground areas as a solution to its climate. However, in general, land use planning has ignored the benefits of harnessing underground space. Technological advances today could enable more facilities to be moved underground and release the surface and aboveground space for activities that require natural ventilation and light.

Private Space Versus Public Space

The role of the spatial planning system is to allocate private property rights while maintaining public values and enabling public services. There is thus a strict division between private space and public space. The system does not easily accommodate shared or joint uses, such as smaller but decent living conditions combined with larger spaces for higher-quality public services. Spatial planning may require reappraisal in relation to the sustainable discourse on what constitutes decent living conditions.

Applications

Paris and Barcelona provide two examples of cities that have applied concepts for sustainable lifestyles in urban development and management strategies. Paris adopted the concept of a 15-minute lifestyle, with an emphasis on proximity, where residents could access their daily needs within a 15-minute radius without the need for a private car.

Barcelona's municipality (Ajuntament de Barcelona) divided sections of the existing urban structure into "superblocks" within which accessibility by vehicles was highly restricted to local residents and priority given to non-vehicular circulation. Each superblock could then re-allocate road space and provide residents with street space for local community urban activities. Barcelona thereby offers an exemplary application of sustainable urban planning and spatial allocation concepts, which is already inspiring further cities to follow it.

Further Reading

Ajuntament de Barcelona. (2016). *Let's fill streets with life – Establishing Superblocks in Barcelona*. Barcelona: Commission for Ecology, Urban Planning and Mobility. Available at: https://ajuntament.barcelona.cat/ecologiaurbana/sites/default/files/en_gb_MESURA%20GOVERN%20SUPERILLES.pdf (accessed: 8 January 2025).

Knaap, G.-J., Nedović-Budić, Z., & Carbonell, A. (Eds.). (2015). *Planning for states and nation-states in the U.S. and Europe*. Cambridge, MA: Lincoln Institute of Land Policy.

Moreno, C. (2024). *The 15-minute city: A solution to saving our time and our planet*. Hoboken, NJ: Wiley.

Nadin, V., Cotella, G., & Schmitt, P. (Eds.). (2024). *Spatial planning systems in Europe: Comparison and trajectories*. Cheltenham: Edward Elgar Publishing. https://doi.org/10.4337/9781839106255.

Vogel, J., Steinberger, J.K., O'Neill, D.W., Lamb, W.F., & Krishnakumar, J. (2021). Socio-economic conditions for satisfying human needs at low energy use: An international analysis of social provisioning. *Global Environmental Change*, 69, 102287. https://doi.org/10.1016/j.gloenvcha.2021.102287.

73 Sustainable Housing

Peter Berrill

Definition

Sustainable housing can be defined as the affordable and safe provision of shelter and domestic services with minimal negative impacts on the environment. Aspects of environmentally sustainable housing include the use of renewable energy, passive provision of thermal comfort and lighting (using naturally available energy inputs), high thermal efficiency, and use of construction materials with low lifecycle environmental impacts. From an economic perspective, sustainable housing should be affordable and retain its value. From a social perspective, sustainable housing should be safe, accessible, and provide a healthy indoor living environment.

History

Fundamental aspects of sustainable housing have been considered and incorporated into the design of housing for millennia (see **Ecodesign**). Prominent examples include design for structural stability, resistance to fire damage, and provision of ventilation to maintain healthy indoor air quality. There is evidence of energy efficiency and passive provision of indoor thermal comfort and lighting in pre-industrial housing, hundreds or thousands of years before design standards existed. Examples include constructing buildings partially in the ground, introducing passively heated or cooled air into buildings, and using materials with large thermal capacity to minimize indoor temperature fluctuations.

Energy efficiency entered building codes and standards in the mid-20th century, particularly after the 1970s energy crises. The existence and stringency of energy standards varies considerably by country and climatic region within countries. Colder regions, for example, tend to require higher levels of insulation. Jurisdictions with building energy efficiency standards typically update their standards to higher levels periodically. In 2020, California's energy efficiency plans stipulated that all new residential buildings must be "zero net energy", meaning that buildings must produce as much clean renewable energy as they consume over the course of a year. Since 2021, new buildings in the European Union must be "nearly-zero energy", a term defined and implemented differently by member states. Improving the energy efficiency of entire building stocks requires paying attention to existing buildings in addition to new construction. In countries with mature (slowly growing) building stocks, strategies to reduce energy demand and environmental impacts will focus primarily on existing buildings (see **Stocks Versus Flows**). The European Union, for instance, announced ambitions for a "renovation wave" in 2020, referring to an increase in the rate and depth of energy renovations. Many (mostly high-income) countries offer subsidies and other incentives for householders and building owners to undertake energy-efficiency renovations.

DOI: 10.4324/9781003584056-79

Voluntary building sustainable certifications (e.g., BREEAM, LEED, HERS) have existed since the 1990s. These award different levels of sustainability performance based on sustainable choices made during building design. "Passive house" buildings embody another approach to sustainable design optimized for energy efficiency.

Standards have recently been developed to measure the "embodied" emissions of new construction, which covers environmental impacts from producing construction materials. In a few jurisdictions (e.g., Canada, California, and some European countries), regulations on the embodied carbon of new construction already exist. Future standards may assess and regulate the "whole-life carbon" of construction, including embodied and energy-related emissions over the building's estimated lifetime.

Different Perspectives

The sustainability of housing can be assessed from different perspectives. The social sustainability of housing can be considered from a household or societal perspective. For individual households, socially sustainable housing would be safe, healthy to occupy in terms of air quality and sanitation, and accessible for different ages and abilities. Resistance to extreme temperatures and acute environmental hazards like earthquakes, flooding, and hurricanes can be included as aspects of housing safety, in addition to basic structural soundness. On a societal level, in addition to being safe, healthy, and accessible, a sustainable housing stock would be affordably available to the entire population, located in regions that are healthy to occupy, and where basic needs and services (education, food, water, healthcare, employment, energy) can be provided (see **Urban Planning and Spatial Allocation**).

The economic sustainability of housing has close connections with social sustainability, particularly at the societal level. For housing to be economically sustainable at the household level, it should be affordable and preserve or grow its value over time. At the societal level, housing should be affordable for all sectors of the population. Tensions can arise between economic sustainability at the household and societal level. Regulations that restrict new construction increase the value of existing housing and the costs of new housing, benefiting incumbent homeowners and landlords at the expense of first-time home buyers and renters. Meanwhile, housing rental markets with rent controls benefit existing tenants at the expense of new tenants (to the extent that controls reduce the supply of new housing). Regulations that reduce housing affordability or availability at the societal level can exacerbate homelessness and housing precarity.

The environmental sustainability of housing can consider impacts beyond those arising from material and energy use during building construction and operation. The location of housing can influence travel behaviors and how people spend their time. Different dimensions of local "urban form" including population density, distance to employment and services, and public transport accessibility can all influence how people travel daily, with clear implications for environmental impacts (see **Sustainable Mobility**, **Social Practice Theory**). Zooming out from the building level to the community or neighborhood level can allow more flexibility in achieving "zero net energy" and lower environmental impacts.

"**Sufficiency**" is a relatively new addition to conceptualizations of sustainability. In buildings, sufficiency predominantly refers to achieving levels of floor space consumption that are "enough" but not "excessive". Suggested values of sufficient residential floor space in the literature range from 15 to 40 m^2 per person. Sufficiency is closely related to the concept of "decent living standards" whereby basic needs for a good life are met for entire populations, and to "**doughnut economics**" where basic needs are met but ecological limits are not transgressed (see also **Fair Consumption Space**, **Consumption Corridors**).

Application

Sustainable housing assessments can be applied at different levels, for different purposes. For a new building, a designer can assess the environmental impacts of alternative designs, ideally from a whole-life perspective. Policies such as energy standards can influence new construction: new homes in many European countries must now use a heat pump as the heating technology. For an existing home, an analyst can assess the multifaceted sustainability outcomes of energy renovation and lifetime extension versus demolition and reconstruction (see **Circular Economy and Society**). This exercise, more common in academic studies than in real-world planning, usually finds it more sustainable to renovate than to demolish and rebuild.

Larger developments, such as plans to deliver hundreds or thousands of new dwellings, can consider the regional environmental impacts of development alternatives (see **Urban Planning and Spatial Allocation**). At a regional or national level, assessments can be made of entire building stocks to identify the relative potential of strategies to reduce environmental impacts for existing and new buildings over a planning horizon of decades. Prospective assessments can incorporate anticipated technological and social changes, assisting comprehensive strategies to maximize the sustainability of the built environment. These are especially important when considering potential requirements to provide many new dwellings to accommodate population growth and reduced household sizes.

Application of sufficiency strategies has not begun in earnest in buildings, but possible approaches include reducing per-capita floor space through constructing smaller new dwellings, internal renovations to create additional dwellings within existing buildings, or encouraging larger household sizes (for a given population). In certain cases, the provision of shared spaces for leisure, storage, or services (e.g., cooking, utilities) can reduce the total space required for all households (see **Product-Service Systems**, **Sharing Economy**).

Further Reading

Berrill, P., Wilson, E.J.H., Reyna, J.L., Fontanini, A.D., & Hertwich, E.G. (2022). Decarbonization pathways for the residential sector in the United States. *Nature Climate Change*, 12(8), 712–718. https://doi.org/10.1038/s41558-022-01429-y.

Hasik, V., Escott, E., Bates, R., Carlisle, S., Faircloth, B., & Bilec, M.M. (2019). Comparative whole-building life cycle assessment of renovation and new construction. *Building and Environment*, 161, 106218. https://doi.org/10.1016/j.buildenv.2019.106218.

Ionescu, C., Baracu, T., Vlad, G.E., Necula, H., & Badea, A. (2015). The historical evolution of the energy efficient buildings. *Renewable and Sustainable Energy Reviews*. https://doi.org/10.1016/j.rser.2015.04.062.

Næss, P., Peters, S., Stefansdottir, H., & Strand, A. (2018). Causality, not just correlation: Residential location, transport rationales and travel behavior across metropolitan contexts. *Journal of Transport Geography*, 69, 181–195. https://doi.org/10.1016/j.jtrangeo.2018.04.003.

Röck, M., Saade, M.R.M., Balouktsi, M., Rasmussen, F.N., Birgisdottir, H., Frischknecht, R., Habert, G., Lützkendorf, T., & Passer, A. (2020). Embodied GHG emissions of buildings – The hidden challenge for effective climate change mitigation. *Applied Energy*, 258, 114107. https://doi.org/10.1016/j.apenergy.2019.114107.

74 Sustainable Mobility

Noel Cass

Definition

A simple definition of sustainable mobility is "traveling with the least possible impact upon the planet". This suggests a hierarchy of preferable transport, from fuel-free "active" walking and cycling, down to private jet-flying or superyacht sailing. The "sustainable mobility *paradigm*" is more specifically related to policymaking and priorities and closely reflects the "A-S-I" hierarchy in recent sustainability studies: *Avoid* (travel demand and car ownership); *Shift* (travel to the most sustainable modes); *Improve* (efficiency of individual modes). Addressing the impacts of car driving and flying is now viewed as a particularly key concern.

History

The first major policy definition of "sustainable mobility" is from the 1992 European Union *Green Paper on the Impact of Transport on the Environment*, which sought to "enable transport to fulfill its economic and social role while containing its harmful effects on the environment". Sustainable mobility thus includes freight, although the most focus has been on passenger transport. Sustainable mobility has a long history subsumed within broad discussions of sustainable development, but as climate change has increasingly dominated policy debates, carbon emissions have risen in importance. That, in turn, eclipsed other impacts of transportation, such as impacts on habitats, air quality, material mining and fossil fuel extraction, as well as road accidents or dividing communities by building roads.

Historically, research and policies for sustainable mobility focused on traffic and congestion. This scope evolved over time to include the "drivers" of individual mobility choices, both psychological and infrastructural (driving and flying), transport infrastructures, mobility systems. The scope has also widened to include the questions of trip purposes (commuting, daily mobility and active long distance leisure travel). The relatively recent shift in framing of why and how people chose transportation modes, from "behavior" to habit, necessity, or social practice shifted the focus of studies and policy options: from information provision and financial (dis)incentives (public transport subsidies, "smarter choices" promotion) to systemic solutions.

A 2019 review of the research literature (see Table 74.1) summarizes the evolution of the studies on impacts of transportation, breaking it up into five "generations" – The most recent interest in demand reduction (around the need to travel) represents the A in the A-S-I approach, and is closely related to **"sufficiency"** studies.

DOI: 10.4324/9781003584056-80

Table 74.1 Five generations of studies on sustainable mobility and their widening scope (adapted from Holden et al., 2019)

Dimension	First 1992–1993	Second 1993–2000	Third 2000–2010	Fourth 2010–2018	Fifth 2013–present
Focus of Research and EU policy	Limit transport volumes	Reduce transport intensity	+ Congestion, equity, competitiveness	+ Decarbonize	+ Reduce travel, demand Mobility Justice
Types of research and policy questions	How to increase the efficiency of different modes of transport?	+ How to manage travel demand?	+ What are different actors' motivations, opportunities, and abilities to change?	+ How to create synergies between environmental effects and the wider social implications on health and inequality?	+ How to reduce the need to travel overall? How to make sure travel reduction is fair – does not exacerbate inequities?
Focus on impacts	Environmental	+ Societal (quality of life)	+ Economic accessibility, distribution (justice)	All dimensions of sustainability	Increased focus on carbon emissions
Categories of mobility	Production travel (work/commute)	+ Reproductive travel (non-work travel by car)	+ Leisure travel (including long-distance travel by car and planes)	+ Shared mobility, autonomous vehicles, and electromobility	+ Immobility, Working from home, virtual meetings, online deliveries
Disciplinary studies	Environmental engineering, planning, transport geography, transport economics	+ Sociology	+ (Social) psychology, anthropology, political science, history, public health	+ Innovation studies, sustainability transitions	Renewed focus from sociology, planning + technology studies,
Theories and methods of study	Environmental impact assessment, quantitative modeling, regression analysis	+ Qualitative analysis (scenario building, scenario analysis?)	+ Case studies, interviews, qualitative modeling, institutional analysis, historical interpretive analysis	+ Multi-Level Perspective for sociotechnical regime transitions, technological innovation systems, big data analysis	+ Social practice theory, mobility as a service

"+" indicates that the focus of the previous generation is broadened to include the marked item

Different Perspectives

The "Mobilities" paradigm, including a journal of the same name, extends consideration of mobilities beyond passenger transport to *movement* as a key characteristic of modern society. In comparison with the well-recognized "socio-technical systems" and multi-level perspectives, which view mobility systems as "regimes" of transport modes and infrastructures, systems of law and regulation, manufacture, fuel supply, and more, the Mobilities paradigm goes further. It includes anthropological and sociological concerns with culture, meaning and identities.

In a related approach, movement and mobility are now understood as driven by **social norms**. This draws attention to what ties people to car driving in particular, which is more complex than simply convenience and access, and includes, for example, geographies of car-friendly suburbs connected to ever-increasing highway networks, the teenage rite of passage of getting a driver's license, the cultural landscape of media where cars and driving are linked with individualism, family life, and freedom (see **Urban Planning and Spatial Allocation, Freedom of Choice, Choice Editing, Attitude-Behavior Gap**). These insights reveal the multi-faceted challenge of encouraging people out of their cars into more sustainable modes of mobility.

Social practice theory (SPT) stresses attention to the materials, skills, and meanings that make up various mobility practices and how they are shared across society rather than being "in people's heads". This shifts the focus to how society coordinates shopping, education, leisure, and work in time and space. The concept of "Autonomobility" combines this more systemic understanding of sustainable mobility with the concerns of mobility justices – stressing that mobility systems should reconcile sustainability with the maximization of freedom, fairness, and fulfilling people's mobility needs.

A "political economy" approach to (un)sustainable mobility ties together the critical study of several of the above aspects: systems of provision (including the powerful car, fossil fuel, and construction lobbies), mobility activities or practices, the energy needs and services being satisfied, and specific technologies involved (see **Political Economy of Consumerism, Foundational Economy**).

Flying and long-distance travel is currently drawing increasing attention, relative to the traditional emphasis on car driving, because of its very large carbon footprint. A recent analysis shows that 3% of plane trips account for 60% of distance traveled and 70% of respective carbon emissions (see **Household Income Versus Carbon Footprint**). There has also been a concern about second-order "rebound" effects as found in other energy research; urbanites drive less than country-dwellers in their daily lives but fly more, owing to their lifestyles and income (see **Rebound Effects**). This again raises the importance of the A-S-I hierarchy of policy approaches – to focus on Avoiding travel before Switching and Improving.

Application

This ever-broadening understanding of sustainable mobility explains the huge number of applications and policy solutions that have been suggested. Currently, the central concerns of the sustainable mobility *paradigm* remain: (i) reducing the need for short-distance travel through the use of **Information and Communication Technology**, telemedicine, or working from home; (ii) encouraging a modal shift to more sustainable modes (especially active modes such as walking and cycling, and public transport); (iii) reducing trip lengths through planning to shorten the distance between places of living, working, education, and so on (see the 15-minute-city concept; also **Urban Planning and Spatial Allocation**); and (iv) encouraging energy efficiency and carbon reduction (especially electrification, with a lesser focus on hydrogen and biofuels from plants). An earlier distinction between "hard" transport measures (i.e., building roads and installing bike paths) and "soft" ones (information, encouragement, subsidies) has been supplemented with a distinction between targeting individual free choices and "**choice editing**" by affecting the available options or systemic change (such as investment in public transport and planning policies) (see **Green Nudging**).

Car-free days or zones and "flight shame" exemplify developments, which could change social norms (see **Social Tipping Points**). But that is not enough. Tackling long-distance and non-commuting travel requires more systemic approaches, including a critical view of the role of the growing menu of options, infrastructures, and technology. These include airport sizes, number and variety of flights, road space for cars, and economic incentives (air miles, tax-exempt jet fuel, company car subsidies). Technological innovation studies stress the potential of drones, light electric vehicles, multi-modality (trips using multiple modes), mobility-as-a-service (purchasing mobility, not individual trips; see **Product-Service Systems**, **Sharing Economy**), and electric micro-mobility (e-bikes, e-cargo bikes, e-scooters, etc.). At the level of major infrastructure investment and planning, sustainable mobility requires a continuing shift from "predict and provide" to demand management; and from simplistic cost-benefit analysis focused time savings and convenience to considering all externalized costs to society, wildlife, biodiversity, and the climate.

Further Reading

Cass, N., & Manderscheid, K. (2010). *Mobility justice and the right to immobility—from automobility to autonomobility*. Presented at the Association of American Geographers Annual Conference, Washington, DC, April. Available at: https://eprints.whiterose.ac.uk/219001/ (accessed: 8 January 2025).

Cass, N., & Manderscheid, K. (2018). The autonomobility system: Mobility justice and freedom under sustainability. In *Mobilities, mobility justice and social justice*. Routledge.

Holden, E., Gilpin, G., & Banister, D. (2019). Sustainable mobility at thirty. *Sustainability*, 11, 1965.

Kemp, R., & Rotmans, J. (2004). Managing the transition to sustainable mobility. In *System innovation and the transition to sustainability: Theory, evidence and policy*, pp. 137–167. Edward Elgar Publishing.

Mattioli, G., Roberts, C., Steinberger, J.K., & Brown, A. (2020). The political economy of car dependence: A systems of provision approach. *Energy Research & Social Science*, 66, 101486.

Rau, H., & Scheiner, J. (2020). Sustainable mobility: Interdisciplinary approaches. *Sustainability,* 12, 9995.

75 Protein Shift

Kristof Rubens and Erik Mathijs

Definition

The protein shift aims to change how we produce and consume food, away from animal-based proteins (both meat and dairy) and toward proteins from other sources. This shift is motivated by the high negative impacts of livestock production and the consumption of animal-based products on the environment, human health, and animal welfare. Potential alternative sources include plants, insects (where culturally accepted), alternative protein sources like single-cell or microbial protein (microalgae, fungi, protein from fermentation [e.g., mycoprotein]), and cultivated meat. "Protein transition" and "protein diversification" are often synonyms for protein shift.

The goal is to meet nutritional needs while staying within ecological boundaries on a global and local level (see **Consumption Corridors**). On average, plant-based and many alternative protein sources have a lower environmental and climate impact than animal-based products (Figure 75.1) (see **Consumption-Based Accounting**). Particularly in high-income countries, current food patterns show high levels of animal-based consumption. The consumption of red and processed meat furthermore often exceeds dietary recommendations, negatively affecting human health. Protein intake in these countries is generally sufficient at a population level and even exceeds nutritional needs.

History

The question of eating or not eating meat or dairy has been part of human life for centuries, for reasons of religion, culture, or specific events (e.g., war). After World War II, the consumption of animal-based products in the Western world dramatically increased, primarily driven by the modernization and industrialization of the agricultural sector. This led to increased production and relatively lower prices for all food products. Once a product for special occasions, meat became a commodity available at any time of the day.

For decades, consumption of animal-based products has been a subject of dietary guidelines, almost exclusively motivated by health. Arguably, the FAO report *Livestock's Long Shadow* (2006) put the detrimental effect of the livestock sector on the environment in the spotlight for the first time. Since then, growing evidence about the climate and biodiversity crises has led to increased policy attention regarding the relationship between food consumption, especially animal-based food products, and its disproportionate effect on land use change, deforestation, water consumption, and animal waste surpluses. This suggests that food system policies should consider the nutritional requirements of present and future generations and which foods we produce, process, and consume.

Since 2010, the FAO has recommended the development of food-based dietary guidelines (FBDGs) that promote healthy diets and consider the impact of global and local agri-food systems on the environment and human health while being culturally and socio-economically appropriate

DOI: 10.4324/9781003584056-81

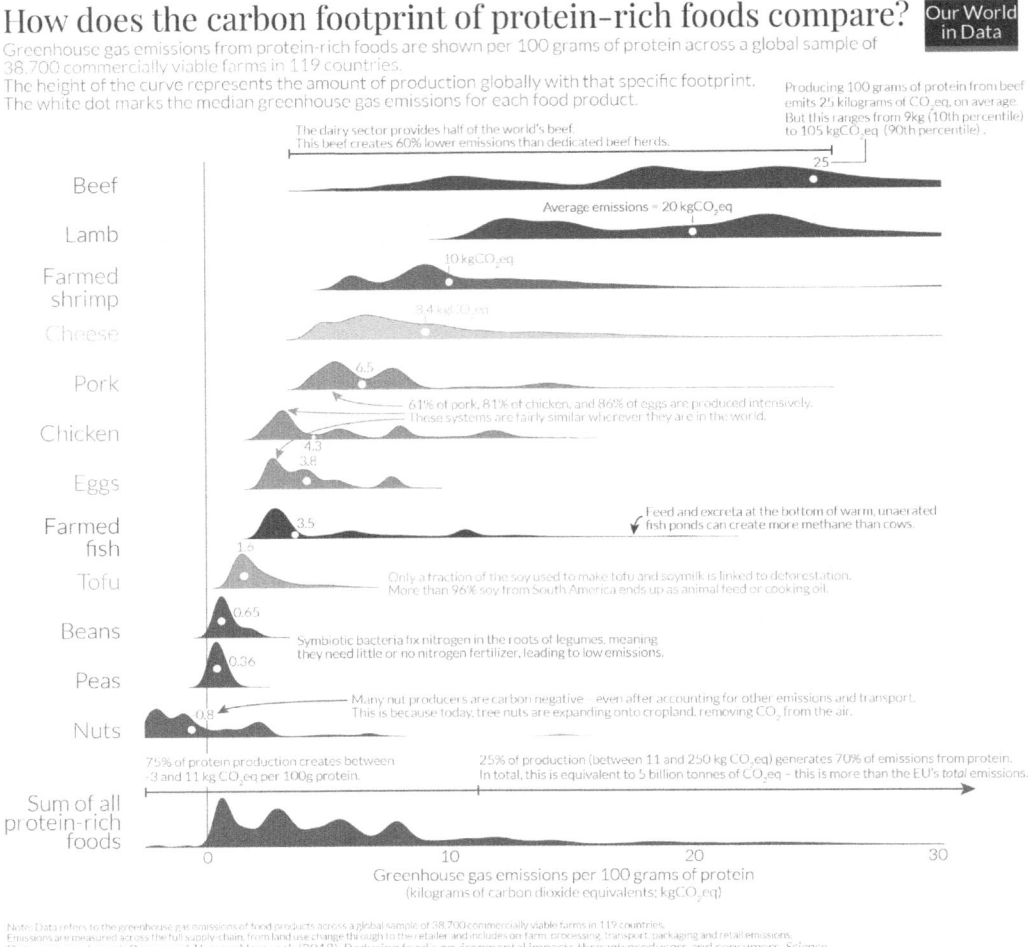

Figure 75.1 Comparison of GHG emissions per 100 grams of protein for different protein-rich foods and their variation

Source: Richie, H. (2020) Less meat is nearly always better than sustainable meat, to reduce your carbon footprint. *Our World in Data*. Available at: https://ourworldindata.org/less-meat-or-sustainable-meat (accessed: 13 June 2025)

(see **Consumption Corridors**). In 2019, the FAO and WHO organized an expert meeting to describe sustainable healthy diets and formulated the Guiding Principles for Sustainable Healthy Diets. Today, other organizations like the IPCC and the World Bank, among others, have recognized that a shift to healthier and sustainable diets, which includes a protein shift, is critical concerning planetary boundaries (see **Doughnut Economics**) and safeguarding well-being.

Attention to the food security aspects of protein has increased, especially in light of attacks on Ukraine by Russia (beginning in 2022). The European Union (EU) is mainly self-sufficient in agricultural products, but a complex picture emerges when looking at protein. The EU critically depends on imports of plant-based protein like soy or grain, mainly sourced for feed, from Argentina, Brazil, and the United States. The EU livestock sector is particularly vulnerable. This has led lawmakers, academia, industry, and NGOs to call for the EU to develop an EU protein strategy.

Different Perspectives

Although sometimes criticized as a reductionist approach to the challenges surrounding food consumption habits, focusing on protein is necessary due to protein's status as an essential macronutrient. Protein (and nutrient) rich, animal-based products are, at the same time, a significant contributor to the environmental burdens of food production. Using protein as a lens brings global food system challenges into sharp focus. By promoting the production and consumption of plant-based proteins for human consumption instead of indirect consumption via animals, the protein shift holds enormous potential for planetary, personal, and animal health. There are, however, also resistances to the protein shift (see Box 75.1).

Box 75.1   Special interest pushback on the protein shift

Resistance to the protein shift can be traced to vested interests in sectors that stand to lose out. The 2024 WHO report "Commercial Determinants of Noncommunicable Diseases in the WHO European Region" described obstructions to reducing meat consumption in Europe. Due to increasing market concentration, ten large global meat companies play a defining role in determining how meat and feed are produced, transported, and traded. This enables these companies to exert influence throughout the supply chain and gain political influence. The meat industry engaged in intensive lobbying against key components of the EU's Farm to Fork strategy, mishandling science and skewing media coverage. These efforts have been successful because proposals put forward as part of the Farm to Fork strategy – such as explicit references to health risks associated with intensive farming, requirements to increase transparency by labeling products, and the ability of EU Member States to impose higher taxes on unsustainable products – have all been delayed or watered down. A proposal to ban the financing of the promotion of red meat was even blocked. In recent years, the EU has invested millions of euros into campaigns promoting beef consumption, including a €4.5 million initiative called Proud of EU beef. The meat industry has also lobbied against initiatives to promote and fund research to develop alternative protein sources. Because of their political power, meat companies have been able to block the development of greener and healthier alternatives.

While the protein shift, as defined here, allows for some animal-based products, some argue that this is only a step toward the ultimate goal of a complete vegetarian (no meat) or vegan (no meat, dairy, eggs, or any animal-based product) lifestyle. Vegetarian and vegan lifestyles do require extra attention to avoid deficiencies in essential nutrients like vitamins (B12, D), minerals (e.g., iron, calcium), and some fatty acids (omega-3), which are less bioavailable in plant-based sources.

The question of which products deserve preference is thus a subject of debate. While it is clear that pulses, nuts, seeds, whole grains, and traditional vegetarian products like tofu, seitan, and tempeh satisfy health and environmental goals, some discussion exists on the risks and benefits of (highly) processed plant-based products. In general, health concerns surrounding ultra-processed food (UPF) consumption are also relevant in the protein shift. On the other hand, these products present an interesting alternative from a **behavioral change** point of view because they resemble products people are used to regarding look, taste, and preparation. Overall, (highly) processed plant-based products remain under scrutiny.

Application

The protein shift is often recommended qualitatively in food-based dietary guidelines (FBDGs): "Eat more plant-based foods than animal-based foods". In some cases, the protein shift translates into a concrete target. The EAT-Lancet Commission calculated a healthy reference diet (with a possible range) in grams per day to align with the UN SDGs and Paris Agreement criteria (Table 75.1).

Some organizations or countries propose a target based on the ratio between plant-based and animal-based protein intake. In Flanders, Belgium, the goal is to reach a 60% plant-based protein intake and a 40% animal-based ratio. In the Netherlands, the target is 50:50.

Initiatives like "Meatless Monday" (USA, 2003), "Meat Free Monday" (UK, 2009), and "Donderdag Veggiedag" (Belgium, 2009) aim to educate and incentivize consumers to stop eating meat for one day a week for environmental, animal welfare and health reasons. Campaigns like "VeggieChallenge" (the Netherlands, 2011) and "Veganuary" (UK, 2014) challenge participants to follow

Table 75.1 Healthy reference diet, with possible ranges, for an intake of 2500 kcal/day

		Macronutrient intake (possible range), g/day	*Calorie intake, kcal/day*
Whole grains	Rice wheat, corn and other	232	811
Tubers or starchy vegetables	Potatoes and cassava	50 (0–100)	39
Vegetables	All vegetables	300 (200–600)	
	Dark green vegetables	100	23
	Red and orange vegetables	100	30
	Other vegetables	100	25
Fruits	All fruit	200 (100–300)	126
Dairy foods	Whole milk or equivalents (e.g., cheese)	250 (0–500)	153
Protein sources	Beef and lamb	7 (0–14)	15
	Pork	7 (0–14)	15
	Chicken and other poultry	29 (0–58)	62
	Eggs	13 (0–25)	19
	Fish	28 (0–100)	40
	Legumes like dry beans, lentils and peas	50 (0–100)	172
	Legumes like soy foods	25(0–50)	112
	Legumes like peanuts	25 (0–75)	142
	Tree nuts	25	149
Added fats	Palm oil	6.8 (0–6.8)	60
	Unsaturated oils	40 (20–80)	354
	Dairy fats (included in milk)	0	0
	Lard or tallow	5 (0–5)	36
Added sweeteners	All sweeteners	31 (0–31)	120

a vegetarian or plant-based diet for one month. Other examples are "Semana sin carne" (Spain, 2017) or "Week zonder vlees" (The Netherlands, 2018), which challenge people for a whole week (see **Behavior Change**).

The Green Protein Alliance (GPA) was launched in 2016 in the Netherlands by a group of retailers, producers, and NGOs to promote plant-based consumption. Similarly, the Flemish Department of Environment and Spatial Development (Belgium) launched a "Green Deal Protein Shift on our Plates" in 2021, together with more than 80 stakeholders. Singapore took the lead in innovative protein production by making the headlines in 2020 as the first country to take cultivated meat to the market. Meanwhile, a movement pushing universities toward 100% plant-based meals kicked off in the United Kingdom in 2021 and has since spread to other countries. In 2023, Denmark was the first country to publish a national action plan outlining how to transition toward a more plant-based food system (see Box 75.2). Local municipalities and states in the United States are spearheading initiatives in schools, hospitals, and other settings.

Box 75.2   World's first national action plan for plant-based foods

"Plant-based foods are the future", the Danish minister for Food, Agriculture and Fisheries wrote in the preface of the "Danish Action Plan for Plant-Based Foods". "This action plan should inspire everyone who works in our food systems and who influences our daily food choices; from the farmer and food producer to the retailer, the canteen provider, the export markets – and of course the consumer on their daily trip to the supermarket". This plan puts forward a specific target ratio and couples it with a dedicated Plant-Based Food Grant of roughly 11.26 million euros (84 million DKK) per year from 2023 to 2030. The grant targets innovative projects and has been a big success in terms of interest. In the first round, there were more than three times as many applications than could be granted. Round two resulted in applications worth three times the available budget (which already doubled). This has led the government to increase the budget for 2024–2026 by more than four million euros per year.

While the protein shift is urgent in high-income countries, the need for a sustainable balance between plant-based and animal-based consumption is also evident in middle-income and low-income countries. This is especially the case as demand for animal-based foods will likely increase in these countries as they become more affluent.

Further Reading

Duluins, O., & Barret, P.V. (2024). A systematic review of the definitions, narratives and paths forwards for a protein transition in high-income countries. *Nature Food*, 5(January), 28–36. https://doi.org/10.1038/s43016-023-00906-7.

Pollicino, D., Blondin, S., & Attwood, S. (2024). *The food service playbook for promoting sustainable food choices*. Report. World Resources Institute, Washington, DC. https://doi.org/10.46830/wrirpt.22.00151.

Pyett, S., Jenkins, W., van Mierlo, B., Trindade, L.M., Welch, D., & van Zanten, H. (Eds.). (2023). *Our future proteins – A diversity in perspectives*. Amsterdam, The Netherlands: VU University Press.

Sutton, W.R., Lotsch, A., Prasann, A. (2024). Recipe for a liveable planet – Achieving net zero emissions in the agrifood system. In *Agriculture and food series*. Washington, DC: World Bank Group.

Willett, W., Rockström, J., Loken, B., Springmann, M., Lang, T., Vermeulen, S., & Murray, C.J.L. (2019). Food in the Anthropocene: The EAT–Lancet Commission on healthy diets from sustainable food systems. *The Lancet*, 393(10170), 447–492. https://doi.org/10.1016/S0140-6736(18)31788-4.

76 Choice Editing

Lewis Akenji and Magnus Bengtsson

Definition

Choice editing is about redesigning the "choice architecture" around lifestyles, focusing on the factors that most influence how people live and seek to satisfy their needs. Through this approach, decision-makers such as businesses and policymakers actively shape the range of products and services available in the marketplace, influencing their competitiveness and attractiveness, and thereby guiding consumer choices. This definition is in line with UNEP's use of the term.

Other uses of the term tend to focus more narrowly on influencing consumer choices, for example through the display and promotion of products in retail outlets. Such measures, which are similar to nudging (see **Green Nudging**), can be part of choice editing as described here, but often in combination with other interventions.

The choices available to consumers are always edited in some way. Choice editing is done by manufacturers, retailers, and service providers when deciding on product and service portfolios, designs, packaging, distribution channels, promotion, and pricing. The criteria businesses use for choice editing are typically based on profitability, available technology, brand image, or market appeal. They usually do not include – or give low priority to – environmental impacts and broader sustainability concerns.

Governments also practice choice editing. It is used, for example, to eliminate unsafe products or services or to incentivize the development of safer alternatives, which may otherwise not be made available. Traditionally, choice editing in public policy has primarily employed the filter of public safety, health, and security. However, considering the current environmental and social crises, governments are increasingly applying choice editing to promote and enable more sustainable ways of living and satisfying human needs.

A key feature of choice editing policies is the coordinated use of combinations of policy tools in the three areas outlined in Box 76.1. Editing out lifestyle and consumption options can be

---

Box 76.1   Elements of choice editing

Choice editing for sustainable lifestyles can be described as a three-pronged approach with the following elements:

- *Editing out*: discouraging high-impact non-essential lifestyle and consumption options or eliminating them from the market. This involves making such options more expensive, less appealing to consumers, and/or harder to access.
- *Editing in*: promoting low-impact alternatives to existing products and services, driving social innovation to ultimately make sustainable options the default. This involves

---

DOI: 10.4324/9781003584056-82

(i) stimulating innovation, (ii) introducing new low-impact options, and (iii) increasing the attractiveness, affordability, and availability of already existing ones.
- *Guaranteeing a social floor for consumption*: ensuring everyone can consume what is necessary to maintain human health, dignity, and social functioning (see **Fair Consumption Space**, **Consumption Corridors**).

controversial (see **Freedom of Choice**). Editing in feasible options and guaranteeing a social floor for consumption can thus be key to gaining public acceptance for editing out policies.

History

There is a long history of governments editing the field of choice for consumers in the public interest. Such policy efforts have ranged from non-intrusive measures, such as awareness-raising campaigns and consumer education, to outright product bans. The latter, which tend to be controversial if not carefully implemented, have been extensively used. For example, in the United States, where since the colonial era, laws prohibiting the sale of adulterated bread and other "unwholesome provisions" were enacted.

Over time, the criteria used for choice editing regulations have widened to include the protection of nature and the environment. For example, in the 1970s, governments worldwide established vehicle emission and fuel economy standards. More recent examples include (i) the phasing out of incandescent light bulbs from retail and wholesale markets, (ii) bans on plastic shopping bags and other single-use plastic packaging, and (iii) sale restrictions on cars with internal combustion engines.

Different Perspectives

Choice editing policies, especially those removing products from the market, can be viewed as paternalistic and heavy-handed. To gain acceptance, they need to be based on scientifically sound and transparent criteria and informed by deliberative processes. This can be seen as a form of collective self-governance – to restrict the availability of options that may be tempting to some consumers, and profitable for manufacturers and retailers, but deemed to have unacceptable impacts.

In democratic societies, government intervention in private consumption choices is often perceived as risky for politicians. Such interventions tend to be framed as restricting individual freedoms, go against neoliberal free-market thinking, and can be seen as a threat to economic growth (see Box 76.2). However, such concerns are based on a particular view of individual freedoms and the role of governments.

Box 76.2 Addressing objections regarding individual freedoms and government overreach

- Pursuing the common good is the responsibility of the political community – those mandated with governing. Governing includes the management of our commons, including the atmosphere and the Earth's life-supporting systems. Thus, the design and implementation

of choice editing policies is a way of guaranteeing the common good, especially when there is a scarcity of said resources or the risk that they may be severely damaged.
- Since the state is responsible for preventing discrimination and protecting individuals against infringements on their freedom by others, it has the right and the obligation to prevent individuals from consuming to such an extent that access to a sufficient quality and quantity of resources is denied to others. Given the significant asymmetries in power existing in the market and politics today, the need to exert this right and obligation to protect freedoms is, in fact, particularly important (see **Political Economy of Consumerism**). This is only reinforced by the growing urgency of the environmental crises. A clearer mandate for democratic choice editing could be part of renewed social contracts, or **ecosocial contracts**, which are increasingly called for by social and environmental advocates.

Choice editing is sometimes criticized for promoting a shift to more efficient and less-polluting products, while not addressing excessive volumes of consumption. However, this is mainly a matter of how the approach is used. Replacing one product with another might indeed fuel "green consumerism" and not achieve meaningful reductions in environmental impact. Hence, when considering alternatives to problematic consumption options, it is important to include not only greener products and other market-based options but also alternative need satisfiers. For example, promoting a shift away from **fast fashion** needs to look not only at how to make clothes less environmentally harmful but also at how the human needs that drive current consumption of fast fashion – for example, social recognition, self-identity, novelty – could be met differently, including through non-material and non-consumptive means (see **Social Norms**, **Social Practice Theory**, **Circular Economy and Society**).

Choice editing is related to approaches based on behavioral economics, such as nudging. However, while nudging aims to influence consumers' purchasing decisions within a given field of choice, choice editing represents a more radical redesign of the marketplace and the choice architecture facing consumers. In this sense, choice editing targets both the demand side (**behavior change**) and the supply of goods and services (provisioning systems, see **Foundational Economy**, **Product-Service-Systems**, **Universal Basic Services**).

Application

Implementing choice editing spans from a narrow focus on removing the worst products to more broadly redesigning choice architectures.

Removing the worst products is best seen with programs like Japan's "Top Runner" energy-efficient appliance program. Each year, the government rates major appliances for energy efficiency, and the top-rated appliances set the mandatory standard for future years, thus forcing the worst-performing models out of the market. This essentially creates a race to the top as there is a clear incentive for companies to make new models more efficient. As a result, in the early 2000s, TVs, air conditioners, and refrigerators became increasingly more efficient.

As a much broader approach, governments and institutions can shift the choice architecture – for example, building sidewalks and bike lanes, and implementing traffic-calming infrastructure (like speed bumps). To reduce car traffic and encourage walking, biking, or public transport use, London introduced the Congestion Charge, which drivers must pay to enter the charge zone in central London. In addition, vehicles that do not meet Ultra Low Emission Zone standards must pay an additional charge to drive in further restricted zones (see **Sustainable Mobility**, **Urban Planning and Spatial Allocation**).

One of the subtlest forms of choice editing is to alter the default options. Limiting the use of public spaces for highways and car parking promotes innovation for more sustainable transport; revising local government zoning laws, and size limits for housing construction, and raising the bar for minimum housing insulation standards defaults toward **sustainable housing**; raising ethical standards for animal farms and mandating reforestation and regeneration of lands previously allocated for cattle and pigs would encourage low-carbon and healthier diets.

Tiered pricing is also an example of shifting choice architecture. By increasing prices according to usage, tiered pricing expands a basic level of access for all but ratchets down consumption as prices increase along with total usage. In Durban, South Africa, for example, the first 750 liters of water per month is free (recognizing that access to water is a basic human right). But as consumption increases, so does the price. The cost of the next 20,000 liters jumps dramatically, and beyond that, the cost doubles again. Tiered pricing could easily be expanded to electricity and heating fuels, which in turn could further incentivize efficiency upgrades and solar panel installations on homes.

Still, on pricing, another approach is to make the least sustainable choices more expensive. Plastic bag charges are a good example. Rather than banning, which draws consumer ire as well as industry lawsuits and workarounds, mandatory charges for plastic bags can also reduce consumption significantly. Gentler changes can also help people get used to a shifting choice architecture. As more people shift to reusable bags to avoid the additional cost, when taking the next step of banning plastic bags, citizens are more comfortable with this further edit, having already gotten used to cultural shifts, such as bringing their own reusable bags.

It is not only governments that can implement significant choice edits. While businesses have mostly used choice editing to sell more products (such as cultivating planned obsolescence; see **Extended Producer Responsibility**), companies can also design products to be longer-lasting, and repairable (see **Ecodesign**), and, through everything from marketing and store design to shelf placement, can encourage more sustainable choices (see **Advertising**, **The Role of Business**). Stores can even take a further step of only stocking sustainable goods, whether removing virgin paper products, selling only sustainably harvested forest products, or selling only sustainably sourced seafood, as many companies have now committed to do. Companies can also shift default options. For example, utilities can make renewable energy the default source of electricity for new customers, or investment companies can make a green portfolio the default, which leads customers to automatically opt for the more sustainable option.

This article draws heavily from two publications:

Akenji, L., & Bengtsson, M. (2022). *Enabling sustainable lifestyles in a climate emergency*. United Nations Environment Programme. Available at: https://hotorcool.org/wp-content/uploads/2023/06/Enabling-Sustainable-Lifestyles_UNEP.pdf (accessed: 8 January 2025).

Akenji, L., Bengtsson, M., Toivio, V., Lettenmeier, M., Fawcett, T., Parag, Y., Saheb, Y., Coote, A., Spangenberg, J.H., Capstick, S., Gore, T., Coscieme, L., Wackernagel, M., & Kenner, D. (2021). 1.5-degree lifestyles: Towards a fair consumption space for all. *Hot or Cool*. Available at: https://hotorcool.org/wp-content/uploads/2021/10/Hot_or_Cool_1_5_lifestyles_FULL_REPORT_AND_ANNEX_B.pdf (accessed: 8 January 2025).

Further Reading

Gumbert, T. (2019). Freedom, autonomy, and sustainable behaviors: The politics of designing consumer choice. In *Power and politics in sustainable consumption research and practice*, pp. 107–123. Routledge.

Maniates, M. (2010). *Editing out unsustainable behavior. State of the world 2010: Transforming cultures, from consumerism to sustainability*. Routledge.

Sustainable Consumption Roundtable. (2006). *Looking back, looking forward: Lessons in choice editing for sustainability: 19 case studies into drivers and barriers to mainstreaming more sustainable products*. Sustainable Development Commission (UK).

77 Green Nudging

Ana Rita Farias and Marta Santos Silva

Definition

Green nudges are subtle interventions aimed at promoting environmentally friendly behaviors and reducing harmful environmental impacts. Unlike restrictive measures, nudges preserve choices while gently influencing behavior through changes in choice architecture (see **Choice Editing**, **Behavior Change**). For example, setting double-sided printing as the default option is a green nudge, specifically a *green default*, which leverages people's tendency to stick with the status quo. Green nudges are recognized as highly effective and are increasingly studied in behavioral economics, playing a significant role in policy discussions in many countries.

History

The rise of nudges in promoting environmental sustainability is tied to the development of behavioral economics in the 1970s, which integrates psychological insights into economic decision-making. Scholars like Herbert Simon, Daniel Kahneman, Amos Tversky, and Richard Thaler shaped this field. Simon's "bounded rationality" challenged the traditional view of humans as fully rational actors, while Kahneman and Tversky's Prospect Theory showed that people are more risk-averse when avoiding losses than when pursuing gains. By framing environmental decisions in terms of avoiding losses (e.g., the monetary and/or environmental costs), rather than in terms of potential gains, policymakers or environmental campaigns can leverage loss aversion to promote more sustainable behaviors. Thaler and Sunstein's nudging framework further advanced strategies to influence behavior without limiting choice (see **Freedom of Choice**). This foundation supports green nudges, which encourage sustainable choices by leveraging human behavior without requiring strict policies.

Different perspectives

Critics argue that nudges can infringe on individuals' privacy rights in two ways: (1) in relation to a lack of consent and (2) regarding independent choice-making.

First, by subtly influencing decisions without explicit consent, they raise concerns about paternalism, transparency, and fears of manipulation. For example, supermarket layouts are often designed to encourage impulse buying, by placing unhealthy snacks near the checkout. While this is a long-established practice, modern digital nudges, such as personalized product recommendations based on browsing data, take this a step further. These suggestions are tailored using sophisticated algorithms, triggering concerns about manipulation and the erosion of individual autonomy

DOI: 10.4324/9781003584056-83

(see **Information and Communication Technology**). This blurs the line between gentle persuasion and manipulative practices.

Second, even when consent is sought and obtained, nudges may compromise an individual's ability to make independent choices. For instance, someone might repeatedly encounter advertisements for energy-efficient appliances after searching for environmentally friendly products online (see **Advertising, Energy Consumption Behavior**). While this nudge toward sustainable purchasing appears beneficial, it can limit the consumer's awareness of the broader market and subtly curate their decision-making environment.

While traditional advertising shapes behavior, modern behavioral targeting using extensive data collection (e.g., browsing history, social media usage) has heightened privacy concerns. Companies like Amazon and Facebook use algorithms to nudge users toward specific actions, often without them realizing the influence.

Conversely, proponents of green nudges argue that modest interventions can yield significant environmental benefits while preserving individual choice. Defaulting to enroll employees in green energy programs, reusable bags in grocery stores, or placing recycling bins in more visible and convenient locations, has proven to increase ecofriendly behaviors.

However, the effectiveness of green nudges is under scrutiny given limited knowledge about the interventions' long-term effects, potential **rebound effects,** and **moral licensing**. For example, studies on reducing energy consumption through feedback systems – such as energy usage comparisons with neighbors – have shown positive short-term results, encouraging households to lower their consumption (see **Energy Consumption Behavior**). However, questions remain about whether these behaviors persist in the long term. There are also potential unintended consequences. For instance, research on moral self-licensing highlights how individuals who perform a positive act, such as recycling, may feel justified in neglecting other environmental responsibilities, like driving less or conserving water. This phenomenon complicates the sustainability impact of nudges, as observed in studies where participants engaged in green behaviors but simultaneously increased their carbon footprint elsewhere.

Debates also surround the acceptability of nudges and the accountability of those implementing them. The UK's "Nudge Unit" has been praised for public health improvements but criticized for lacking transparency, prompting calls for clearer ethical guidelines and transparency in nudge use to protect individual autonomy.

Application

Green nudges have successfully promoted sustainable behaviors, especially when combined with policy interventions like monetary incentives. Informational nudges, such as **ecolabels** and light-switch reminders, encourage energy conservation and reduce food waste, while measures like Clear Bag Policies boost recycling rates. Subtle tactics, such as strategically placing recycling bins or using visual cues like "watching eyes" around recycling points can effectively guide pro-environmental behavior. Promoting vegan food options (see **Protein Shift**) and setting green energy defaults have also shown success.

Green nudges can complement traditional regulations by encouraging voluntary environmental compliance. Cities use nudges like real-time bus schedules and colored sidewalks to promote public transit and walking. However, critics argue that green nudges focus on "weak" sustainable consumption, addressing small changes while ignoring systemic issues behind environmental degradation. This reliance on nudges can lead to **consumer scapegoatism**, where individuals bear the burden rather than broader systemic reforms. To effectively combat climate change and other environmental challenges, more profound and equitable transformations are necessary. This includes

Table 77.1 Pros and cons of green nudges (GNs)

Pros	Cons
Environmental benefits: GNs can improve the environment with minimal effort (e.g., reducing energy consumption and waste).	*Manipulation concerns*: Nudges may manipulate individuals by influencing their decisions without their explicit consent.
Preservation of choice: GNs guide individuals toward more sustainable choices while preserving their autonomy.	*Transparency issues*: By relying on subtlety, GNs can lead to a lack of transparency and deception.
Cost-effectiveness: Implementing GNs tends to be less costly and less intrusive than implementing regulations.	*Short-term impact*: Studies suggest that behavior changes may not be sustained over time.
Non-paternalistic approach: GNs often aim to benefit society and the environment without being overtly paternalistic.	*Erosion of privacy*: The rise of behavioral targeting and data-driven nudging raises concerns about the erosion of privacy and autonomy.
Social acceptability: Green nudges tend to be more socially acceptable than other interventions because they align with public interest in environmental conservation.	*Unintended consequences*: Some nudges may lead to moral self-licensing or consumer scapegoatism, placing the sustainability burden on individuals rather than systemic changes.

addressing the structural inequities that exacerbate environmental harm and disproportionately affect marginalized communities (see **Carbon Inequality, Climate Justice**). Acknowledging the need for deeper systemic change alongside the use of green nudges ensures a more comprehensive approach to sustainability, one that balances individual actions with broader social, economic, and environmental reforms.

Table 77.1 summarizes the debate surrounding the use and effectiveness of green nudges, highlighting their advantages on the one hand ("Pros") and the disadvantages on the other ("Cons").

Further Reading

Clot, S., Della Giusta, M., & Jewell, S. (2022). Once good, always good? Testing nudge's spillovers on pro-environmental behavior. *Environment and Behavior*, 54(3), 655–669. https://doi.org/10.1177/0013916 5211060524.

Santos Silva, M. (2022). Nudging and other behaviourally based policies as enablers for environmental sustainability. *Laws*, 11(1), 9. https://doi.org/10.3390/laws11010009.

Schubert, C. (2017). Green nudges: Do they work? Are they ethical? *Ecological Economics*, 132, 329–342. https://doi.org/10.1016/j.ecolecon.2016.11.009.

Sunstein, C.R., & Reisch, L.A. (2014). Automatically green: Behavioral economics and environmental protection. *Harvard Environmental Law Review*, 38, 127. https://doi.org/10.2139/ssrn.2245657.

Thaler, R.H., & Sunstein, C.R. (2021). *Nudge: Improving decisions about health, wealth, and happiness*. Yale University Press.

78 Ecolabeling

Maike Gossen and Johann M. Majer

Definition

Ecolabels are informational tools that indicate the environmental performance of a product or service. Manufacturers use ecolabels to communicate environmental information to other businesses and consumers. Ecolabels are usually based on a voluntary certification, which provides information on the extent to which a product or service meets predefined criteria such as energy efficiency, resource conservation, emissions reduction, and waste minimization. The aim is to reduce the information asymmetry between manufacturers (who know a lot) and consumers (who know relatively little) about the environmental impact of a product so that consumers can make more environmentally friendly choices (Noblet & Teisl, 2015). There are different types of ecolabels, depending on the specific scope (from a wide range of criteria to a single criterion) and the type of certifying institution (from third-party governmental labels to self-certified labels by companies).

History

In the 1960s and 1970s, public concern about sustainable consumption and production patterns grew (e.g., through the publication of the Club of Rome's *Limits to Growth* in 1972). Originally, ecolabeling schemes were intended as an alternative to government regulations, to enable consumers to make more sustainable choices and reduce the market share of products with poor environmental performance. In 1978, the world's first government ecolabel was introduced as an environmental protection measure in Germany: the so-called Blue Angel. The environmental label started with six product groups, including CFC-free spray cans, quiet lawnmowers, and reusable bottles.

In Agenda 21, participants in the UN Earth Summit in Rio de Janeiro (1992) described environmental labeling as an important instrument for promoting sustainable consumption and production by making it easier for consumers to buy resource-efficient and environmentally friendly products. Since then, ecolabels have proliferated. In 1992, for example, the European Commission decided to introduce the EU Ecolabel for the entire European market. At the same time, however, criticism of the actual effectiveness of ecolabels in reducing the environmental and social impact of consumption has steadily increased. According to the Ecolabel Index (2024), there are currently 456 ecolabels informing consumers from 199 countries about products and services in 25 sectors, such as fashion, tourism, and food. Figure 78.1 shows these milestones in the history of ecolabeling.

The International Organization for Standardization (ISO) provides standards to ensure the reliability and credibility of environmental claims and declarations. It distinguishes between three types of environmental labels (see Figure 78.2). Type I labels are verified by a third party, that is, an independent organization evaluates and certifies a product's environmental claims (e.g., Nordic Swan Ecolabel). Type II labels set standards for self-declared environmental claims that do

DOI: 10.4324/9781003584056-84

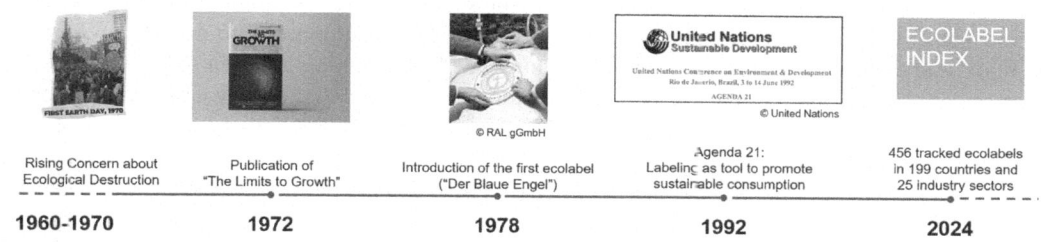

Figure 78.1 Milestones in the history of ecolabeling

Source: Own illustration, with permission by Club of Rome, Blauer Engel, and Nordic Swan

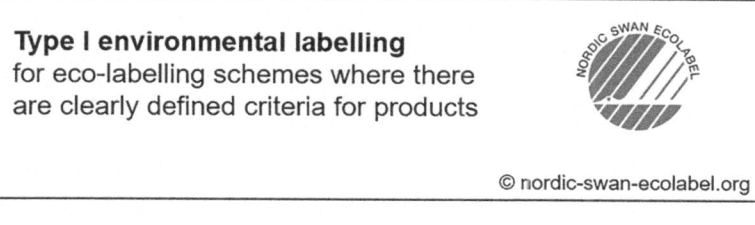

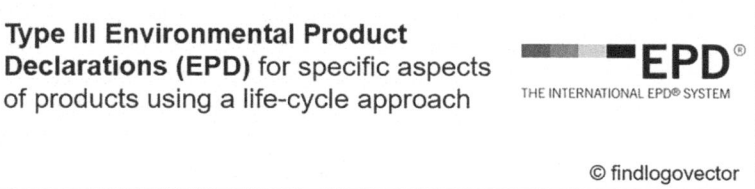

Figure 78.2 ISO standards for environmental labels

Source: Own illustration based on International Organization for Standardization. (2019). *Environmental labels*. Geneva: ISO

not require third-party verification (e.g., recycling label). These green claims are made directly by manufacturers or producers and are often scrutinized by consumer protection bodies for their potential to mislead consumers. Type III, on the other hand, is mainly a standard for the reporting of environmental data of products between companies (e.g., EPD).

Different Perspectives

More recently, the term sustainability label has come into use, referring to labels that focus on the environmental and/or social impact of a product in a broader sense. The effectiveness of sustainability labels varies greatly depending on contextual and personal factors such as age, education, and income. Research suggests that sustainability labels can positively influence consumer behavior and help translate people's attitudes toward sustainable consumption into actual purchasing behavior, that is, narrowing the **attitude-behavior gap**. For example, a recent study has shown that compared to positive and negative labels, graded labels (i.e., traffic light labels such as the Nutri-Score in the food sector) are most effective in guiding consumers to make more sustainable product choices (Thøgersen et al., 2024).

With the increasing number and heterogeneity of sustainability labels, criticism of labeling practices has further increased, to the point that they are considered by skeptics as **greenwashing.** A recurring debate revolves around the reliability and rigor of the criteria for sustainability labels. Critics argue that some labels prioritize marketability over actual environmental or social impact. This problem has been exacerbated by various private companies introducing their own self-declared sustainability labels (i.e., green claims), which often exaggerate the sustainability benefits of products. Concerns about greenwashing underscore the importance of rigorous certification processes, reliable data, and transparent communication to maintain the credibility and integrity of labeling schemes and certification institutions. In general, labeling programs need to be carefully designed to improve transparency, credibility, and accessibility and to increase consumer trust and engagement (Majer et al., 2022).

Another challenge is the proliferation of multiple sustainability labels within the same product category, leading to confusion and skepticism among consumers and hindering informed purchasing decisions (see e.g., **Choice Paralysis**). Researchers, governments, and companies have begun to consider integrated labeling approaches, which combine multiple environmental dimensions into one overarching meta-sustainability label. This may better address information asymmetry between producers and consumers, reduce information overload, and increase product comparability (Torma & Thøgersen, 2021).

Despite these efforts, sustainability labels often fail to achieve their primary goal of significantly reducing the negative environmental and social impact of consumer goods. One of the unintended consequences of sustainability labels is the **rebound effect**, whereby environmental impact "savings" from purchasing sustainable products in one domain lead to increased consumption in other domains (see **Moral Licensing**). Rather than encouraging genuinely sustainable practices, such as not buying products at all, these labels may maintain or exacerbate current unsustainable levels of consumption by inadvertently increasing consumers' desires or spreading green consumerism.

To overcome these obstacles, some scholars even suggest reversing the logic of sustainability labeling: mandatory non-sustainable labels for less sustainable options (i.e., negative labels, such as those used for tobacco), while sustainable products are given priority and easier market access. This change could better reflect the systemic and complex nature of sustainability challenges and promote strong rather than weak sustainability.

In summary, the potential of sustainability labels in addressing broader systemic issues such as overconsumption and ecological overshoot is limited, as the mere provision of product information is only one of many possible **behavior-change** interventions (e.g., social comparison, **choice editing**, incentives). Labeling can only be effective in changing consumer **norms** and habits in conjunction with government intervention, technological developments, and systemic change promoted by long-term policies (Horne, 2009).

Application

In addition to consumers, there are a variety of stakeholders with different interests in the sustainability labeling landscape. Producers try to gain a competitive edge through certification by balancing costs against the market benefits. In the certification process, they often face the technical challenge of calculating the product-specific environmental (e.g., carbon emissions) and social impacts (e.g., labor conditions) of products with complex or diverse supply chains. Retailers try to influence consumer choice by promoting products with sustainability labels to satisfy demand and improve brand reputation. Business customers and public procurers can prioritize sustainable products and services and set standards for environmental and social responsibility through their purchasing decisions. Policymakers advocate for regulatory frameworks that incentivize the introduction of sustainability labels and encourage market proliferation. Certification organizations strive for credibility by applying strict standards and involving stakeholders to ensure transparency.

Sustainability labeling as an information-based tool to promote sustainable consumption must be subject to political initiatives and regulations to be effective. Important developments in this area (e.g., the EU Green Claims Regulation) aim to combat **greenwashing** and ensure the integrity and credibility of sustainability labels. Such regulations underline the importance of reliable third-party certification, reliable data, and strict enforcement mechanisms to protect against misleading environmental claims and promote consumer trust.

At the same time, increasing digitalization and big data are enabling new tools (e.g., the European Digital Product Passport [DPP]) to improve transparency, traceability, and disclosure of information along the entire value chain. In addition, the Product Environmental Footprint (PEF) will standardize the measurement of environmental impacts and create a more consistent and transparent basis for sustainability claims in the EU. Such initiatives can provide consumers with comprehensive digital profiles of products, including environmental performance data, supply chain information, and end-of-life considerations. Machine learning models can be conducive to communicating high standards of sustainability in online environments at scale (see **Information and Communication Technology**). These technical trends may significantly impact the potential and pitfalls of sustainability labeling in the coming decades.

Further Reading

Horne, R.E. (2009). Limits to labels: The role of eco-labels in the assessment of product sustainability and routes to sustainable consumption. *International Journal of Consumer Studies, 33*(2), 175–182. https://doi.org/10.1111/j.1470-6431.2009.00752.x.

Majer, J.M., Henscher, H., Reuber, P., Fischer-Kreer, D., & Fischer, D. (2022). The effects of visual sustainability labels on consumer perception and behavior: A systematic review of the empirical literature. *Sustainable Production and Consumption, 33*, 1–14. https://doi.org/10.1016/j.spc.2022.06.012.

Noblet, C.L., & Teisl, M.F. (2015). Eco-labelling as sustainable consumption policy. In L. Reisch & J. Thøgersen (Eds.), *Handbook of research on sustainable consumption*. Cheltenham, UK: Edward Elgar Publishing. https://doi.org/10.4337/9781783471270.00031.

Thøgersen, J., Dessart, F.J., Marandola, G., & Hille, S.L. (2024). Positive, negative, or graded sustainability labelling? Which is most effective at promoting a shift towards more sustainable product choices? *Business Strategy and the Environment, 33*(7), 6795–6813. https://doi.org/10.1002/bse.3838.

Torma, G., & Thøgersen, J. (2021). A systematic literature review on meta sustainability labeling – What do we (not) know? *Journal of Cleaner Production, 293*, 126194. https://doi.org/10.1016/j.jclepro.2021.126194.

79 Advertising

Timothy de Waal Malefyt

Definition

Advertising motivates consumption. Its primary purpose has been viewed as persuading individual consumers to purchase products or services by offering, apparently, more telling benefits than rival products. This casts consumption as a rational and evaluative process that the consumer ultimately controls. However, highlighting the symbolic, hedonic, and often irrational elements of people and consumption, advertising has evolved into a more multi-media approach. It uses multiple communication channels to engage with consumers in versions of storytelling and narratives, projecting positive lifestyles and symbolic images that include user-generated publicity to build relations with consumers.

Advertising is one of the most powerful cultural forces at work in the contemporary world, affecting not only consumption but many aspects of human social behavior. In the United States, where over 70% of global spending on advertising takes place, about $300 billion is spent annually. The digital share of this, at about 60% of the total, has been expanding rapidly in recent decades.

History

Advertising historically dates to ancient civilizations in which merchants used papyrus, for instance, or painted advertisements on prominent surfaces. In the modern era, advertising flourished with the rise of the industrial revolution. In the 19th century, it became a major economic force, initially through magazines and newspapers, and later through radio, television, and, most recently, electronic media.

Advertisements of goods were initially based on product characteristics and their unique advantages. In the 1920s and 1930s, advertising methods began focusing on appealing to the psychology of consumers. Edward Bernays was the first to broadly employ symbolism in advertising. In his book *Propaganda*, he described how the "invisible" people can draw on psychological research to shape the thoughts and values of the masses. His success in "selling desire" made him a highly sought-after consultant to industry and government; even Hitler adopted his methods for spreading political propaganda.

In the 1940s, cultural theorists Theodor Adorno and Max Horkheimer condemned the increased influence of advertising on consumption and the accompanying expansion of mass production, as leading to the passivity of consumers and the commodification of culture. They called advertising a "hegemonic tool" that can easily evolve into ideological control. In the late 1950s, Vance Packard's *The Hidden Persuaders* posited that advertising subliminally evoked consumer desire through hidden or embedded messages, such as inserting "Drink Coca-Cola!" in split-second film frames, ostensibly prompting movie viewers to buy concessions.

DOI: 10.4324/9781003584056-85

Recent studies suggest that consumers are neither passive recipients nor cultural dupes of advertising messages. Consumers actively co-create and help produce consumption to foster human relations and identity formation. For instance, daily shopping by mothers for their families can be viewed as a devotional ritual or labor of love (Miller, 1998); advertising images can be appropriated by consumers for their own ends, such as forming ideologies that create youth subculture identities or launch national identity debates, such as Molson Canadian's beer commercial, "I am Canadian" (MacGregor, 2003). Advertising continues to evolve with cultural norms, and on social media, using influencers. Examples include healthy living and "body positivity" in Dove's campaign for "real beauty"; or racial injustice in Nike's ads with American football player Colin Kaepernick. Advertisers appropriate and reuse these populist innovations, repeating the cycle, and thus resisting social change.

Different Perspectives

On the positive side, advertising can inform consumers about helpful products or new medical treatments, or alert them to public service announcements. It can stimulate economic activity and contribute to consumer well-being (Stafford & Pounders, 2021). As a semiotic enterprise of the firm and culture at large, it shapes the consumption experience and the creation of ideals to aspire to in lifestyles and values (see **Social Norms**, **Values and Consumption**). Advertisements become part of our cultural discourse and the exchange of ideas in society, such as Nike's slogan, "Just do it", or L'Oreal's, "Because you're worth it". They reflect and instigate how people acquire and share cultural knowledge about products and services, and encourage consumers to become co-creators of meaning creation.

On the negative side, advertising can promote unnecessary consumption, drive rampant materialism, and affect people's health and lifestyles (including eating habits). By inflaming desire, it also increases dissatisfaction and arouses anxiety (see **Hedonic Treadmill**, **Choice Paralysis**). It operates as a "panacea and a pandemic [advertising] giveth and it taketh away" (Sherry, 2008, 88). Consumers receive thousands of messages through multiple media channels, with little regard for their best interests or a product's usefulness (Malefyt & Morais, 2017). For example, advertising of ultra-processed foods has been linked to a rapid increase in obesity, especially among children, which led to regulatory restrictions on such adverts in several European countries and Canada (though the United States largely relies on self-regulation by industry). Direct advertising of medicinal drugs to patients (illegal in most industrial countries, except in the United States) has been linked to increased use of drugs with few benefits.

Advertising's effects are indeed questionable. Nevertheless, it has become a powerful economic and cultural force, providing employment for millions and deeply embedded in institutions and culture. For example, public service organizations depend on private advertising dollars for significant portions of their operating budgets. Public radio in the United States is a case in point: 38% of the 2023 budget of the National Public Radio Corporation came from corporate sponsors. While these messages are built around NPR's social mission, the practice nonetheless supports businesses ultimately seeking to increase sales.

Application

Various forms of advertisements are neither all good nor all bad in an absolute sense. Rather, they are morally ambiguous. Promoting an art exhibit is desirable while showcasing fur for fashion is callous; advertising to children is reprehensible, but advocating dolls that support racial or ethnic identity can uplift young girl's self-esteem and pride (Malefyt & Morais, 2017).

Advertisers can and sometimes do help to educate consumers about responsible consumption, recycling, and waste reduction (see **Education for Sustainable Consumption**). They can and sometimes do advertise healthy foods and eating habits instead of ultra-processed foods. Still, although increasing, "purpose-driven marketing" represents only a small portion of the total. Patagonia, whose advertisements focus on outdoor lifestyles and long-lasting, repairable, and recyclable products "that give back to the Earth as much as they take", famously launched a "Don't Buy This Jacket" campaign in 2011. Ironically, the ad increased jacket sales, yet it also established the company as a leader in green marketing and prompted other clothing brands, like **fast fashion**-oriented H&M, to initiate recycling and clothing return programs (see **Extended Producer Responsibility**, **Product Returns and Right of Withdrawal**, **Greenwashing**).

The practice of mass advertising, which in the United States is a form of constitutionally protected commercial free speech, is unlikely to diminish any time soon. However, certain types of advertising may be blocked through organized campaigns modeled on the anti-tobacco campaigns of the 1980s and 1990s. The public health community campaigns to restrict the advertising of pharmaceuticals (in the United States) and ultra-processed foods. Some cities follow the model set by São Paulo, Brazil, which in 2006 banned all advertising in outdoor public spaces (Clean City Law).

Meanwhile, environmental activists call for bans on ads for fossil fuel products and carbon-intensive services (e.g., cars, flights, and cruises). Starting in 2025, The Hague in the Netherlands will implement such a policy, followed by other cities, like Sheffield, Edinburgh, and Amsterdam. France is the first European country to ban advertising for all energy products related to fossil fuels and recently prohibited ultra-fast fashion advertising. By measuring spending on advertisements for carbon-intensive products and services, the European Climate Neutrality Observatory tracks progress toward the EU goal of climate neutrality by 2050.

Further Reading

MacGregor, Robert M. (2003). I am Canadian: National identity in beer commercials. *Journal of Popular Culture; Oxford*, 37(2), 276–286. https://doi.org/10.1111/1540–5931.00068.

Malefyt, T.d.W., & Morais, R.J. (2017). Advertising anthropology ethics. In *Ethics in the anthropology of business*, pp. 104–118. Routledge. https://doi.org/10.4324/9781315197098-7.

Miller, D. (1998). *A theory of shopping*. Cambridge: Polity Press.

Sherry, J.F., Jr. (2008). The ethnographer's apprentice: Trying consumer culture from the outside in. *Journal of Business Ethics,* 80(1), 85–95. https://doi.org/10.1007/s10551-007-9448-7.

Stafford, M.R., & Pounders, K. (2021). The power of advertising in society: Does advertising help or hinder consumer well-being? *International Journal of Advertising*, 40(4), 487–490. https://doi.org/10.1080/0265 0487.2021.1893943.

80 Greenwashing

Panayiota Alevizou and Claudia E. Henninger

Definition

Greenwashing is a form of selective or deceptive communication whereby an organization presents itself as being more environmentally friendly than it actually is, exaggerating or misrepresenting the extent or impact of its environmental initiatives and practices. Greenwashing capitalizes on a growing societal interest in sustainable organizations, products, and services, without necessarily implementing genuine sustainable business models or mitigating the organization's ecological and social impact. In essence, greenwashing enables organizations to portray themselves as environmentally conscious and gain the perception of being good corporate citizens, without making substantial changes to their operations.

Greenwashing is a concept that encompasses various aspects of corporate operations and product life cycles. It is a dynamic phenomenon that co-evolves with market trends and consumer preferences and involves multiple stakeholders, making it difficult to establish a universally accepted definition. Greenwashing comes in different forms, including intentional (see Box 80.1) or unintentional, direct or indirect, factual-based or "aspirations"-focused, or visual and/or textual. It can use multiple media to spread misinformation, which can result in delaying and sabotaging global priorities and progress toward the Sustainable Development Goals (SDGs). Many platforms focused on sustainability collect and disseminate examples of greenwashing (e.g., ClientEarth, Greenwashingindex, Earth.org, or The Sustainable Agency).

Box 80.1   Examples of intentional direct greenwashing

Nature-inspired packaging and branding: Using symbols, visuals, and language associated with nature on products or services can give the impression that they are made from natural materials or cause less environmental harm, even if no evidence is provided to support these claims.

Irrelevant claims: Labeling a product as "free from" certain chemicals when those chemicals are not typically used in that type of product, creates a misleading impression of environmental superiority.

Overemphasizing a minor ecofriendly feature or improvement: Highlighting a small environmentally friendly aspect of a product, such as recyclable packaging, while ignoring a more significant environmental impact of the product, such as the use of harmful ingredients or unsustainable production and distribution processes.

DOI: 10.4324/9781003584056-86

History

The term "greenwashing" was coined in the 1980s by environmental activist Jay Westerveld in an essay inspired by the hotel industry's "save the towel" movement. Hotels encouraged guests to reuse and reduce towel use to conserve resources, but Westerveld argued that this was more about reducing costs than genuine environmental concern. The essay sparked a broader awareness of deceptive marketing practices where companies made unsubstantiated or exaggerated claims about their environmental efforts.

Greenwashing gained momentum in the 1990s, as sustainability became a significant global concern. Companies across various industries began to recognize the potential of appearing to be *sustainable,* and engaged in marketing this. However, many of these sustainability-inspired claims were superficial or outright misleading, resulting in consumer skepticism and calls for greater transparency and accountability. Additionally, the rise of social media and increased consumer awareness have further pressured companies to substantiate their claimed environmental credentials, making it harder to "claim sustainability", without facing backlash (see **Boycotting and Buycotting, Subvertising**). To that end, regulatory bodies, consumer protection groups, and NGOs have introduced guidelines for organizations engaging in sustainability claims and for consumers trying to decipher such messages.

Governmental efforts to combat greenwashing practices have intensified globally. In 2024, the European Parliament formally approved a new directive requiring EU member states to apply stricter rules against misleading environmental marketing claims made by companies. Similarly, in 2023, the Australian Competition and Consumer Commission published guidance on environmental claims, while the US Federal Trade Commission updated its Green Guides to deal with a rising number of greenwashing cases.

East Asian countries also appear to be following suit with similar regulatory approaches. These actions by governments and regulatory bodies across different regions indicate that greenwashing has become a widespread issue transcending companies, industries, and geographical boundaries. Yet, the implementation and enforcement of these regulations have been inconsistent across different regions.

Different Perspectives

Communicating sustainability has been associated with various advantages for organizations. For instance, organizations communicate sustainability to enhance their legitimacy by demonstrating their commitment to, for instance, the UN's SDGs (see **The Role of Business**). This type of communication can build trust with the public and regulators and lead to greater market performance, as consumers increasingly favor more sustainable options. Additionally, communicating sustainability efforts can provide a competitive advantage and differentiation. Finally, showcasing sustainability initiatives attracts stakeholders, including investors, employees, and partners, who increasingly prioritize environmental and social governance (ESG) criteria, thus ensuring long-term collaboration.

From a corporate perspective, communicating commitments to sustainability can be seen as increasingly challenging given potential allegations of greenwashing. Historically, there have been few legal repercussions for greenwashing, largely because consumers are often unaware of the practice and its underlying science. Additionally, the absence of a universally accepted definition and standardized auditing systems further enables companies to exploit this practice. Despite the risk of backlash, many organizations still are keen to communicate their commitment to sustainability, as in many cases the perceived benefits of being seen as a sustainable organization by the public often outweigh the risks of such accusations.

A growing number of companies engage in "green hushing" or "brownwashing": the practice of remaining silent about their sustainability practices and, as such, avoiding market and stakeholder pressures. These practices may, however, also have negative impacts on firms, such as lower financial performance and reputation. Consequently, companies face a dilemma (and paradox) when it comes to effectively communicating their sustainability efforts.

Application

Greenwashing comes in different forms, ranging from product-level greenwashing to corporate promises for future sustainability commitments. In the late 2000s, TerraChoice, a consulting firm, released a seminal report titled *The Six Sins of Greenwashing*, which brought the issue to the forefront of attention for policymakers, organizations, and academics, sparking discussions and receiving extensive media attention. Greenwashing as both a practice and study has evolved greatly since the time of the report and organizations have become more *sophisticated* in their approach to greenwashing. Greenwashing constantly evolves in both concept and practice, thus posing ongoing challenges for consumers and policymakers trying to keep up and respond.

Greenwashing can be direct and indirect. Organizations engage with *direct greenwashing* when they intentionally practice it on the level of a product, brand, or corporate image. The most prominent forms of direct greenwashing involve the use of misleading **advertising** claims and deceptive labeling. They often employ visual imagery (e.g., nature scenes, animals) and sustainability-inspired language (e.g., "ecofriendly", "natural") to create an illusion of environmental responsibility. Claims may be vague, unsubstantiated, and/or false. Direct greenwashing can also manifest through corporate statements or visions that set overly ambitious future goals and commitments (Box 80.1).

Indirect greenwashing, on the other hand, can occur when a company is associated with suppliers, distributors, and investors who may engage in questionable practices; or with **ecolabeling** and certification organizations that have been criticized for having lower auditing standards and criteria compared to others. This may be *intentional* or *unintentional*. *Intentional* indirect greenwashing companies may, for instance, deliberately omit their associations with questionable third-party organizations.

Greenwashing is a complicated phenomenon that needs to be carefully investigated and managed to avoid delaying progress toward sustainability goals. Looking ahead, tightening regulations may lead organizations who are greenwashing to eventually "walk the talk", but also guide them to substantiate sustainability claims. To gain and maintain market trust, organizations may view sustainability communications as integral to their ethos and core value system, fostering transparency and accountability while steering clear from the "sins of greenwashing".

Further Reading

Bowen, F. (2014). *After greenwashing: Symbolic corporate environmentalism and society*. Cambridge, England: Cambridge University Press.

Delmas, M.A., & Burbano, V.C. (2011). The drivers of greenwashing. *California Management Review*, 54(1), 64–87. https://doi.org/10.1525/cmr.2011.54.1.64.

Greer, J., & Bruno, K. (1996). *Greenwash: The reality behind corporate environmentalism. Reality behind corporate environmentalism*. New York: Apex Press.

Montgomery, A.W., Lyon, T.P., & Barg, J. (2024). No end in sight? A greenwash review and research agenda. In *Organization and environment*, Vol. 37, Issue 2. https://doi.org/10.1177/10860266231168905.

Seele, P., & Gatti, L. (2017). Greenwashing revisited: In search of a typology and accusation-based definition incorporating legitimacy strategies. *Business Strategy and the Environment*, 26(2), 239–252. https://doi.org/10.1002/bse.1912.

The steel bumper is heavier, but overall it requires less energy (and thus emissions) than the aluminum version, due to the high energy inputs required in aluminum production. It is also cheaper in production. From a production point-of-view, the steel version is therefore the best ecological choice.

However, in the use phase, the weight of the bumper has a large impact on fuel efficiency and carbon emissions. When all the calculations are done, aluminum emerges as the optimal ecological choice.

History

Approaches to dealing with environmental issues inherently differ from country to country, and from continent to continent. The Western world plays a significant role in both the problem and the solution because of its early industrialization and policy impacts.

The emergence of the grassroots environmental movement in the 1960s brought environmental awareness to the masses (see **Social Movements**, **Grassroots Innovation**). In the 1970s, in response to the oil crisis, the focus was mainly on reducing energy consumption. In the early 1980s, the discovery of hazardous waste dumps directed attention to waste management, largely by incineration. Guided by the principle of "cradle-to-grave", end-of-pipe techniques made their appearance to comply with government regulations seeking to reduce air emissions and other

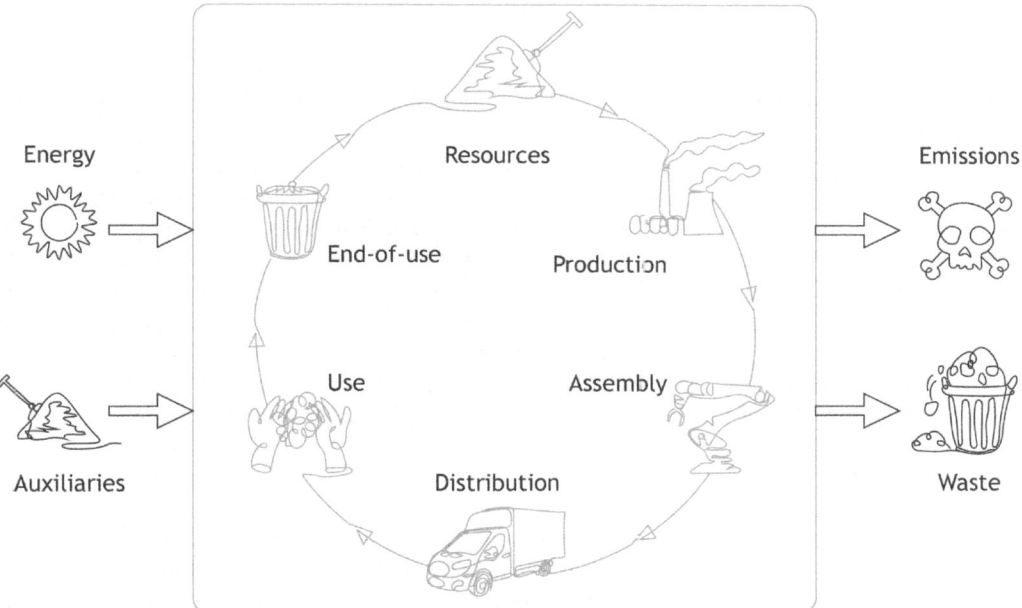

Figure 81.1 Life-cycle thinking

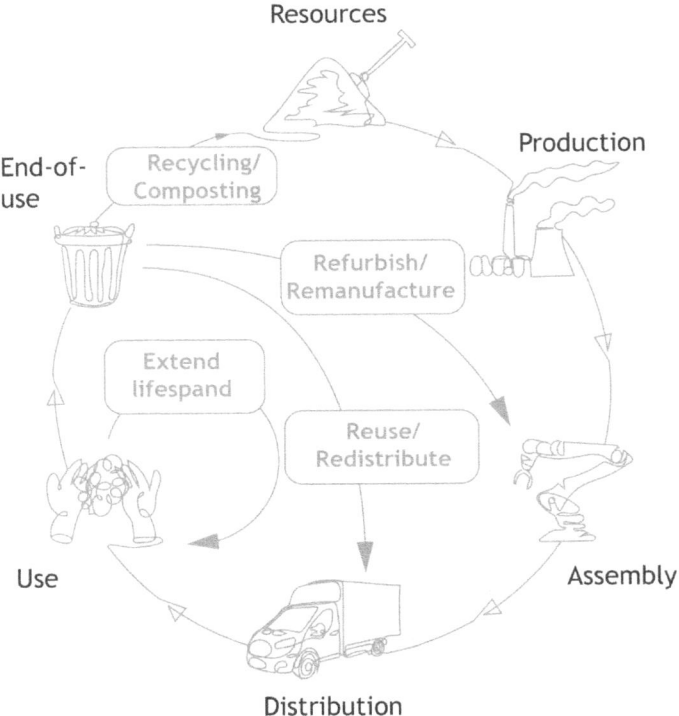

Figure 81.2 The link between ecodesign and product life cycles

Source: By authors

discharges to the environment. Overall, the focus was on harm reduction and process-orientation, not product-orientation.

In the 1990s, attention shifted upstream to product-integrated environmental care. The terms "ecodesign", "Design for Environment", "environmentally friendly design", "ecoefficiency", and others were given more and more substance. In the beginning, ecodesign was mainly focused on designing products made from waste, bio-based materials, or the development of non-fossil-energy-consuming products such as electrical products based on solar panels. In the 2000s, the ecodesign-based approach became increasingly formalized and standardized, supported by the popularization of the cradle-to-cradle concept. The European Ecodesign Directives and the European Green Deal, as well as the elaboration of the circular economy model by the Ellen Mac-Arthur Foundation and others, gave a major boost to the wide adoption of ecodesign.

Different perspectives

Ecodesign is not a specific extra layer in the design process, but an integral part of it. It is sometimes misused as the "green sauce" on top of the product, with many such examples emerging in packaging (see **Greenwashing**). Some replace single-use recyclable plastic packaging with complex paper-based non-recyclable packaging. Elsewhere, we regularly see the claim of plastic-free packaging, while it is made from bio-based plastic. Ecodesign efforts alone cannot guarantee sustainability; other stakeholders should adapt as well, including consumers in their behavior and decision-making by policymakers (see **Behavior Change, Co-Benefits of Climate Policy**). There are many incorrect interpretations and confusion regarding the ecodesign approach. In Box 81.32, we illustrate the most important such misconceptions.

Application

The practical implementation of ecodesign is based on the use of guidelines, often portrayed as "Design for X"-guidelines. X stands for the different design strategies, such as "Design for Disassembly", "Design for Repair", etc. These are usually connected to one or more life cycle phases. Some examples are presented in Table 81.1.

Box 81.2   Common misconceptions about the ecodesign approach

Misconception: Ecodesign ensures sustainability.

Correction: Ecodesign only focuses on the ecological aspect of sustainability. Of course, social and economic issues relevant to sustainability are also taken into account during the design process, but these are not the main targets of ecodesign.

Misperception: Ecodesign is mainly for energy-consuming or energy-producing products.

Correction: Ecodesign applies to all kinds of products in all different sectors: consumables, consumer goods, or capital goods.

Misperception: Ecodesign and circularity are synonyms.

Correction: Design for circularity is part of ecodesign. Design for circularity focuses on closing the loop, while Ecodesign also focuses on the environmental impact of each step of the life cycle of the product (input energy and materials and output of waste and emissions.

Misperception Ecodesign means selecting ecological materials.

Correction: Ecodesign is more than material selection. The ecological impact of a product is defined by much more than the material alone. Aspects like physical connections, business models, product architecture, looks, etc. are all part of ecodesign as well (see **The Role of Business**). Furthermore, the environmental impact of a material is often only evaluated for the production phase. The material choice has consequences for all life cycle phases, so should be evaluated accordingly (e.g., see Box 81.1).

Misperception: Ecodesign is making products from bio-based materials.

Correction: Bio-based materials are often portrayed as the holy grail in sustainable product design. Some definitely have sustainable potential, but bio-based materials also have environmental issues. No fundamentally ecological material exists; every material has its impact. It is all about the sustainable use of materials.

Misperception: Ecodesign and circularity are mostly focused on the recyclability of the materials.

Correction: Ecodesign supports prolonging the use phase of the product. Recycling has its limits and comes at a price, both economically and ecologically. Also, there may be an important difference between theory and practice (see **Attitude-Behavior Gap**).

Misperception: The environmental impact and circularity can be fully calculated.

Correction: The environmental impact and circularity cannot be narrowed down to a quantitative figure. Life Cycle Assessment (LCA) studies incorporate life cycle thinking but remain based on many estimations, assumptions, and general data. Also, inherently different types of impacts are matched through weighting factors. In the end, the LCA results have a big uncertainty range and can be manipulated. This does not mean that the designer should not use quantitative numbers at all to evaluate impacts but be aware of complex comparisons. Either way, the designers should use qualitative methods, which are based on ecodesign guidelines (Design for X guidelines).

Table 81.1 Examples of design for X and ecodesign

Product name	Product type	Design for X	Intervention
Slimbox	Packaging machine & software	Logistics	Made-to-measure cardboard packaging to reduce packaging waste production and CO_2 emissions while optimizing costs
Frame.work	Modular laptop	Longevity	Laptop with replaceable and upgradeable parts architecture, allowing for longer use
Juunoo	Room separator walls	Disassembly	Interior wall system that can be disassembled and reassembled so reuse is possible
The New Raw	3D-printed furniture	Recycling	3D-printed large objects made from recycled plastic, also in mono-material
Atlas Copco Reman	Compressors & generators	Remanufacturing	Business model where they remanufacture compressors and generators, reducing the need for new resources
Bounty Select-a-size	Paper towels	Sustainable behavior	Additional perforations let the user choose exactly the amount (s)he needs
Signify Light as a Service	Corporate lighting	Energy Efficiency	The vendor provides and maintains energy-efficient lights, and optimizes the lighting scheme
Fjallraven Greenland Wax	Wax for outdoor clothing	Maintenance	This company sells materials for the care and maintenance of clothes, so they last longer

The main challenge in ecodesign is to find a compromise between the minimization of the environmental impact and the other product requirements. For instance, optimizing energy use may induce a larger use of critical materials (see **Rebound Effect**, **Energy Overshoot**, **Stocks Versus Flows**) or may weaken the functionality of the product. Proper ecodesign finds the right balance. As this is always context-dependent, we must stress that there is no universal ecological solution or material. There is always some form of impact. Ecodesign limits this impact and enables the transition to the circular economy, as stated in the EU Green Deal and the Ecodesign for Sustainable Products Regulation.

Further Reading

Bakker, C., & Den Hollander, M. (2014). *Products that last: Product design for circular business models*. Amsterdam: BIS Publishers.

Braungart, M., & McDonough, W. (2002). *Cradle to cradle: Remaking the way we make things*. New York: Search Knowledge.

Circular design: Open source list of book recommendations. (2024). Available at: https://www.circulardesign. it/publications/ (accessed: 8 January 2025).

Ellen MacArthur Foundation. (2024). Available at: www.ellenmacarthurfoundation.org (accessed: 8 January 2025).

Van Doorsselaer, K., & Koopmans, R.J. (2020). *Ecodesign: A life cycle approach for a sustainable future*. Munich: Hanser.

82 Extended Producer Responsibility

Jessica Stubenrauch, Maike Demandt, and
Joachim H. Spangenberg

Definition

Extended producer responsibility (EPR) is a product-oriented approach in environmental policy. While producers have a moral responsibility for the environmental impacts of products, up until recently the legal responsibility has been different in two ways: it is time-limited, and it is usually restricted to guaranteeing the functioning of a product, regardless of the impacts of that functioning and the necessary post-consumer waste management. Consequently, EPR extends (a) responsibilities over time and (b) guarantees beyond mere functionality.

The OECD (2001) defines EPR as an environmental approach that extends the producer's physical and/or financial responsibility for a product to the post-consumer stage of a product's life cycle. There are two intentions behind this: firstly, the direct mandatory effect that EPR shifts the responsibility for taking back, transporting, recycling, or disposing of products from municipalities and consumers directly to the producer. This allows the costs of waste management to be internalized in the product cost, encouraging consumers to choose more environmentally benign goods. Secondly and indirectly, EPR aims to encourage producers to integrate environmental considerations into the design of their products – often a logical consequence of EPR, as producers want to minimize the full costs of a product.

In this respect, EPR is related to the concept of **ecodesign** while they are not the same. Ecodesign is the systematic integration of environmental aspects into product design and development. This requires environmental aspects to be considered at the earliest stages of product development to minimize negative environmental impacts throughout the product's life cycle, including product design and supply chain. EPR, on the other hand, stresses the end-of-life impacts.

History

The concept was first defined in a report to the Swedish Ministry of the Environment by Thomas Lindqvist in 1990, as a response to the growing volumes of waste generated by increased consumption. Several other institutions, such as the OECD (2001) or the European Commission, published guidelines in subsequent years in which the concept was also further elaborated (see Box 82.1). EPR is linked to other concepts, such as Corporate Social Responsibility (CSR), **Circular Economy and Society**, and Environmental, Social, Governance (ESG) criteria, as its goals align with the principles underpinning these frameworks.

The concept was first implemented in 1991 in Germany, where it was introduced by the packaging ordinance under the circular economy law. Other European countries, such as Austria, the Netherlands, France, and Sweden, followed suit and introduced similar obligations for packaging manufacturers. Regarding the second dimension of EPR, in 2005, the European Commission enacted a directive setting out requirements for the ecodesign of energy-using products. This

DOI: 10.4324/9781003584056-88

directive has been further developed and currently includes over 30 product categories and criteria that go beyond energy efficiency, such as durability or recyclability. In addition, the EU Corporate Sustainability Due Diligence Directive (CSDDD) defines the social and environmental obligations of manufacturers along the value chain.

In the United States, efforts have been made to introduce EPR policies at the national level; however, these initiatives have primarily materialized at the state level. Notably, the product categories encompassed by EPR systems in the United States diverge from those in European countries, as they focus on waste streams such as thermostats and paints, and an EPR system for packaging was not implemented until 2021. Canada unveiled a national action plan for EPR in 2009, aiming to enhance waste prevention strategies and to embed the principle within Canadian waste policy frameworks. Since the early 2000s, EPR systems were also introduced in the Asia-Pacific region and South America. South Africa and Kenya passed EPR laws for packaging, among other things, in 2020. In addition, several countries (including the United States, Japan, and Australia) have adopted legislation similar to the EU Ecodesign Directive, thus focusing not only on the end-of-life phase but also on other stages of the product life cycle. However, internationally, there have also been multiple efforts to water down the implications of such legislative developments.

Box 82.1. Three examples of EU legislation supporting the EPR

Waste Framework Directive (WFD)

The directive introduces an EPR with extended waste prevention requirements and mandatory minimum recycling rates for municipal waste. Member States are encouraged to take legislative or non-legislative measures to implement the EPR and to ensure the acceptance of returned products and waste at the end of their life. This involves the subsequent management of waste and the financial responsibility for these activities. Measures may include an obligation to provide publicly available information on the extent to which the product is reusable and recyclable. The Directive thus places a strong emphasis on the implementation of the first dimension of the EPR.

Ecodesign for Sustainable Products Regulation (ESPR)

The Regulation entered into force in July 2024. It aims to improve the circularity of products and to ensure that they last longer, use energy and resources more efficiently, are easier to **repair** and recycle, contain fewer hazardous substances, and incorporate more recycled content. This is to be achieved, for instance, by measuring carbon and environmental footprints and providing access to data and information throughout the product value chain through a digital product passport or a product label. Rapid implementation of the Regulation would therefore significantly support the design dimension of the EPR.

Corporate Sustainability Due Diligence Directive (CSDDD)

The Directive implements comprehensive due diligence processes to mitigate negative human rights and environmental impacts in value chains within and outside Europe. It enforces business decisions that consider human rights and the climate and environmental impacts of the entire production process (see **The Role of Business**). This strengthens the incentive for management to minimize negative impacts from the outset, that is, already in the design phase.

Different Perspectives

Although many aspects of EPR are already considered in different regulatory measures, in practice it is often reduced to a financial tool for the waste treatment of products. For example, most European Producer Responsibility Organizations for packaging do not reward circular design and thus do not incentivize manufacturers to change their design. While economic and environmental criteria are inherent to EPR regulations, social criteria (like bans on child labor) have been added by the CSDDD for the production process, but do not affect the design dimension of EPR. Since the design of the products affects what consumers can and will do with them, it is important to also consider behavioral implications when establishing the design dimension of an EPR system (see **Behavior Change**, **Attitude-Behavior Gap**, **Social Practice Theory**).

While current EPR regulations are a significant step toward responsibility for life cycle impacts, four kinds of improvement deserve to be highlighted to make EPR effective under 21st-century conditions:

i. *Beyond discounting:* With increasing knowledge of long-term damages (e.g., from per- and polyfluoroalkyl substances [PFAS] and other 'eternity chemicals"), and the future cost they will cause (like the "eternity cost" of mining), it becomes obvious that damage assessments based on cost discounting ignore environmental realities. As polluting facilities do not necessarily live long enough to pay for the damage they have caused, upfront payments into restoration funds could be a condition for licensing.

ii. *Understanding constraints:* While in the past, relative scarcities could be balanced by the market (every demand is fulfilled given sufficient purchasing power), this will not be the case in a resource-constrained economy, as the projected demand for critical raw materials vastly exceeds the known resources. EPR requires a rethinking of the goods produced, and their contribution to the well-being of society, beyond a few high consumers: what to build instead of private jets?

iii. *Accepting boundaries:* The EU has defined its aim as human well-being within the Planetary Boundaries (see **Doughnut Economy**). However, products accumulate in the environment, so transgressing them is not an individual, but a cumulative collective act. Collective responsibility could be defined by revitalizing the bubble concept, known in air pollution regulation since the 1970s. Rather than limiting the emissions from individual sources, the concept attributes accountability to all producers under the bubble. EPR would then imply collective limits, instead of individual limits per firm, with obligations allocated to individual firms based on regional carrying capacity and company performance.

iv. *Responsible corporate citizenship:* beyond the frequent focus on Corporate Social Responsibility (CSR), citizenship – a concept from the 1990s – includes all activities of a company intended to improve societal welfare. It should be revitalized and extended to cover accountability for corporate actions trying and/or succeeding to influence political decisions and social processes.

While being an important tool, EPR cannot be a stand-alone policy; to effectively reduce consumption, it must be part of an integrated policy mix with technical regulations, environmental taxes, corporate social citizenship, and others (see **Personal Carbon Allowance**, **Co-Benefits of Climate Policy**, **Ecosocial Contract**). For political and market competition reasons, flexible but binding global environmental governance is needed.

Application

Despite still-growing waste volumes, EPR has been considered an effective instrument for the end-of-life management of a wide range of products. However, it is mainly developed for industrial countries and application to countries of the Global South requires adaptations and effective institutions for enforcement. Moreover, EPR is often implemented as a recycling scheme, instead of upstream interventions and for rethinking the entire value chain.

Apart from packaging, the EU prescribes mandatory EPR systems for other product categories, including end-of-life vehicles and waste electrical and electronic equipment. Some European countries have regulations for other product categories as well, such as tires or medical products. It is important to note that there is no universal principle for all product groups and all countries. The design of an EPR system for medicines is not transferable to textiles – nor is an EPR system in Spain, say, applicable to another country. Consequently, there must be individually adaptable models for each product type and region, see Box 82.2 for an example.

Box 82.2   Implementing EPR in practice

EPR for packaging with a focus on ecomodulation: Citeo France

Within the French EPR system for household paper and packaging, the non-profit Producer Responsibility Organisation Citeo undertakes the respective duties for around 28,000 producers. In exchange, the producers have to pay a contribution and provide a declaration on the amount of products they have put on the French market.

The total contribution a producer has to pay depends not only on the amount of products and materials used but also on the ecodesign and a company's efforts to promote circularity. The calculation for the individual contribution includes

1. *A weight-based contribution*, which is based on the material used. The fee indicators for the various materials reflect the costs of collecting, sorting, and processing the respective material at the end-of-life.
2. *A contribution per unit*, which aims to promote the reduction of various components to simplify sorting and processing.
3. *Bonus or penalty* options, which are introduced to reduce packaging at source and eliminate components hindering recycling, among other things. Bonuses are granted, for instance, if refill options are offered with a weight reduction of at least 33% compared to the original product, or if the packaging is reusable. Penalties are imposed, e.g., if the material contained reduces the quality of recycling.
4. *Incentives*, that is, financial relief, are granted if the packaging contains recycled material.

These contributions incentivize applying ecodesign principles to the products and are used to finance research and development projects for ecodesign, solutions for collection, re-use, recycling, as well as recovery.

Further Reading

Ekvall, T., Hirschnitz-Garbers, M., Eboli, F., & Śniegocki, A. (2016). A systemic and systematic approach to the development of a policy mix for material resource efficiency. *Sustainability*, 8(4), 373. https://doi.org/10.3390/su8040373.

European Commission. (2014). *Development of guidance on extended producer responsibility (EPR)*. Available at: https://ec.europa.eu/environment/pdf/waste/target_review/Guidance%20on%20EPR%20-%20Final%20Report.pdf (accessed: 8 January 2025).

Grabs, J. (2023). Business accountability in the Anthropocene. *Environmental Policy and Governance*, 33(6), 615–630. https://doi.org/10.1002/eet.2081.

Macdonald, K. (2023). Accountability in the Anthropocene: Activating responsible agents of reform or futile finger-pointing? *Environmental Policy and Governance*, 33(6), 604–614. https://doi.org/10.1002/eet.2084.

OECD. (2001). *Extended producer responsibility: A guidance manual for governments*. Paris: OECD. https://doi.org/10.1787/9789264189867-en (accessed: 8 January 2025).

83 Product Returns and Right of Withdrawal

Marta Santos Silva

Definition

The Right of Withdrawal is a consumer protection mechanism that allows consumers in the EU to return goods purchased online within 14 days of receipt, without needing to provide a reason. This right provides an exception to the general contractual principle that contracts are binding once concluded. It enables consumers to inspect and test products before deciding whether to keep them or cancel the purchase, thus terminating the contract.

The right to return products, as well as to withdraw from a purchase agreement, is a cornerstone of modern consumer protection, reflecting evolving standards of fairness in commerce. The history of this right can be traced through a series of legal, social, and economic shifts, each spurred by consumer dissatisfaction and the growing complexity of market transactions, particularly in the online era.

History

Before the 20th century, the ability to return products was largely determined by the discretion of retailers. In many economies, the principle of *caveat emptor* ("let the buyer beware") dominated commerce, placing the burden of assessing product quality entirely on consumers. Transactions were often final, and the possibility of returning defective or unsatisfactory goods depended on the goodwill of individual merchants rather than a structured legal framework.

In the early 20th century, growing consumer dissatisfaction with unfair business practices led to reforms that laid the groundwork for return policies and withdrawal rights. With the rise of consumer awareness, businesses began to recognize the importance of corporate social responsibility (CSR), acknowledging that ethical practices in sales and returns were essential to building consumer trust. In Europe, by the mid-20th century, consumer protection organizations, particularly in the United Kingdom and Germany, advocated for stronger rights to ensure that consumers could return defective or unsatisfactory goods. These early policies established the idea that consumers should have a remedy when products failed to meet reasonable expectations. They marked a shift toward CSR as companies increasingly viewed responsible return policies as part of their commitment to fair treatment and transparency.

The concept of a Right of Withdrawal – allowing consumers to change their minds about a purchase within a set period – first emerged in the context of distance selling. As mail-order catalogs grew in popularity throughout the mid-20th century, new consumer protection challenges arose. Unlike in-person shopping, distance selling removed the buyer's ability to physically inspect goods before purchase. To address these concerns, governments began to introduce "cooling-off" periods, during which consumers could return products without penalty.

DOI: 10.4324/9781003584056-89

One of the earliest examples emerged in Germany's 1950s regulations on doorstep sales, which allowed consumers to cancel purchases made at their homes within a short period. This cooling-off period provided essential protection against high-pressure sales tactics and misleading offers, recognizing that consumers in these situations were at a disadvantage and more susceptible to coercive or impulsive buying decisions.

The rise of e-commerce in the late 1990s and early 2000s spurred a new era of consumer protection legislation, particularly around the Right of Withdrawal. Online shopping, which exploded with the advent of major marketplaces like Amazon and eBay, presented unique challenges for consumer protection (see **Money**). The inability to inspect goods firsthand before purchase, coupled with cross-border transactions, made it easier for consumers to fall victim to fraud, misrepresentation, or simply dissatisfaction with a product's quality. A new wave of consumer protection laws followed.

In the European Union (EU), the landmark 2011 Consumer Rights Directive (2011/83/EU) codified the right to withdraw from a purchase made online, by phone, or through other remote means within 14 days (Articles 9 to 16 of the Consumer Rights Directive, CRD). This provided consumers with a critical safeguard, allowing them to return items without having to provide a reason, ensuring their right to a refund.

Different Perspectives

The right of withdrawal is a subject of ongoing debate, particularly concerning its environmental impact and the ethics of consumer behavior. While its primary purpose is to protect consumers, there is growing concern that it may inadvertently encourage unethical practices and contribute to both environmental and economic unsustainability. The rise of e-commerce in Europe, coupled with the ease of returns, has contributed to a "chronic" return culture, where consumers often engage in impulsive or compulsive buying, followed by regret-driven returns, as shown by Santos Silva & García-Micó (2024). This trend is compounded by practices like "wardrobing" – returning items after brief use for a social event or photo – and "bracketing" where multiple sizes are ordered and those that don't fit are returned.

Such behaviors have environmental and financial costs. Returns increase transportation emissions and generate waste from packaging materials. Businesses face the expense of storing, reselling, or disposing of returned goods, with costs often passed on to all consumers, leading to higher prices. Furthermore, the financial strain of providing free returns impacts all buyers, including those who avoid returns for sustainability reasons, as they too bear the hidden costs of these policies.

From a sustainability perspective, the right of withdrawal is thus a double-edged sword. It enhances consumer confidence by addressing power imbalances in business-to-consumer (B2C) relationships, mitigating risks, and preventing buyer's remorse. However, critics argue that the ease of returning goods fosters excessive consumption and waste, particularly in environments designed to encourage impulse buying.

Recent studies suggest that the convenience of returns in online shopping may indeed contribute to a culture of overconsumption and even addictive shopping behaviors. By removing the finality of purchase decisions, lenient return policies can create a sense of "low-stakes" buying, where consumers deliberate less when making purchases. Scholars also argue that this ease of return fosters a "buy-and-try" mentality, where the temporary ownership model diminishes a consumer's attachment to the goods purchased, reinforcing a disposable culture that values short-term gratification over sustainability (see **Fast Fashion**).

These findings indicate that, while consumer protections are necessary, an overreliance on easy return policies may unintentionally encourage more frequent and impulsive purchases, contributing to a cycle of excessive consumption and waste that undercuts sustainability efforts. This dynamic thus compels policymakers and businesses to strike a balance between upholding consumer rights and implementing sustainable practices that minimize environmental impact and economic waste.

Application

Several alternatives have been proposed to mitigate the negative impacts of excessive returns. One approach is to expand the list of exceptions to the Right of Withdrawal, particularly for products that cannot be resold as new, as well as single-use items not covered by the Single Use Plastics Directive. This could help reduce the volume of returns and their associated environmental costs. Another proposed solution involves offering consumers a choice between two contracts: one without a withdrawal right at a lower price, and another with a withdrawal right at a higher price (Karampatzos & Ilic, 2023).

Introducing a "pre-cooling off" period before a purchase is finalized could give consumers additional time to reflect on their decisions, potentially decreasing impulse purchases and reducing the subsequent need for returns. Stricter enforcement of return policies, such as mandating that consumers cover shipping costs for items they send back, could discourage unnecessary returns and encourage more deliberate purchasing decisions. In addition, there is growing interest in leveraging technology to minimize returns. Tools like virtual fitting rooms (immersive retail), augmented reality (AR) previews, and comprehensive product reviews could enable consumers to make more informed choices upfront, reducing the likelihood of returns (see **Information and Communication Technology**).

Educating consumers on the environmental impact of returns can further encourage responsible shopping practices, aligning consumer behavior with broader sustainability goals (see **Education for Sustainable Consumption**).

Further Reading

Karampatzos, A., & Ilic, N. (2023). Law and economics of the withdrawal right in EU consumer law. *Review of Law and Economics*, 19(5). Available at: https://www.degruyter.com/document/doi/10.1515/rle-2022-0076/html (accessed: 14 August 2024).

Rekati, P., & Van den Bergh, R. (2000). Cooling-off periods in the consumer laws of the EC member states: A comparative law and economics approach. *Journal of Consumer Policy*, 23(4), 371 ff. https://doi.org/10.1023/a:1007203426046.

Santos Silva, M., & García-Micó, T.G. (2024). Cooling-off hot deals. A plea for green sludge in electronic consumer contracts. In *Routledge handbook on private law and sustainability*. Routledge.

Terryn, E., & Van Gool, E. (2020). The role of European consumer regulation in shaping the environmental impact of e-commerce. *EuCML*, 3, 89–101. https://doi.org/10.2139/ssrn.3732911.

Wood, S.L. (2001). Remote purchase environments: The influence of return policy leniency on two-stage decision processes. *Journal of Marketing Research*, 38(2), 157–169. https://doi.org/10.1509/jmkr.38.2.157.18847.

84 Information and Communication Technology

Georgina Guillen-Hanson, Heli Hallikainen,
Nannan Xi, and Juho Hamari

Definitions

Information and Communication Technologies (ICT) manage and communicate data in an increasingly interconnected world. They include a series of technologies and services such as computers, mobile devices, communication networks, the Internet, cloud computing, data analytics, and various other tools that continuously shape business practices and consumers' behaviors.

Digital technologies offer a significant potential for enhancing efficient production, reducing resource use, facilitating the exchange of goods and services, and promoting knowledge creation and social interactions. These technologies can contribute to decoupling economic growth from resource consumption and foster sustainable lifestyles. However, ICT can also lead to the overuse of resources through overproduction and compulsive consumption, technostress, social isolation, being detached from physical reality and nature, and mental overload. Ethical concerns such as privacy violations, fraud, empathy loss, and inequality also arise. While these issues do not necessarily result from ICT itself, they are linked to the purposes for which it is developed, and adopted – creating a dilemma for sustainable living.

ICT also creates manifold socio-environmental pressures, such as the extraction of raw materials, increasing energy and infrastructural demands, CO_2 emissions, complex supply chains, planned obsolescence, and e-waste generation, among other issues. Despite the widespread adoption of technologies like mobile phones, a growing digital divide persists between those who can afford access and those who cannot. Digital lifestyles are also unequal, with increasingly blurred lines between the real and the online worlds, leading to diverse malaises, physical, mental, and social, calling for digital literacy education, stringent data protection laws, and also learning how to be "offline". Furthermore, the pervasive use of digital technologies poses potential threats to traditional cultures and human development. Addressing these challenges requires a balanced approach that harnesses the benefits of digital technologies while mitigating their negative impacts on society and the environment.

History

Arguably, media consumption and its influence on our lifestyles date back to the first printing press in the 15th century. Its development accelerated rapidly toward the end of the 20th century when the computing industry took off and led to greatly accelerating consumption. From using radio and television for product placement and the presentation of aspirational lifestyles to the proliferation of e-commerce and the increasing sophistication of data collection processes to personalize

DOI: 10.4324/9781003584056-90

advertising, technology keeps shaping everyday decisions, behaviors, attitudes, **social norms,** and social dynamics alike.

The speed and volume of data processed and mined through personal devices have led to an explosion of online services, such as e-commerce and social media, which, besides displacing conventional consumer-seller-producer interactions, have created entirely new commercialization and information channels. These have been used to drive consumption, empower (and manipulate) consumers, and raise awareness about underlying sustainability issues. The provided information induced anxiety about the global impact of increased consumption, but at the same time encouraged global movements to curb production and shift lifestyles. Since the 2010s, the Internet of Things (IoT) has significantly contributed to more efficient use of resources at the household level; a decade later, emerging virtual technologies such as machine learning, artificial intelligence, blockchain, and the metaverse have opened new opportunities for humans to engage in sustainable consumption practices in a fully digitized world, extending beyond physical reality.

Different Perspectives

ICT is a double-edged sword. Solutions touted as environmentally friendly have other, less desirable, features and socio-economic implications. For example, artificial intelligence (see Box 84.1), which could enable more efficient use of resources, also perfects consumer targeting with personalized **advertisements** that could lead to more mindless consumption and **rebound effects.** Robotics, which allows faster, more efficient production and services, can potentially replace human labor, increase unemployment, and widen poverty gaps.

Box 84.1   Artificial Intelligence

Artificial Intelligence (AI) – applications and technologies programmed to think and learn like humans, simulating human intelligence in machines. Its application in robotics, particularly in the service industry, has significantly improved the efficiency in decision-making, mainly in terms of customer satisfaction and improving decision quality via automation, personalization, and humanization. These non-human agents, whether embodied as physical robots, chatbots, or virtual assistants, have seamlessly integrated into the fabric of daily consumption life.

The Internet of Things (see Box 84.2) aims to reduce energy and material consumption. Virtual technologies (see Box 84.3) reshape multiple realities (e.g., virtual reality, augmented reality, extended reality, mixed reality) and promise to dematerialize consumption and enable sensorial experiences. However, these technologies may increase the dependency on gadgets, sensors, and demand for new products – again requiring new resources. Consumption experiences in virtual realities are expected to reduce the need for physical goods and substitute them with experiences. Yet, access to these worlds demands other types of consumption choices, skills, physical abilities, and, of course, the literacy and infrastructure to make everything possible.

Box 84.2   Internet of Things

The Internet of Things (IoT) describes any sensor-enabled device that collects and transmits data through internet connections. It is also a framework that bridges the physical and virtual worlds through new applications and services on the Internet, creating an inclusive, "smart" information system that helps regulate consumption (e.g., energy) and improve efficiency. Its applications include healthcare, manufacturing, transportation, agriculture, smart cities, industries, and commerce.

Box 84.3   Virtual technologies and the metaverse

Virtual technologies comprise the devices and platforms that simulate environments and stimulate the senses to interact with digitally generated content. The two core virtual technologies are Virtual Reality (VR) – which recreates new realities to immerse human senses via different stimuli – and Augmented Reality (AR) – which overlays information (e.g., images, videos) onto the physical environment. Other technologies include Augmented Virtuality (AV), which superimposes content on top of physical realities, Mixed Realities (MR), which allows real-time coexistence and interactions between physical and digital objects, and Extended Reality (XR), which encompasses all the virtual and real human-machine interactions. Besides their hedonic and fun applications, these technologies are being increasingly applied by the utility and business sectors.

The **metaverse** consists of these artificially generated, modified, diverse, and extended realities built on the convergence of virtual technologies with visual, linguistic, spatial, aural, and gestural compositions (multimodality) and human representations (avatars).

ICT is also changing the notions of ownership, safety, and even economic systems, for example, blockchain technology (BT), known mainly for cryptocurrency trading, is also used for patient data management, supply chain transactions, voting systems, and legal agreements. While more literacy for understanding its use and applications is needed, its socio-environmental impacts in the mid-and long-term remain an open research area.

Applications

The case of ICT-enabled efficiencies to address environmental concerns is clear: more efficient production processes convey less use of resources and waste reduction. However, the intricacies of ICT include other aspects with considerable environmental impacts, such as planned obsolescence and low device recyclability rates, the exploding demand for energy to enable these technologies, and adequate infrastructure, to name a few. On the social front, automation translates into job losses and greater demand for high-skilled workers. Other concerns include data ownership, manipulation, safety, digital literacy, cyber-bullying, management of intellectual property, and recognition of artistic creations. Nowadays, solutions like the Metaverse (see Box 84.4) are seen as enablers of sustainable lifestyles because of their potential, for example, to assess life cycles and derive solutions for reducing material consumption and to simulate environments where the consequences of decisions can be experienced (e.g., living in a world 3°C hotter) thus encouraging lifestyle shifts in the "real" world.

Besides having the potential to enable more convenient consumption practices, ICT often enhances consumption's emotional and behavioral aspects through Gamification and Digital Sensory Marketing (see Box 84.4), approaches that have shown their potential to support individual efforts across several lifestyle domains. While gamification is predominantly used in environmental efforts, such as reducing energy consumption at the household level or promoting ecofriendly driving, its application is expanding into social domains. Here, it aims to influence diverse consumption-related behaviors – from providing information and encouraging more informed, conscious decisions to fostering overconsumption through playful environments designed to keep users coming back for more. Similarly, Digital Sensory Marketing enables the possibility to experience the impacts of lifestyle choices, thus enabling better-informed decision-making processes. Both approaches are highly persuasive, and when applied to incentivize consumption, it can lead to more wasteful lifestyles as the purchase of cheap, low-quality products is only a tap away.

Box 84.4   Exemplary applications of technology

Gamification – "an intentional process of transforming any activity, system, service, product, or organizational structure into one which affords positive experiences, skills, and practices similar to those afforded by games . . . the gameful experience" (Hamari, 2019: 1)

Digital Sensory Marketing (DSM) – integrates new technologies into a multisensory experience, engaging the consumers' five senses (namely visual, olfactory, auditory, gustatory, and tactile), gathering information from our daily surroundings, and generating physical sensations that impact consumers' emotional, cognitive and social perception and experiences as well as behavior and decision-making.

Overall, digital technologies have the potential to reduce people's general need to move from one place to another while enabling "feeling" and experiencing products and places as if they were real, albeit in a digital realm. Despite the promising narratives of ICT to enhance more sustainable consumption, their actual development and application constitute an oxymoron where innovation and profitability clash with societal well-being and environmental impacts. To address this situation, all stakeholders should engage in responsible practices, such as considering technology's unintended impacts, the transparent use of data, and developing technology-free habits to enjoy in our daily lives.

Further Reading

Hargittai, E. (2018). The digital reproduction of inequality. In D.B. Grusky & S. Szelényi (Eds.), *The inequality reader*, 2nd ed. New York: Routledge. https://doi.org/10.4324/9780429494468.

Hamari, J. (2019). Gamification. In G. Ritzek & C. Rojek (Eds.), *The Blackwell Encyclopaedia of Sociology*. John Wiley and Sons.

Pellegrino, A., Stasi, A., & Wang, R. (2023). Exploring the intersection of sustainable consumption and the Metaverse: A review of current literature and future research directions. *Heliyon*. https://doi.org/10.1016/j.heliyon.2023.e19190.

Petit, O., Velasco, C., & Spence, C. (2019). Digital sensory marketing: Integrating new technologies into multisensory online experience. *Journal of Interactive Marketing*, 45(1), 42–61. https://doi.org/10.1016/j.intmar.2018.07.004.

Stephanidis, C., Salvendy, G., Antona, M., Chen, J. Y. C., Dong, J., Duffy, V. G., Fang, X., Fidopiastis, C., Fragomeni, G., Fu, L. P., Guo, Y., Harris, D., Ioannou, A., Jeong, K. (Kate), Konomi, S., Krömker, H., Kurosu, M., Lewis, J. R., Marcus, A.,. . . Zhou, J. (2019). Seven HCI grand challenges. In *International journal of human–computer interaction*, Vol. 35, Issue 14, pp. 1229–1269. Informa UK Limited. https://doi.org/10.1080/10447318.2019.1619259.

Xi, N., Chen, J., Gama, F., Korkeila, H., & Hamari, J. (2024). Virtual experiences, real memories? A study on information recall and recognition in the metaverse. In *Information systems frontiers*. Springer Science and Business Media LLC. https://doi.org/10.1007/s10796-024-10500-2.

85 Consumption-Based Accounting

Jukka Heinonen

Definition

Consumption-based accounting (CBA) is an approach for calculating ecological and social impacts related to a particular process or product, based on its full lifecycle, including the end-use, and attributing them to the final consumer. The result is called a consumption-based footprint. This approach is complementary to so-called territorial or production-based accounting, which attributes the impacts to the location where they occur or to the producer. For instance, if a particular material good is manufactured in one location and exported elsewhere, CBA attributes its impacts to the end user, whereas territorial accounting would attribute them to where it was produced. In CBA, the entity in question can be anything from an individual to an organization, city, nation, region, or the globe. On a global level, consumption-based and territorial/production-based impacts are the same.

The endpoints measured in CBA can include carbon, biodiversity, nitrogen, water, energy, and land-use footprints in the ecological domain. Social footprint endpoints have been more scarce, but include, for example, global slavery footprint. Carbon footprints, based on carbon dioxide emissions, are the most commonly used metric. Often, a carbon footprint also includes other greenhouse gases, such as methane, normalized to carbon equivalents based on their atmospheric warming potential per unit volume. In such cases, carbon footprint becomes a *de facto* greenhouse gas or climate footprint.

History

CBA emerged in the late 20th century with the first consumer carbon footprint estimations. Its widespread adoption took place in the early years of the millennium, first in academia, then followed by other organizations. In 2005, British Petroleum (BP) introduced a major campaign to popularize consumption-based carbon footprints, though this move has been widely criticized as a cynical attempt by the company to emphasize the impacts of personal consumption and lifestyles on global carbon emissions ("blaming the victim") (see **Consumer Scapegoatism, Greenwashing**). The campaign achieved that goal.

The consumption-based carbon footprint assessment of 79 member cities of the C40 Cities Climate Leadership Group in 2018 was an important milestone in the diffusion of CBA outside academia. The Intergovernmental Panel on Climate Change (IPCC) has also started emphasizing the importance of the consumption-based perspective. The 2022 Sixth Assessment Report (AR6) devoted a lot of space to mitigation options based on reduced consumption, with the Working Group III Co-Chair Priyadarshi Shukla stating in a press release that "*having the right policies, infrastructure and technology in place to enable changes to our lifestyles and behaviour can result in a 40–70% reduction in greenhouse gas emissions by 2050. This offers significant untapped potential*".

DOI: 10.4324/9781003584056-91

On a national level, the Nordic countries have been at the forefront in recognizing the importance of the demand side – especially household consumption – in driving up global carbon emissions. In 2023, the Nordic Council of Ministers declared that

> Nordic consumption must change if [the sustainable development goals] are to be achieved . . . Nordic consumption has a huge environmental and climate footprint in other parts of the world. The Nordic Council of Ministers is working to turn this around and make the Nordic Region the most sustainable region in the world.

However, even in the Nordic countries, there is a lack of serious commitment to take responsibility for the environmental impacts caused by them outside their territories.

Different Perspectives

CBA is an umbrella term for two different allocation/responsibility principles. If a territorial or areal principle is followed, the impacts are allocated based on consumption within a certain location regardless of who within the area or territory is consuming. For example, if a resident of country A travels to country B on a leisure trip, all their consumption in country B would be allocated to the consumption-based carbon footprint of country B. Vice versa if a residence or personal principle is adopted, the impacts are allocated based on residency, regardless of where consumption happens. Using this principle, the consumption by the resident of country A on a leisure trip to country B would be counted in the consumption-based carbon footprint of country A. These definitional differences mean that when looking at geographic areas – such as a city or a nation – the territorial/areal principle shows higher impacts where tourists and commuters are responsible for an important share of all demand for goods and services. In turn, it is lower where there are no tourists or where the residents commute away for work, school, and errands. On the other hand, the residence/personal principle shows how the impacts follow personal consumption, even when that consumption takes place outside the home location.

CBAs are often divided into three main domains: personal/private consumption, governmental consumption, and capital production. With the territorial/areal principle, it is clear that the boundaries of the assessment coincide with the area in question, plus imports and minus exports. With the residence/personal principle, the scope is different. The personal/private consumption component includes only the activities of the residents of the area in question and excludes those of visitors to the location. The personal/private consumption component therefore consists of the consumption of residents and visitors within the area in question if using the territorial/areal principle, but of the consumption of residents within and outside the area in question if using the residence/personal principle.

Governmental consumption always serves a particular location (e.g., green spaces, safety and waste collection services) and, therefore, benefits both residents and visitors. This means that governmental consumption is very difficult to include in the residence/personal principle. Using the residence/personal principle, governmental consumption would partially consist of local governmental consumption, but also partially of governmental consumption in all the locations visited outside the home location. It would also include governmental consumption serving the production of goods and services consumed by the individuals in question.

Similarly, with the territorial/areal principle, capital production can be allocated to the location in question using the same allocation method as with governmental consumption. With the residence/personal principle, allocation of capital production is again difficult.

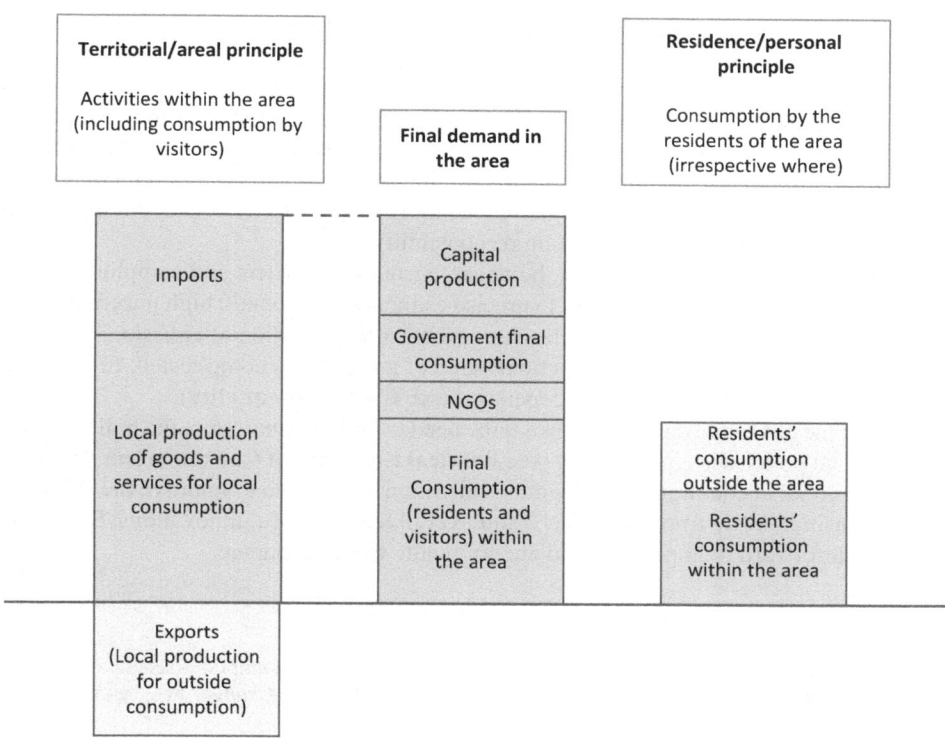

Figure 85.1 Comparison of territorial/areal accounting principle and residence/personal principle in CBA

Source: Adapted from Heinonen et al. (2020)

Due to these allocation problems, the scopes of the two CBA approaches typically become very different from one another, and therefore a reader should not compare the results when the allocation principle has not been the same.

Figure 85.1 shows how the territorial/areal principle captures the global impact of the area in question, and how that matches with final demand in the area, whereas the residence/personal principle captures the global impact caused by the residents of the area in question.

Application

While CBA has not gained much popularity in policymaking, there are signs of change. Accordingly, Sweden recently announced a target to become the first nation to include the impacts caused by their societies outside their territories in their carbon neutrality targets. Sweden and Finland have also started publishing national greenhouse gas emissions accounts based on CBA to inform citizens and policymakers. In Finland, a recent national CBA analysis concluded that with the right policies, consumption-based carbon footprints could be reduced by 50% by 2035. Iceland is trying to map its so-called spillover effects, that is, the effects occurring beyond its territory, but induced by its economy.

On a city level, in the United States, Portland, Oregon, and several other cities have adopted consumption-based greenhouse gas accounts as companions to their territorial inventories.

Portland's CB Inventory in 2011 showed that the city's consumption-based emissions are 40% greater than the emissions based on territorial accounting.

Another application of CBA is the use of mobile apps that provide consumers with information about the carbon footprint of their purchases at the point of transaction. In the future, CBA could be very useful for allocating responsibility among nations for developing effective climate policies and for setting long-term targets, based on the emissions driven by each country's consumption. Accounting with metrics other than greenhouse gases – for example, biodiversity footprints – would be also valuable for allocating responsibility.

The adoption of CBA in policymaking, however, encounters barriers and complications. First, current assessment techniques do not lead to precise estimates and contain high uncertainty, which may create political tensions. Second, in the cases where a large portion of emissions is imported, controlling the impacts outside the jurisdictional area of a country is complicated, for example, if the country of origin is highly fossil-fuel dependent (see **Carbon Inequality**).

Moreover, if the CBA analysis points toward the need to curb consumption, the political willingness to adopt such policies is typically low (see **Political Economy of Consumerism**, **Degrowth**). In contrast, territorial accounting naturally draws attention to the need to improve the efficiency of production, transition away from fossil fuels, and overall technological innovations. Such policies are more straightforward to implement and are politically easier to pursue.

Further Reading

Afionis, S., Sakai, M., Scott, K., Barrett, J., & Gouldson, A. (2017). Consumption-based carbon accounting: Does it have a future? *Wiley Interdisciplinary Reviews: Climate Change*, 8(1), e438. https://doi.org/10.1002/wcc.438.

Heinonen, J., Ottelin, J., Ala-Mantila, S., Wiedmann, T., Clarke, J., & Junnila, S. (2020). Spatial consumption-based carbon footprint assessments – A review of recent developments in the field. *Journal of Cleaner Production*, 256, 120335. https://doi.org/10.1016/j.jclepro.2020.120335.

Ivanova, D., Stadler, K., Steen-Olsen, K., Wood, R., Vita, G., Tukker, A., & Hertwich, E. (2016). Environmental impact assessment of household consumption. *Journal of Industrial Ecology*, 20(3), 526–536. https://doi.org/10.1111/jiec.12371.

Malik, A., McBain, D., Wiedmann, T., Lenzen, M., & Murray, J. (2018). Advancements in input-output models and indicators for consumption-based accounting. *Journal of Industrial Ecology*. https://doi.org/10.1111/jiec.12771.

Wilting, H., Schipper, A., Bakkenes, M., Meijer, J., & Huijbregts, M. (2017). Quantifying biodiversity losses due to human consumption: A global-scale footprint analysis. *Environmental Science & Technology*, 51(6), 3298–3306. https://doi.org/10.1021/acs.est.6b05296.

86 Personal Carbon Allowance

Tina Fawcett and Yael Parag

Definition

Despite global efforts to decarbonize energy supply, carbon emissions keep rising. Personal carbon allowances (PCAs) is an umbrella term for mitigation policies that aim to limit carbon emissions by allocating a carbon budget to individuals, encouraging them to change their energy consumption patterns, and engaging them in emissions reduction (see **Behavior Change** and **Energy Consumption Behavior**). Unlike carbon mitigation policies which tend to focus on energy production, PCA focuses on energy consumers and is sometimes described as a downstream cap-and-trade policy. Several PCA policy designs have been proposed. These vary in their combination of system boundaries, the energy services covered (heating, power, mobility, flights), geographical scale, trading rules, and the allocation and surrender of allowances (Figure 86.1). Fawcett and Parag have shown how different options can be combined to create varied PCA designs including cap and share; tradable consumption quotas; carbon rations; tradable transport carbon permits; personal carbon trading; and individual carbon quotas.

The underlying assumption about how PCA works is that emission reductions would arise through three interconnected mechanisms influencing individual behavior, that is, economic, cognitive, and normative mechanisms (Figure 86.2). In the economic mechanism, the carbon "price" on emissions would incentivize people to choose low-carbon activities and goods. Introducing a virtual "carbon currency" and the scarcity of carbon units resulting from a shrinking cap would also encourage people to mentally account for their carbon usage and promote lower-carbon choices.

In the cognitive mechanism, the carbon cap would spark new conversations about carbon and climate in society. This increased visibility of carbon and heightened awareness of individual impacts on the climate would prompt people to opt for lower-carbon activities.

In the normative mechanism, the underlying principles of PCA – such as environmental limits, fair shares of emissions, societal climate responsibility, and social solidarity – would establish new **social norms** regarding acceptable behavior, thus encouraging widespread adoption of low-carbon lifestyles.

History

The idea of personal carbon allowances emerged from concepts developed independently in the 1990s by two British researchers, David Fleming and Mayer Hillman. With his economics expertise, David Fleming developed the idea of "domestic tradable energy quotas" as part of an economy-wide mechanism that included carbon auctions for firms. He believed this would reduce carbon emissions more cost-effectively and equitably than carbon taxation. Mayer Hillman, a transport and social policy researcher, was inspired by Aubrey Meyer's proposal for global "contraction

DOI: 10.4324/9781003584056-92

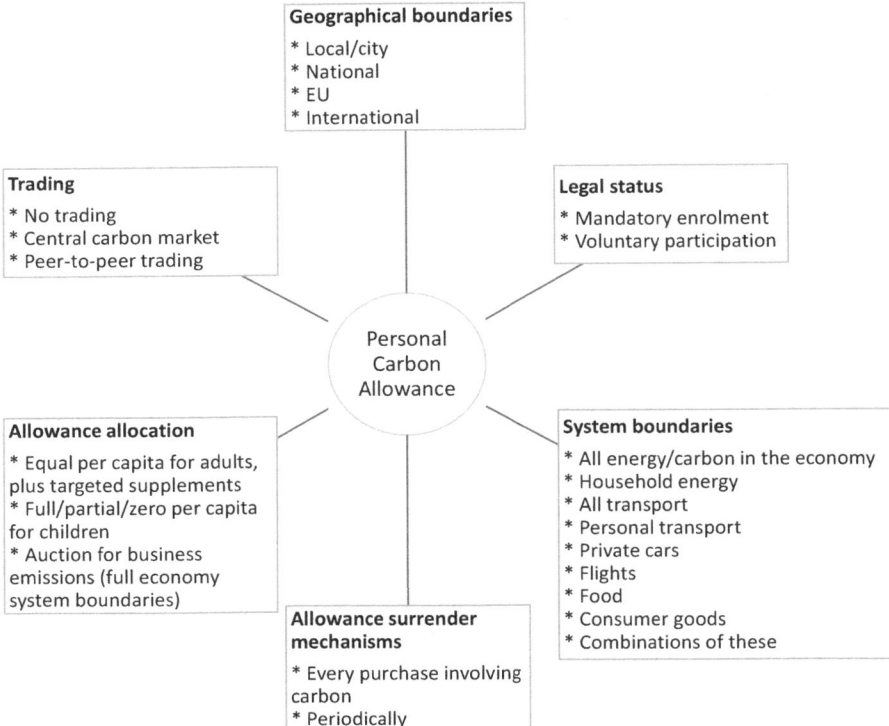

Figure 86.1 Key design elements of a personal carbon allowance policy

Source: Authors' own production, drawing from earlier work – Akenji, L., Bengtsson, M., Toivio, V., Lettenmeier, M., Fawcett, T., Parag, Y., Saheb, Y., Coote, A., Spangenberg, J.H., Capstick, S., Gore, T., Coscieme, L., Wackernagel, M., & Kenner, D. (2021) *1.5 degree lifestyles: Towards a fair consumption space for all*. Berlin: Hot or Cool Institute

and convergence" of carbon emissions, where nation state emissions per capita first converged and then contracted to stay within safe limits. He was also influenced by Britain's experience of food rationing in World War II. Hillman called his idea "carbon rations" and saw it as a national route to global contraction and convergence. International researchers developed the central ideas further, with different policy designs emerging.

In 2006, the idea attracted high-level political interest in the United Kingdom, resulting in government-sponsored studies in 2007–2008, including analysis of equity impacts and costs. The UK government department Defra concluded that PCA would be expensive and complex to introduce and was "an idea ahead of its time". Subsequently, there was limited further official interest in PCA. The only country that has conducted a PCA pre-feasibility study is the United Kingdom.

Different Perspectives

PCA have always engendered divided opinions about its likely effectiveness, cost, fairness, and public acceptability. In part, this is because its impacts depend on the details of policy design and implementation (Figure 86.1). The absence of a detailed or operational PCA scheme creates ambiguity. Disagreements also emerge from competing understandings of fairness, **climate justice**, mechanisms for change in relation to energy use (Figure 86.2), personal responsibility, and

81 Ecodesign

Max Schoepen and Karine Van Doorsselaer

Definition

Ecodesign is an integral part of the design process, used to minimize the environmental impact of a product throughout its entire life cycle. The life cycle of a product is the complete path that a product takes, from raw material extraction to end of use. For each phase, inputs (auxiliary materials and energy) and outputs (waste and emissions) need to be considered and minimized (Figure 81.1). Overall, the designer seeks an optimal compromise between all of the product aspects (ecological, technical, human-centered, economical), as the environmental impact will always exist. There is no such thing as a one-fits-all solution, with each case having its own optimal solution.

Ecodesign facilitates the main principles of the circular economy by slowing down, narrowing, and closing resource loops (see **Circular Economy and Society**). During the design process, the principles such as reuse, **repair**, refurbishment, and recycling are taken into account (Figure 81.2). Ecodesign also supports the implementation of new circular business models such as product-as-a-service (see **Product-Service Systems**).

We identify three key aspects of ecodesign:

1. The first aspect is thinking in *functionality, not materiality*. The designer starts from the functional fulfillment of the product. Why design an improved shampoo bottle, for instance, when a block of soap has the same functionality, without the need for packaging and transport of water? In the same context, we see the emerging model of product-as-a-service. Many people need mobility, not a car. Any vehicle that is owned and maintained by the manufacturer will be designed and used in a different manner than one that is sold to the customer.
2. The second key aspect is *life cycle thinking*. The example in Box 81.1 explains this approach.
3. The third key aspect is *value chain cooperation*. Cooperation in the entire economic system is essential to the transition to a sustainable society (see **Role of Business in Sustainable Consumption**). This means cooperation between existing and new partners, between colleagues and competitors, and so on. The designer should interact with all involved parties and facilitate their contribution to sustainability.

Box 81.1   Life cycle thinking as an approach to ecodesign

Let us take the simplified choice between a steel and an aluminum car bumper, looking at ecological impact:

DOI: 10.4324/9781003584056-87

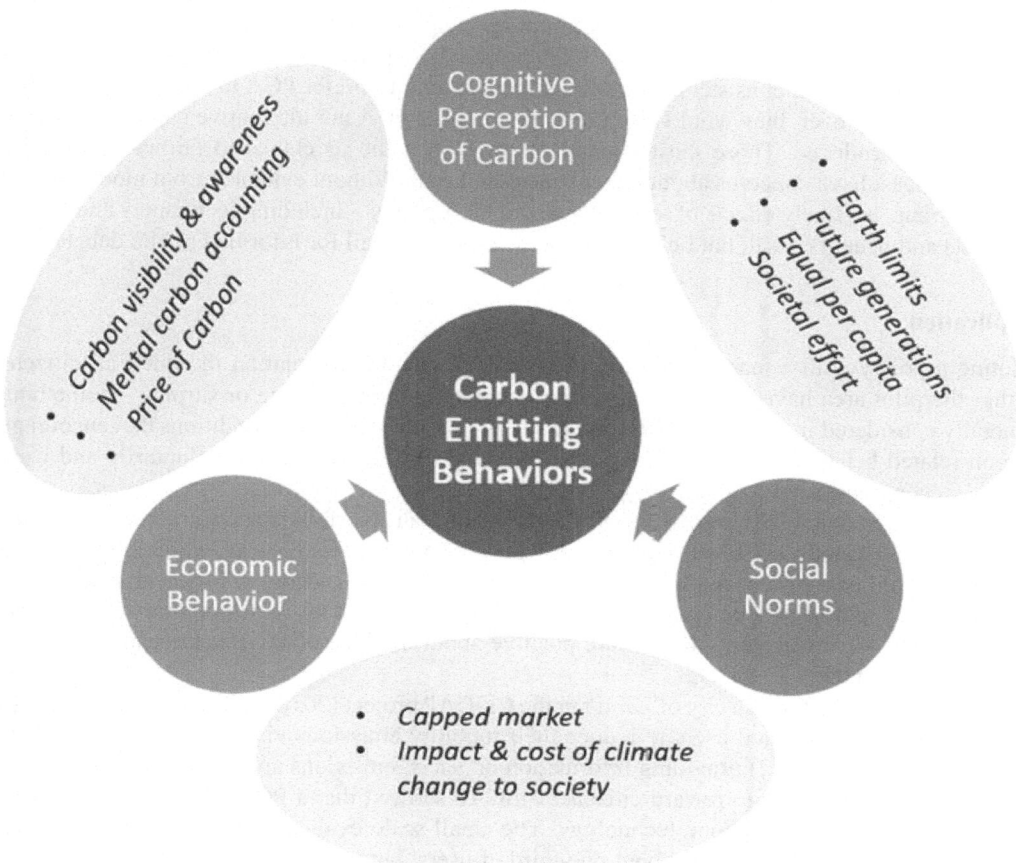

Figure 86.2 Mechanisms through which PCA may influence carbon-emitting behaviors

Source: Authors' own production, drawing from earlier work – Akenji, L., Bengtsson, M., Toivio, V., Lettenmeier, M., Fawcett, T., Parag, Y., Saheb, Y., Coote, A., Spangenberg, J.H., Capstick, S., Gore, T., Coscieme, L., Wackernagel, M., & Kenner, D. (2021) *1.5 degree lifestyles: Towards a fair consumption space for all.* Berlin: Hot or Cool Institute

freedom (see **Freedom of Choice**, **Carbon Inequality**, **Household Income Versus Carbon Footprint**, and **Consumer Scapegoatism**).

Regarding fairness, PCA policies suggest each individual is permitted to generate the same amount of carbon emissions. However, equal carbon emissions do not equate to equal energy use, let alone equal access to energy services. For example, different sources of heating or transportation energy vary in their carbon intensity (emissions per kWh), and renewable energy can be emissions-free – meaning low emissions can result from high energy use and vice versa. Further, an equal per capita emissions allocation does not reflect the variation in people's energy service needs. Questions raised include how vulnerable households or those with higher energy needs would be treated, what would happen if households ran out of allowances, and the consequences of a particularly cold winter. These questions can only be answered through detailed policy design, which could include allowance for these, and other, circumstances. An implemented PCA policy is likely to look much more complex than the simple principles on which it is based.

Research on public opinion regarding PCA proposals, whether conducted via surveys, interviews, or focus groups, consistently shows that people tend to prefer PCA to alternatives such as a carbon tax. However, they would also prefer that neither PCA nor alternative capping or taxing policies are introduced. Those most positive about PCA value its claims to fairness, and those who like it least have concerns about effectiveness and cost. Without evidence from modeling and research trials, the likely effects of an implemented PCA policy – including its winners and losers, and costs and benefits – will not be understood in sufficient detail for informed public debate.

Application

Piloting a policy with a mandatory carbon cap entails creating a situation in which all citizens within the pilot area have emission caps and allowances are in shortage or surplus – something generally considered impossible. Therefore, trials cannot fully create the conditions that encourage carbon-related behavioral change. The pilots to date have been undertaken voluntarily and were limited in geography and scope.

The most significant PCA research trial was the Australian "Norfolk Island Carbon/Health Evaluation" (NICHE) study, established in 2011; 350 people registered for the trial, which encompassed food, household energy use, and travel, and incorporated a carbon accounting system, rewards for participation, and feedback on emissions. Results showed a switch away from motorized transport (>20% reduction). Participants were more positive about the idea of PCA at the end of the trial than at the beginning.

More recently, the Finnish city of Lahti ran the CitiCAP project (2018–2021), a voluntary incentive system that encouraged users to reduce their mobility emissions via (1) financial incentives through carbon pricing; (2) providing information on users' emissions and reduction options; and (3) an online marketplace to reward citizens. CitiCAP showed that a PCA for mobility could be implemented with mobile phone technology. The small-scale evaluation indicated self-reported transport emissions reductions by about one-third of users, but with overall rejection of a mandatory mobility PCA.

When the concept of PCA was introduced in the 1990s and gained policy traction in the United Kingdom (from 2006 to 2010), transportation and heating systems – the major sources of direct personal emissions – primarily relied on fossil fuels and carbon-intensive electricity generation. Since then, the electricity system has been decarbonizing and, in many countries (especially Europe and North America), the transition to electric mobility and heating is being supported by dedicated policies (see **Energy Overshoot**). Given the complexity of introducing PCA, its relevance for the transition to a net zero-carbon economy is open to question.

Most of the current interest in PCA concerns mobility, the sector where energy use is still rising, decarbonization is progressing slowly and individual choices play a major role (see **Sustainable Mobility**). Various Personal Mobility Carbon Allowance (PMCA) schemes have been proposed, particularly in China. Recent Chinese research includes studies on the design of trading systems, public willingness to participate, explored via choice experiments and other methods, and learning from voluntary personal carbon transport schemes in particular cities. Some PMCA proposals would allow users to trade carbon credits in a market, influencing their travel decisions within set carbon budgets. If combined with good quality, affordable public transport and other low-carbon infrastructure for walking, cycling, and EV charging, PCA might prove a useful addition to the policy mix (see **Urban Planning and Spatial Allocation**).

Further Reading

Bothner, F. (2021). Personal carbon trading – Lost in the policy primeval soup? *Sustainability*, 13, 4592.

Bristow, A.L., Wardman, M., Zanni, A.M., & Chintakayala, P.K. (2010). Public acceptability of personal carbon trading and carbon tax. *Ecological Economics*, 69(9), 1824–1837. https://doi.org/10.1016/j.ecolecon.2010.04.021.

City of Lahti. (2022). *CitiCAP – Citizens' cap and trade co-created*. Available at: https://www.lahti.fi/en/files/citicap-final-report/ (accessed: 20 December 2024).

DEFRA. (2008). *Synthesis report on the findings from Defra's pre-feasibility study into personal carbon trading*. UK Department for Environment Food and Rural Affairs. Available at: www.defra.gov.uk (accessed: 20 December 2024).

Fawcett, T., & Parag, Y. (2010). An introduction to personal carbon trading. *Climate Policy*, 10, 329–338. https://doi.org/10.3763/cpol.2010.0649.

87 Co-Benefits of Climate Policy

Eva Alfredsson

Definition

The Intergovernmental Panel on Climate Change (IPCC) initially distinguished between side effects specific to climate change policy (ancillary benefits) and side effects of any policy (co-benefits). However, it has since 2018 used "co-benefits" for both, defined as "a positive effect that a policy or measure aimed at one objective has on another objective, thereby increasing the total benefit to society or the environment" (IPCC, 2022).

Climate-related co-benefits can be grouped into Type 1 co-benefits, that is, those arising from policies aiming at decreasing greenhouse gas emissions (GHG); Type 2 co-benefits, that is, those arising from other types of policies, not directed at climate protection; and Type 3 co-benefits, that is, those with multiple objectives (see Table 87.1). The scientific literature is heavily dominated by studies focused on Type 1 co-benefits.

In this entry, we will focus mainly on type 1 co-benefits.

History

The concept of climate policy co-benefits appeared in the literature in the early 1990s. As mentioned, the IPCC used to differentiate these co-benefits based on the policy they originated from. Climate mitigation policy has, within the wider economic literature, historically been regarded as a cost. This started to change more seriously with the 2006 Stern Review, which underlined not only co-benefits but also the costs of climate inaction. It concluded that the benefits of strong and early interventions outweigh the economic costs of not acting. According to the Review, without action, the overall costs of climate change will be equivalent to losing at least 5% of global gross domestic product (GDP) each year, now and forever. Mitigation costs, on the other hand, were estimated to cost around 1–7% of global GDP per year.

Since then, numerous cost-benefit analyses have been published, assessing the quantitative impacts of various policies on the economy as a whole. Meanwhile, the previously used concepts "double dividend", "ancillary benefits", "no-regret", and "win-win-policies" have been replaced with "co-benefits" in the rapidly growing literature.

Different Perspectives

Co-benefit studies cover a range of topics, stretching from air quality to soil and water quality, diet and physical activity, biodiversity, economic and organizational performance, and energy security. The topics can be grouped into economic, ecological, social, and institutional (see Figure 87.1)

DOI: 10.4324/9781003584056-93

Table 87.1 Categorization of climate-related co-benefits

Type of climate policy co-benefits	*Examples*
Type 1 (GHG reduction policies)	Improved air quality following carbon taxation and reduced use of fossil fuel combustion. Enhanced biodiversity or soil fertility due to carbon sink restoration in forests.
Type 2 (other types of policies)	Reduced GHG due to, e.g., water consumption reduction policies, or power plant shutdowns to enhance energy security by substituting oil imports with wind power development.
Type 3 (multiple objectives)	Decreased carbon dioxide emissions and fewer road accidents following traffic safety measures.

Source: Karlsson et al. (2020)

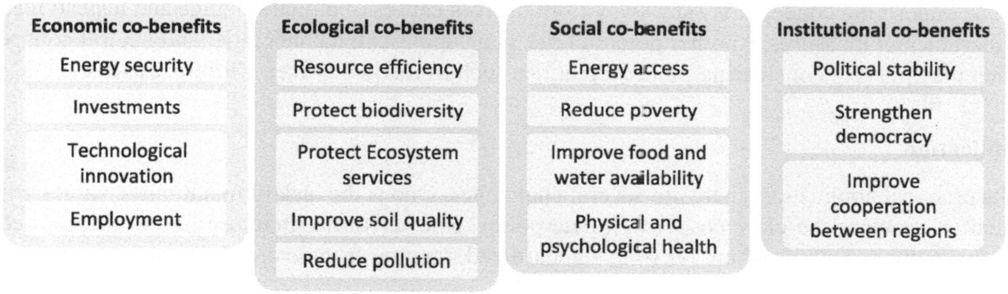

Figure 87.1 Topic areas of co-benefits

Source: Mayrhofer and Gupta (2016)

In areas that have been extensively studied, such as air quality, the climate co-benefits – like lives saved – are often economically valued similarly to the costs of mitigation. In many cases, the benefits are, however, significantly larger. As such, improved air quality from climate mitigation policies has substantial positive effects on health. Studies, like that of Lelieveld et al. (2023), estimate that 5.13 million (ranging from 3.63 million to 6.32 million) excess deaths per year globally are linked to ambient air pollution from fossil fuel use, which could be largely avoided by phasing out fossil fuels.

As for improved water and soil quality, farming practices that promote carbon sequestration can remove carbon dioxide from the atmosphere, while simultaneously improving soil quality and productivity. Studies estimate that by the end of the 21st century, atmospheric carbon dioxide levels could be significantly lowered, while aboveground productivity could be increased.

Biodiversity co-benefits, for example, stem from reduced deforestation, afforestation, and improved forest management. Co-benefits in terms of a healthier diet and physical activity have wide reach and importance. Reducing climate impact from diets while improving people's health may help motivate climate policy. A transport system that encourages active mobility, walking, and cycling can similarly improve public health (see **Sustainable Mobility**, **Urban Planning**

and Spatial Allocation). Many mitigation strategies in the transport sector would have various co-benefits, including air quality improvements, health benefits, equitable access to transportation services, reduced congestion, and reduced material demand (Karlsson et al., 2023).

Within the energy sector several mitigation options, notably solar and wind energy, electrification, and energy efficiency are generally supported by the public and have environmental co-benefits, including improved air quality (see **Energy Consumption Behavior**, **Energy Overshoot**).

Most studies analyzing co-benefits in terms of improved economic performance focus on the existence of a double dividend, that is, improved economic efficiency when environmental tax revenues are used to reduce distortionary taxes. These studies confirm that carbon taxes are beneficial in terms of both GDP and reduced carbon dioxide emissions (Karlsson et al., 2023). Another economic co-benefit is reduced costs for energy supply and energy security.

In academia, a critique against the co-benefit concept is that it lacks identifiable boundaries (Mayrhofer & Gupta, 2016) and scholars often use the concept without defining it.

While net effects are positive, there are also cases of goal conflicts and examples of reverse effects, that is, where climate policy may increase poverty and reduce biodiversity. It is thus important to analyze all potential effects of climate policy to facilitate effective designs.

The underlying concept may shy away from the root causes of climate change and instead focus on incremental measures. Another potential risk is an apolitical, technocratic, and neutral context, which may prevent considering political realities such as inequality (see **Climate Justice**).

Application

In its practical application, there are several challenges. One is the lack of quantitative values with which to compare the clear costs of climate policy. The scientific literature has so far not been helpful, as the metrics used are far from standardized, ranging from avoided deaths, percentage of mitigation costs, and monetary value per ton CO_2e.

Another challenge concerns governance, which often does not consider the wide societal costs and benefits, lack of policy integration (considering several goals simultaneously when designing policy), and fragmented institutional regimes with multitudes of isolated ministries that deal with particular problems.

Generally, there is an increased recognition of the costs of climate change, as they materialize in the direct consequences of extreme weather events on infrastructure and livelihoods, and effects on harvests. The EU, for example, is actively tracking economic losses from weather- and climate-related extremes. The linking of climate risk to financial risks by the Network of Central Banks and Supervisors for Greening the Financial System (NGFS) has been influencing financial institutions to strengthen their climate risk management. At COP 29, an agreement was reached to increase the climate finance to "developing countries" from $100 billion to $300 billion annually by 2035, which was, however, deemed widely unacceptable. While long-term economic consequences are starting to be taken more seriously, the commitments must yet live up to the promise of effectively addressing the global environmental crisis and supporting countries of the Global South.

Meanwhile, the costs and benefits are rarely explored systematically and even less frequently considered in real-world climate policy decision-making. Sweden serves as a striking example; despite its long history of international climate leadership and readily available knowledge on, for example, air pollution co-benefits, such benefits have deliberately been excluded from decision-making processes.

Further Reading

IPCC. (2022). Summary for policymakers. In *Climate change 2022: Mitigation of climate change*. Contribution of Working Group III to the Sixth Assessment Report of the Intergovernmental Panel on Climate Change [P.R. Shukla, J. Skea, R. Slade, A. Al Khourdaji, R. van Diemen, D. McCollum, & M. Pathak].

Karlsson, M., Alfredsson, E., & Westling, N. (2020). Climate policy co-benefits: A review. *Climate Policy*, 20, 292–316. https://doi.org/10.1080/14693062.2020.1724070.

Karlsson, M., Westling, N., & Lindgren, O. (2023). Climate-related co-benefits and the case of Swedish policy. *Climate*, 11, 40. https://doi.org/10.3390/cli11020040.

Lelieveld, J., Haines, A., Burnett, R., Tonne, C., Klingmüller, K., Münzel, T., & Pozzer, A. (2023). Air pollution deaths attributable to fossil fuels: Observational and modelling study. *BMJ (Online)*, 383, e077784–e077784. https://doi.org/10.1136/bmj-2023-077784.

Mayrhofer, J.P., & Gupta, J. (2016). The science and politics of co-benefits in climate policy. *Environmental Science & Policy*, 57, 22–30. https://doi.org/10.1016/j.envsci.2015.11.005.

Afterword

The wide range of thought and practice covered by the 87 essays in this book make it abundantly clear that consumer society is a complex system of production-consumption. Its many elements are intertwined and mutually dependent, including the structure of the economy and the financial system, major societal institutions, and the built environment as well as social practices, behaviors, beliefs, dominant values, cultural understandings, political ideologies, power centers, and individual and collective identities. This complexity makes the system resilient in the face of challenges and resistant to any rapid changes.

This complex system underwent explosive growth in the United States after World War II. Driven by business, media indoctrination, and government propaganda, it quickly spread to other parts of the world. Economic growth fueled by consumerism brought unprecedented material prosperity to the Global North, so that any reservations about the headlong race into consumerism were outshone by new gadgets and cars, and the sounds of endless construction: of houses, roads, suburbs, oil refineries, and much more. Even the rebellious decade of the sixties did not slow that down.

But we have long known that consumerism in high-income countries has been paid for by borrowing from the future of our children and their children's children, and by exploitation of the low-income countries, which supplied cheap labor, fossil energy, food, and minerals. It resulted in massive ecological degradation, much of it in distant locations.

The extraction of wealth by the haves from the have-nots, necessary for maintaining high standards of living and societal peace among the haves, continues. Billions of people around the globe, who were kept from sharing the fruits of consumer society, suffer from its consequences: poverty, unemployment, political instability, violent conflicts, pollution, and climate-related natural disasters. Disenfranchised and prospectless youth the world over are restless. They are abandoning traditions and are attracted to the lifestyles they see on the screens of their TVs and mobile phones. But to enable them to satisfy their basic needs, including consuming more, wealthy societies must consume less.

Despite the high price paid for the affluence of a minority of the global population, delivering well-being is just as elusive today as ever. Social solidarity, democracy, and widespread satisfaction with life in the high-income countries are in decline. Instead, we are witnessing a concentration of political and economic power, mistrust in people and institutions, declining political participation, and alarming rates of mental and physical ill health. The younger generations in high-income countries are anxious about the future. Their elders' prescriptions sound hollow to them: work hard, even at the cost of spending time with family and friends; compete with your neighbors and friends; accumulate as much wealth and stuff as you can; and do not rock the boat.

We write this Afterword at a time of heightened social divisions and shifts further toward the political right in major global power centers. This includes the polarization of wealth and political discourses, international isolationism, and a decline in climate mitigation and social safety net policies. Global billionaires are grabbing political power in their countries while having more solidarity with each other than with their own compatriots.

In the meantime, advanced technologies continue to drive consumption, using artificial intelligence and social media. In the United States, commerce on social media platforms has more than doubled in the last three years. Most purchases are made by the 18–44 age cohort, who also spend the greatest amount of time online. Other large economies in Europe, China, Australia, and others follow closely behind. The rate of increase in demand for energy surpasses the growth rate in renewable energy supply.

Many visions of a less consumerist future have been published, ranging from abstract to more specific, place-based, and culture-specific. That body of literature is burgeoning. But the question of *how* to transition toward a different organizing principle of social and economic life has been less explored, and even less supported by empirical data.

As we noted in the Introduction (section 6), the cause of sustainable consumption is a political orphan. In a related scenario, what would happen, for instance, if younger generations collectively reject a consumerist way of life, extreme individualism, and competition? What if they were to look to other cultures and value systems for inspiration? This is not outside the realm of possibilities. The Fridays for Future international climate movement demonstrated that young people can make themselves heard and affect government agendas. Going forward, they could become the engine of changing social norms.

In this volume, we provide a common language for inspiring champions of change around the globe. We translate often-abstract concepts into their sphere of implementation. Our hope is that the vocabulary will create a political space for a discourse about consumption and lifestyles; and be used to grow a coherent movement that rejects consumerist lifestyles.

Contributors

Saamah Abdallah is program lead for sustainable wellbeing at the Hot or Cool Institute in Berlin. He has worked in the fields of well-being and beyond GDP since 2006, having co-developed the Happy Planet Index and worked on several other alternative indicators of progress including the Index of Sustainable Economic Welfare and Eurostat's Quality of Life indicators.

Alberto Acosta is a grandfather and economist with a degree from the University of Cologne, Germany. He has been a university professor in several countries, including as visiting professor at the University of Florida, and a Judge of the International Tribunal for the Rights of Nature.

Panayiota Alevizou is a lecturer in socially responsible marketing at Sheffield University Management School. She holds a PhD in Sustainability Labeling, and her research explores sustainability labeling as well as sustainable production and consumption practices across various industries.

Eva Alfredsson is a researcher at Uppsala University with a PhD in human geography. Eva is also an analyst working for the government in the area of economic growth and sustainability. Her research focuses on how to transition to a sustainable green economy, sustainable consumption and production, green structural change, green investments and sufficiency.

Erik Assadourian is a sustainability researcher, writer and communicator. Over 16 years with the Worldwatch Institute (2001–2017), Erik co-authored more than a dozen books, co-directed two book projects, and directed four, including Transforming Cultures: From Consumerism to Sustainability. Erik is also the director of The Gaian Way (gaianway.org), an ecospiritual organization founded in 2019 that works to reconnect people with the living Earth they are part of and depend on.

Tessa Avermaete is the founder of Run and Harvest. Tessa holds a PhD in applied biological sciences. She is a board member of Groene Kring, the young farmers in Flanders, and was co-chair of the Global Food Security conference 2024. Based on over 20 years of research in the domain of sustainable food systems, Tessa founded Run and Harvest in 2025, which aims at facilitating the constructive dialogue on food and farming between different actors in society.

Julia Backhaus is a research associate at the Living Labs Incubator of RWTH Aachen University, where she studies and supports the use of living labs for the co-production of knowledge in research and education. Julia has a background in the liberal arts (BSc) and science and technology studies (MPhil) and is currently finishing her PhD on societal transformation processes at the Maastricht Sustainability Institute of Maastricht University.

Lindsay Barbieri is a cofounder of DegrowUS and an assistant research professor and science-policy lead for the Global Land Programme at the University of Maryland College Park. With her research focus at the intersection of land, society, and climate, Bar works to deepen scientific understandings of land system science and inform just and sustainable pathways of climate action from local to global scales.

Richard Bärnthaler is a lecturer in ecological economics at the Sustainability Research Institute (SRI) and co-lead of the research group "Economics and Policy for Sustainability" at the University of Leeds. His research interests include social provisioning, foundational economy, sufficiency as a social organizing principle, sustainable well-being and ecosocial policies, the political economy of climate change, and de-/postgrowth.

Beatriz Barros is an activist and PhD candidate in anthropology at Indiana University, Bloomington, working for environmental and social justice. She is concerned with the environmental impacts of wealth and the links between climate change and economic inequality.

Magnus Bengtsson is an international sustainability expert advising governments and organizations on policies addressing consumption, production, and circular economy. He is the policy director at the Berlin-based Hot or Cool Institute, a sustainability advisor to a group of Japanese corporations, a lecturer at Toyo University, and a consultant to international organizations in Southeast Asia.

Peter Berrill is an assistant professor in the Industrial Ecology department (CML-IE) at Leiden University, where he teaches Material Flow Analysis. Peter's research focuses on the building, transport, and energy sectors and employs a variety of methods and tools including material flow and stock analysis, life cycle analysis, machine learning, and other statistical models. He is particularly interested in sufficiency and location-based strategies for mitigation of environmental impacts.

Soumyajit Bhar is the assistant dean of admissions and outreach and a founding faculty member at the School of Liberal Studies, BML Munjal University. With over 15 years of experience in environmental and sustainability research, he holds a PhD in sustainability studies from ATREE. His work explores the socio-psychological drivers of conspicuous consumption in India and its environmental impacts, while also engaging with broader questions of philosophy and ethics, particularly those related to environmental issues.

Sam Bliss teaches ecological economics and studies non-market food practices at the University of Vermont. He is also a co-founder of DegrowUS, co-organizer of Food Not Bombs Burlington, and a part-time shepherd at Rock Point Commons.

Roxana Bobulescu holds a PhD in economics and works as an associate professor at Grenoble Ecole de Management, France. Her research and teaching experience covers areas such as economic theory, history of economic thought, international economics, industrial organization, critical pedagogy, and ecological economics. Currently, she is studying the paradigm of "degrowth" from a multidisciplinary perspective and teaches the course "Sustainable Degrowth" to masters' students.

Bernd Bonfert is an assistant professor at the Department of Territories and Sustainable Development, EM Normandie Business School. His research lies at the intersection of political economy and transformation studies, as he investigates the governance models, political strategies, and role of citizen participation in spreading socio-ecological innovations such as the commons.

Valerie Brachya is a former lecturer in environmental planning and sustainability policy at Tel Aviv University and Hebrew University and a former research associate at the Jerusalem Institute for Policy Research. She was a founding member of the Environmental Protection Service in Israel and former Deputy Director General of the Ministry for Environmental Protection.

Aleksandra Burgiel-Szewc is an associate professor at the University of Economics in Katowice, Poland. She holds an MA in management and a PhD in economics. Her transdisciplinary research covered socially determined consumption, psychological determinants of consumer choices, and new consumer trends. Her recent studies refer mostly to sustainable consumption as well as the sharing economy and access-based consumption as alternative modes of needs satisfaction.

Kate Burningham is a sociologist in the Department of Sociology, University of Surrey, UK. Her research focuses on the social construction of environmental issues, environmental inequalities, and sustainable lifestyles. She is co-director of CUSP (Centre for the Understanding of Sustainable Prosperity), where she leads research on social and psychological understandings of "the good life", and is a co-investigator in the ACCESS Network https://accessnetwork.uk, which aims to raise the profile of environmental social sciences in tackling climate and environmental problems.

Martin Calisto Friant is the Global Value Chains Lead at Circle Economy Foundation and lectures on circular economy at the University of Amsterdam. His current work focuses on analysing the social and environmental justice implications of circularity transitions through a decolonial and degrowth lens, as well as developing policies to foster inclusive socio-ecological transformations at the local, national and international levels.

Noel Cass is a research fellow in energy demand behavior in the Institute for Transport Studies, University of Leeds, UK. His research has been almost exclusively on the social science and sociology of energy and everyday life, from work on nuclear waste management and renewable energy, to building design and office work, and transport and mobility studies including everyday travel, modal shift, EVs, autonomous vehicles, and electric (cargo) bikes.

Anna Coote is principal fellow at the New Economics Foundation (NEF). A leading analyst, writer, and advocate in social policy, she has written widely on social justice, sustainable development, working time, public health policy, and equality. She directed the Social Guarantee initiative (2021–2024) to develop the theory and practice of universal basic services (UBS). She was Commissioner for Health with the UK Sustainable Development Commission from 2000 to 2009.

Nilo Coradini de Freitas holds a PhD in Organization Studies, and from August 2025, he is an assistant professor in Management at the American University of Paris. Nilo has written a Master of Science dissertation relating the oeuvre of Ivan Illich to the field of Organization Studies (2019, PPGA/UFRGS, Brazil). With Neto Leão, he has also translated Ivan Illich's In the Mirror of the Past to Portuguese (N-1 Edições, 2024). Nilo has also co-authored papers using the concept of conviviality in Cadernos EBAPE.BR, Farol – *Journal of Organization Studies and Society*, and *Management and Connections Journal*.

Luca Coscieme leads the Sustainable Lifestyles program at the Hot or Cool Institute. He has been working for more than a decade in activating policymakers and other actors for enabling sustainable societies. In his current role, he develops and manages international Science-Policy projects considering aspects of inequality, and social and economic stability.

Duncan Crowley is an Irish architect exploring community-led ecocities through an action research PhD in Lisbon, Portugal. A climate activist for a quarter century, his work builds bridges between different communities, languages, movements, and peoples. He works with ECOLISE on the FLIARA and WAVE projects. Previously, he was part of UrbanA.

Callie Dance is a doctoral student at Idaho State University with a master of public administration. She is currently a grant coordinator for Idaho State University Institute of Rural Health, where she manages a Public Health AmeriCorps grant that focuses on boosting the workforce and organizational capacity for nonprofits that provide mental health and well-being services.

Katia Dayan Vladimirova is the founder of Post Growth Fashion Agency and an affiliated researcher at the University of Geneva, where she works on the topics at the intersection of fashion consumption, policy, and sufficiency. Her main research interest is systemic change toward a post-growth future, through the lens of fashion.

Elena Dawkins is a research fellow at the University of Surrey. She holds a doctorate in planning and decision analysis, with a specialization in environmental strategic analysis, from KTH Royal Institute of Technology in Sweden. Her research focuses on low-carbon transitions, sustainable lifestyles, carbon and environmental footprints, and policy for sustainable consumption and production systems.

Rico Defila is attorney at law. He is deputy leader of the research group Inter-/Transdisciplinarity, which is affiliated to the Social Transitions Research Group (STR) of the Department of Social Sciences at the University of Basel. His research covers the theory and methodology of inter- and transdisciplinarity, sustainable consumption, and quality of life.

John de Graaf is a documentary filmmaker, author, and frequent public speaker in the United States and Europe. His film *Affluenza* reached 15 million viewers on PBS and his book *Affluenza: The All-Consuming Epidemic* has sold nearly 200,000 copies in 12 languages. He has produced nearly 50 award-winning documentaries, most for public television.

Maike Demandt completed her master of science in sustainability management from the University of Wuppertal. In her master's thesis, she looked at possible extended producer responsibility models in the domestic mattress industry. In her work at the Wuppertal Institute's Circular Economy Department, her focus is on circular economy concepts for the automotive and the textile industries.

Timothy de Waal Malefyt (PhD anthropology, Brown) is a clinical professor of marketing at Gabelli School of Business, Fordham University. Previously, he held executive positions as VP, director of consumer insights at BBDO and D'Arcy advertising agencies, where he used cultural approaches to develop brand strategies and consumer insights for corporate clients.

Antonietta Di Giulio holds a PhD in philosophy. She is the head of the research group Inter-/Transdisciplinarity, which is affiliated to the Social Transitions Research Group (STR) of the Department of Social Sciences at the University of Basel. She is a founding member of SCORAI Europe. Her research and teaching interests cover inter- and transdisciplinary research and teaching, transformative research as well as sustainable consumption.

Marc Dijk is an assistant professor of Sustainability Transformations and research coordinator at the Maastricht Sustainability Institute, Maastricht University. His main research interests are sustainable innovation and policy, sustainability assessment and societal transformations toward sustainability. Marc is currently working on action-research projects in Living Labs focused on learning, upscaling, and social exclusion (SUMMALab, EmbedterLabs), and previously SmarterLabs.

Brett Dolter is an associate professor in the Department of Economics, University of Regina, where he teaches courses in climate change policy, cost-benefit analysis, ecological economics, and microeconomics. Brett's research is focused on modelling the costs and consequences of climate and energy policy.

Angela Druckman is emerita professor of sustainable consumption and production at the University of Surrey. Angela's research focuses on investigating avenues to more sustainable lifestyles. She takes a holistic, systems-based approach that encompasses supply chain analysis combined with understandings of individual and societal behaviors.

Fabián Echegaray is founder and director of Market Analysis, a market and opinion research company in Brazil. He is the current president of the Latin American chapter of WAPOR (World Association for Public Opinion Research). He holds a PhD from the University of Connecticut and is author of articles, chapters, and books on topics of political culture, electoral behavior, and sustainable consumption.

Megan Egler is an ecological economist interested in the political economy of extraction, energy transition, and desirable futures in historical fossil fuel regions. She cofounded DegrowUS in 2018 and has organized with the group ever since, and is currently a postdoctoral fellow at the University of Victoria, where her work supports community-led energy transitions.

Patrick Elf is an associate professor of sustainable business at the Centre for Enterprise, Environment and Development Research (CEEDR), Middlesex University, and co-investigator at the ESRC-funded Centre for the Understanding of Sustainable Prosperity (CUSP). Patrick's research focuses on investigating avenues for organizational change, sustainable business models as well as the mechanisms facilitating behavior change approaches toward the adoption of more sustainable lifestyles.

Roberto Falanga is an Italian sociologist. He earned a master's degree in psychology at the University La Sapienza of Rome in 2009 and a PhD degree in sociology (Democracy in the XXI Century) at the University of Coimbra in 2013. He is currently assistant research professor at the Institute of Social Sciences at the University of Lisbon (ICS).

Ana Rita Farias is a senior researcher and an assistant professor at Hei-Lab: Digital Human-Environment Interaction Lab. Professor Farias has held research positions and served as a guest researcher in various psychology and business and economics departments, including Utrecht University, University of Wisconsin, Católica Lisbon Research Unit in Business and Economics, and the Center for Economics and Finance at the University of Porto.

Tina Fawcett is a senior researcher and an associate professor in the Energy Program at Oxford University's Environmental Change Institute. Her research focuses on how and why we use energy, as individuals and organizations, and explores pathways and policies to reduce energy demand and carbon emissions.

Mariëlle Feenstra is co-founder and scientific director of 75inQ, the Institute on gender and energy promoting more visibility for diversity in the energy transition. After 15 years as a policy advisor for municipalities in the Netherlands, she returned to academia with her PhD project "Gender Just Energy Policy: engendering the energy transition in Europe", graduating in 2021 at University Twente, the Netherlands. She has been working on gender and energy since 2000.

Kuishuang Feng is an Ecological Economist in the Department of Geographical Sciences at the University of Maryland, College Park. Dr. Feng's research addresses urgent environmental and climate policy issues, including the distributional effects of environmental policies across regions and income groups, social and environmental inequality, and the implementation of the Paris Agreement and Sustainable Development Goals.

Daniel Fischer is professor of sustainability education and communication at Leuphana University, Lüneburg, where he also holds the UNESCO Chair in Higher Education for Sustainable Development. His research explores how more sustainable ways of living can be facilitated in education, with a special emphasis on teacher education.

Jared Berry Fitzgerald is an assistant professor of sociology at Oklahoma State University. His research broadly explores the political economy of environmental change and sustainability. Specific areas of research include working time, inequality, human well-being, climate change, water resources, and energy consumption.

Doris Fuchs is the director of the Research Institute for Sustainability (RIFS) at GFZ Helmholtz Centre for Geosciences in Potsdam and Professor of Sustainable Development at the University of Münster. Her research, teaching, and transfer activities focus on sustainability governance, sustainable consumption, and the sustainability-democracy nexus in particular, often centering on concepts and issues such as strong sustainable consumption, consumption corridors, power, and participation.

Jacob Gordon's background is in environmental journalism, the technology industry, sustainable building, and racial-equity-focused mental health services. His current studies at the Harvard Kennedy School of Government focus on the safe and humane deployment of emerging technologies.

Maike Gossen is a research associate at the chair of economic education and sustainable consumption at the Technische Universität Berlin, Germany. Maike's research interests focus on sustainable consumption and behavior change as well as sufficiency and sufficiency-promoting business strategies.

Ian Gough is visiting professor in CASE (Centre for the Analysis of Social Exclusion) and an associate of the Grantham Research Institute on Climate Change and the Environment, both at the London School of Economics. He is the author of numerous publications including *The Political Economy of the Welfare State* (1979); *A Theory of Human Need (1991): Global Capital, Human Needs and Social Policies* (2000); and *Heat, Greed and Human Need: Climate Change, Capitalism and Sustainable Wellbeing* (2017).

Mary Greene is an assistant professor of sustainable consumption at the Environmental Policy Group, Wageningen University. She specializes in critical geographical and sociological social practice research on (un)sustainable consumption. Focusing on the intersection of consumption with broader systems of provision, Mary explores everyday dynamics of change across energy, mobility, food, and circular society transformations.

Georgina Guillen-Hanson is a postdoctoral researcher working at the intersection of sustainability, gamification, and human behavior. At Tampere University's Gamification Group, she has been exploring the implications of gamification as a responsible research and innovation approach to advance sustainability transitions and translate scientific research outcomes into everyday language.

Juho Hamari is a professor of gamification at the Faculty of Information Technology and Communications, Tampere University. He leads the Gamification Group and the Research Centre of Gameful Realities. Dr. Hamari's group is focused on multidisciplinary and multimethodological research on gamification research holistically, including gamification, game-based learning, motivational systems, and extended realities (VR, AR, XR), situated in domains striving for ecological, economical, and social sustainability.

Ian Hamilton is the environmental discourse officer for the United States Bahá'í community. In this role, he seeks to promote principled dialogue among like-minded organizations with a view to facilitating a shift to sustainable development. Before joining the US Bahá'í community, Ian worked in climate finance for both the World Bank and African Development Bank.

Heli Hallikainen is an assistant professor at the Business School of the University of Eastern Finland with a tenure track on digital marketing and analytics. She coordinates UEF SenseLab, and her research topics include, e.g., consumer behavior, digital marketing, sensory marketing, and green consumption.

Anders Hayden is an associate professor in the Department of Political Science at Dalhousie University, with an emphasis on environmental politics. He is particularly interested in the concept of sufficiency and related post-growth ideas and initiatives, including sustainable consumption, work-time reduction, "beyond GDP" measurement, and the well-being economy.

Teresa Heath is an associate professor with habilitation in marketing at the University of Minho, prior to which she lectured at the University of Nottingham. She is a member of the research group NIPE. Her research focuses on transformative marketing and its intersection with ethics, sustainability, and consumption.

Jukka Heinonen is a professor of sustainable built environment at the Faculty of Civil and Environmental Engineering, University of Iceland. He also has a Docent position at Aalto University. He specializes in the challenges of sustainable development in community structures and lifestyles.

Claudia E. Henninger is a reader in fashion marketing management in the Department of Materials, University of Manchester. Her research interest is sustainability in the fashion industry, with a specific focus on the circular economy and end-of-life of garments.

Anna Horodecka's current research focuses on sustainable consumption and development, applying a multidisciplinary perspective – economic (including heterodox), ethical, philosophical, and sociological. She received three MA degrees in different disciplines and held scholarships from DAAD, Bavarian Ministry, and Renovabis. She holds two PhDs in economics and theology and a habilitation in economics and finance and is an associate professor at the Institute of International Economic Policy at the Warsaw School of Economics.

Jiayu Huang is a graduate student in sociology at Boston College. Her research interests are in environmental sociology, consumers and consumption, and cultural sociology, with a particular interest in sustainable food movements.

Amy Isham is a lecturer (assistant professor) in psychology at Swansea University, where she founded and leads the Sustainable Wellbeing Research Group, and a research fellow at the Centre for the Understanding of Sustainable Prosperity (CUSP). Amy's work explores the relationship between human well-being and engagement in ecologically sustainable lifestyles.

Diana Ivanova is a lecturer in environmental and climate governance at the School of Earth and Environment at the University of Leeds. As part of her Marie Skłodowska-Curie Individual Fellowship (ShaRe, 2019–2024), she has researched the social and environmental impacts of household, community and stranger sharing, including the sharing economy.

Melanie Jaeger-Erben is professor of sociology of technology and the environment at Brandenburg University of Technology and guest professor at the Technology for Circular Economy Center at University of Aalborg. Melanie does research in systems and practices of sustainable consumption and production, socio-scientific technology research, social innovation, and sustainability transformation.

Petr Jehlička is senior researcher at the Department of Ecological Anthropology at the Institute of Ethnology of the Czech Academy of Sciences. His research is located in agri-food and environmental studies and revolves around everyday environmentalism and sustainable food consumption at the intersection of formal and informal food economies.

Fatemeh Jouzi is a junior researcher at Lappeenranta-Lahti University of Technology. Her research in sustainability science focuses on conceptual aspects of sufficiency, such as money and time. She has an academic background in physics and the philosophy of science.

Shahzad Khan Durrani is a PhD scholar at the School of Economics, Beijing Institute of Technology. His areas of interest include energy economics, environmental economics, social and behavioral theories, and sustainable development policies.

Joseph Koetsier is a retired senior lecturer in cross-cultural educational studies and lifelong learning from Groningen University, the Netherlands, and University of the Western Cape, South Africa. He did field work and consultancies in India, Sri Lanka, Mozambique, and Suriname.

Vuyiswa Lamfiti runs the Phakama Africa campaign of the Ubuchule Resource Centre NPC. The center aims to reset the collective mindset to Ubuntu and self-reliance, by connecting with core values epitomized through the lives of African pioneers and legends (historical and living).

Eleftheria Lekakis is an associate professor in media and communications at the Faculty of Media, Arts, and Humanities, University of Sussex. Her research focuses on communication, promotional culture, and politics.

Jarkko Levänen is an associate professor of sustainability science at Lappeenranta–Lahti University of Technology. His research focuses on new socio-ecological models and sustainability management in the areas of circular economy and climate change mitigation. Jarkko's interests include sectoral transitions, new business models, innovation processes, and novel policy approaches in various socio-economic and cultural contexts.

Lassi Linnanen is a professor of environmental economics and management at Lappeenranta–Lahti University of Technology. His research over the years has covered a variety of topics, including sustainability transitions of energy, food and textile systems, sustainable innovations, corporate responsibility strategies, life cycle assessment, environmental and energy policy, and sufficiency as a solution for polycrisis.

Wenling Liu is an associate professor at the School of Economics, Beijing Institute of Technology. She got her doctoral degree in environmental economics and policy from Wageningen University, the Netherlands. Her research area is sustainable consumption theory and practice, with a focus on the formation mechanisms and transformations of household or individual behavior, as well as the interactions between behavior and the environment.

Sylvia Lorek is a researcher and policy consultant for sustainable consumption. She holds a PhD in consumer economics and has a lecturer position at the University of Applied Science in Münster. As chair of the Sustainable Europe Research Institute Germany e.V. she provides research for national and international organizations and institutes and is engaged in various national and international networks on sustainable consumption.

Stephan Lorenz is an "außerplanmäßiger" (non-scheduled) professor of sociology and project leader at the institute of sociology, Friedrich Schiller University of Jena. His main fields of research include environmental and sustainable sociology with a focus on consumption, technologies, infrastructure, food, water, poverty, and social exclusion. His entry on Freedom of Choice results from many years of research, but the writing of it was made possible by current project funding from the German Federal Ministry of Education and Research (grant number 03ZU1214NA).

Jessika Luth Richter is an associate senior lecturer at Lund University in Sweden. She researches policies and initiatives enabling sustainable consumption through a circular economy, including extended producer responsibility (EPR) policies, ecodesign policies, green procurement, circular business practices, circular city, and climate initiatives.

Donna Lybecker is a professor of political science at Idaho State University where she teaches classes on international relations, comparative politics, and environmental politics and policy. Her research looks at environmental policies and governance and the role of narrative in both policymaking and public opinion.

Johann M. Majer is postdoctoral fellow at the Department of Social, Organizational, and Economic Psychology, University of Hildesheim, Germany. Johann's research has two main focuses spanning across scales: sustainable consumption and behavior change at the individual level; the resolution of transition conflicts and social change at the societal level.

Tamar Makov is an associate professor and head of the Circular Economy Lab at the Department of Management, Ben-Gurion University of the Negev. Combining Data-Science approaches with behavior experiments and Industrial Ecology tools, she investigates how consumer behavior affects the potential to address social and environmental challenges through alternative provisioning systems.

Mikael Malmaeus is associate professor of environmental analysis at IVL Swedish Environmental Research Institute. He has wide research interests including sustainable macroeconomy, green transitions, and future scenarios.

Teresa Marat-Mendes is a full professor in architecture and urbanism at ISCTE, at the Department of Architecture and Urbanism and a senior researcher at DINÂMIA'CET. Teresa's research activity is centered on the following areas: architecture and urbanism, urban morphology, urban metabolism, urban design, urban sustainability, food systems, water and agriculture in urban planning, and Portuguese architecture and urban planning in the 20th and 21st centuries.

Aitor Marcos is an assistant professor of business economics at the University of the Basque Country. He is also affiliated with the Wrigley Institute for Environment and Sustainability at the University of Southern California. As a behavioral-ecological economist, he investigates the norms, beliefs, preferences, and structural factors that shape resource and energy demand. His recent work has focused on the limitations of promoting individual behavior change to transform sustainable consumption patterns.

Lisa Mastny is a senior associate and communications strategist at OneEarth Living. She provides research, editing, and communications support to a wide range of nonprofit and international organizations focused on sustainable development, including OneEarth Living, UNEP, the International Renewable Energy Agency, REN21, SLOCAT, and Food & Water Watch.

Manu V. Mathai is the director of research, data, and impact at WRI India. He was previously a university professor for 12 years, when he taught in the United States, Japan, and India. He studies the political economy of development as expressed in energy infrastructure choices and its implications for finding justice on a shared and finite planet. The content and opinions expressed in the chapter are those of the author and are not necessarily endorsed by, or do not necessarily reflect the views of, WRI India.

Erik Mathijs is director of SFERE (Sustainable Food Economies Research Group) and professor of agricultural and resource economics at the Department of Earth and Environmental Sciences, KU Leuven. His research focuses on the practices, metrics, and policies fostering the transformation of the agricultural and food system toward sustainability and resilience.

Giovanna Micarelli is professor of anthropology at the Pontificia Universidad Javeriana of Bogotá and guest researcher at the Center for Social Studies of the University of Coimbra. Her ethnographic research, primarily centered in Amazonia, is dedicated to bridging academic inquiry with support for indigenous peoples' rights and recognition. She undertakes collaborative research on indigenous alternatives to Western development, intercultural health and education, territorial protection, food sovereignty, and the commons.

Mirja Mikkilä works as an associate professor (tenure track) of business economics of forestry at the Department of Forest Sciences, University of Helsinki. She holds also an adjunct professor (Docent) position in Sustainability Science, LUT University. Her expertise covers sustainable management and utilization of natural resources, sustainable systemic transition, environmental economics, natural resource economics, corporate responsibility, and sustainability.

Jihoon Min is a research scholar at the International Institute of Applied Systems Analysis (IIASA) in Laxenburg, Austria. He has been working on issues related to poverty, inequality, and well-being especially around formulating decent living standards and their connection to energy, material, and climate problems.

Oksana Mont is a professor of sustainable consumption governance. She conducts transdisciplinary and international research on sustainable business models, sustainable consumption, and sustainable consumption policy. The current topical focus of her work is post-growth business, the sharing economy, and sustainable lifestyles.

John Mulrow is a degrowth researcher and activist based in Chicago, Illinois. He is a cofounder of DegrowUS, executive director of the Degrowth Institute, and adjunct assistant professor of environmental and ecological engineering at Purdue University. His work is focused on improving environmental assessment methodology through perspectives that prioritize social equity and limits to growth.

Lindsay Naylor is a feminist political geographer who has worked with indigenous communities in the Americas for over a decade. She uses qualitative methods to understand and explain injustices across space. She is the author of the award-winning book *Fair Trade Rebels: Coffee Production and Struggles for Autonomy in Chiapas*.

Beyza Oba is a lecturer and researcher in organization studies at Istanbul Bilgi University. She carried out her doctoral study on business ethics at İstanbul University. She has researched and published, with a critical management perspective, on topics such as the hegemonic struggle of employer unions, the role of trust in industrial districts, strategies in and around open-source software communities, and gender diversity practices of Turkish companies. She currently works on alternative consumer cooperatives and food politics.

Robert Orzanna is a long-term affiliate at the Sustainable Consumption Research and Action Initiative (SCORAI). He completed a master's degree in sustainable development at Utrecht University, the Netherlands, and since then contributed to several projects focused on communicating sustainable consumption and degrowth to the wider public.

Zeynep Özsoy has built her academic career in Organization Studies for more than 25 years. She is currently the acting Dean of the Business School at Altınbaş University. She has written about corporate governance practices of Turkish listed companies, boards of directors, gender diversity, and alternative organizations.

Yael Parag is a professor of energy policy. She is the head of the energy track and former vice dean of Reichman University's School of Sustainability. She holds a BSc degree in biology, and an MA and PhD in social science. Her research focuses on energy consumer behavior, consumer integration into the smart energy grid via smart technologies, prosumer markets, demand flexibility, energy security, local and community governance of energy, public acceptance of energy generation technologies, microgrids, personal carbon trading, and more.

Susan Paulson serves as professor at the University of Florida's Center for Latin American Studies and previously taught sustainability studies during five years in Europe and ten years in South America. Her research on human-environment relations in Andean and Amazonian communities, and beyond, explores dynamics of gender and masculinities.

Jenny Pickerill is a professor of environmental geography at the University of Sheffield, England. Her research focuses on inspiring grassroots solutions to environmental problems and on hopeful and positive ways in which we can change social practices. She has published books and articles on themes around ecohousing, eco-communities, social justice, and environmentalism.

Veronica Pizziol is a research fellow at the Economics Department of the University of Bologna. She obtained her PhD in economics at IMT School for Advanced Studies Lucca. Her main research interests lie in experimental and behavioral economics, with a particular focus on cooperation, pro-sociality, and norms.

Raquel Rebouças is a professor of graphic design at the Federal Institute of Paraiba, Brazil. She conducts research on consumer behavior and social and cultural dimensions of design.

William E. Rees is an ecologist, ecological economist, professor emeritus, and former director of the University of British Columbia's School of Community and Regional Planning. He researches global ecological trends with a special interest in psycho-cognitive barriers to corrective societal change. Prof Rees is the originator and co-developer (with his graduate students) of "ecological footprint analysis" (EFA). He has authored hundreds of peer-reviewed and popular articles on EFA and humanity's overshoot predicament.

Andrew M.M. Reeves is a postdoctoral fellow in ecological macroeconomics and metrics, the Faculty of Environmental and Urban Change, York University and a member of the International Ecological Footprint Learning Lab. Originally a PhD astrophysicist studying galaxy formation, he now conducts research in ecological economics, with particular interests in macroeconomic modeling, degrowth, and Global North – Global South dynamics.

Lucia A. Reisch is the inaugural director of the El-Erian Institute for Behavioural Economics and Policy at Cambridge Judge Business School in the UK and a professor of behavioural economics and policy for sustainable development. Her research focuses on the theory and application of behavioral insights to promote behavioral change in individuals and organizations to promote sustainability.

Kristof Rubens is a policy advisor at the Department of Environment & Spatial Development of the Flemish government in Belgium. His work revolves around environmentally responsible food consumption and sustainable food systems. He has a keen interest in bringing together and developing insights on the environmental impacts of food consumption and behavioral strategies to influence consumption patterns.

Jennifer D. Russell is an associate professor in the Department of Sustainable Biomaterials at Virginia Polytechnic Institute & State University. Her research into circular economy and sustainability explores the flows of materials within economic consumption-production systems, and strategies for slowing and closing material loops.

Marlyne Sahakian is associate professor of sociology at the University of Geneva, where she brings a sociological lens to consumption studies and sustainability. She coordinates national and European research projects on energy, food, and well-being, often working with inter- and transdisciplinary teams.

Marta Santos Silva (PhD in law) is a postdoctoral researcher at the Research Centre for Justice and Governance at the University of Minho. She completed executive training in Behavioral Economics and Nudging at the University of Chicago Booth School of Business and served on the EC's Expert Group on Liability and New Technologies. Dr Santos Silva's contributions to this book are funded by the contract-program "Stimulus to Scientific Employment – Institutional" CEECINST/00157/2018 by the Fundação para a Ciência e a Tecnologia.

Benedikt Schmid holds a doctorate in geography from the University of Luxembourg and is currently a postdoctoral researcher at the chair of geography of global change at the University of Freiburg. His research focuses on the interplay between community-led organizations, ecosocial enterprises, and local institutions in post-growth-oriented transformations.

Max Schoepen is an applied researcher and designer. He holds a master's degree in industrial design engineering and an interdisciplinary PhD in industrial design engineering and health sciences (Ghent University). He's an educational assistant for product development students at the University of Antwerp.

Ulf Schrader is a professor of economic education and sustainable consumption and director of the School of Education at Technische Universität Berlin. He is trained as an economist and social scientist and has done research mainly on (education for) sustainable consumption and on corporate responsibility. His teaching focuses on teacher training for the school subject

Arbeitslehre (Work Studies), which integrates economic education, technical education, consumer education, and vocational orientation with a focus on sustainable development.

Adrian Smith is a professor of technology and society based at the Science Policy Research Unit of the University of Sussex, where he leads research projects and contributes to postgraduate teaching and professional training. Adrian has over 20 years' experience researching grassroots innovation initiatives in Latin America, Europe, and India.

Ana Maria Soares is an associate professor with habilitation at the School of Economics, and Management and Political Science, University of Minho (EEG-UMinho), and a member of iBMS – Centre for Research in Business, Markets & Society, University of Minho, Braga, Portugal. Her research interests include voluntary simplicity, the impact of virtual assistant technologies, social networks, and marketing for higher education.

Kate Soper is emeritus professor of philosophy and former researcher at the Institute for the Study of European Transformations at London Metropolitan University. She has published widely on environmental philosophy, aesthetics of nature, theory of needs and consumption, and cultural theory. Her study on "Alternative hedonism and the theory and politics of consumption" was funded in the ESRC/AHRC "Cultures of Consumption" Programme (www.consume.bbk.ac.uk).

Lucie Sovová is assistant professor at the Rural Sociology Group of Wageningen University in the Netherlands. Her work revolves around alternative economic thought and practice, particularly in relation to food and agriculture. She is interested in food provisioning practices that are not only more environmentally friendly, but that also create distinctive socio-economic relations geared toward well-being of more-than-human communities, care and solidarity.

Joachim H. Spangenberg is a senior scientist at the Juelich Research Centre and co-chairs the Steering Committee of the Ecosystem Services Partnership. He recently served on the German Bioeconomy Council and as a member of the Scientific Committee of the European Environment Agency. EEA, Copenhagen, and co-chairs the Steering Committee of the Ecosystem Services Partnership. Earlier, he served on and participated in UN and OECD expert groups on sustainable consumption and on the sustainable development indicators underpinning Agenda 21.

Jared Starr is a sustainability scientist, ecological economist, and associate director of the Energy Transition Institute at the University of Massachusetts Amherst. His research analyzes how economic inequality – particularly at the very top of the income distribution – shapes the distribution of environmental benefits and harms through the global economy.

Katarzyna Stasiuk is an associate professor at the Institute of Applied Psychology at the Jagiellonian University in Krakow, Poland. She specializes in social psychology and the psychology of consumer behavior. Recently, her interests have focused on sustainable consumption issues and the application of consumer psychology in addressing legal problems. She has many years of practical experience in conducting qualitative and quantitative market research.

Jennie C. Stephens is a professor of climate justice at the Irish ICARUS Climate Research Centre at Maynooth University and lives in Dublin, Ireland. She is the author of Climate Justice and the University: Shaping a Hopeful Future for All (Hopkins University Press, 2024) and Diversifying Power: Why We Need Antiracist, Feminist Leadership on Climate and Energy (Island Press, 2020).

Jessica Stubenrauch is a postdoctoral researcher in the Department of Environmental and Planning Law at Helmholtz Centre for Environmental Research, Leipzig. Her research focuses on sustainable land use in line with international climate and biodiversity targets from a legal and governance perspective. Her particular interest lies in the implementation of sustainable agriculture and forestry systems.

Bianca Stumbitz is a senior research fellow at the Centre for Enterprise, Environment and Development Research (CEEDR) at Middlesex University Business School in London. Her key research interests include equity, diversity, and inclusion in the workplace, with an emphasis on gender- and work-family-related issues.

Sahra Svensson-Hoglund is an interdisciplinary PhD candidate at Virginia Polytechnic Institute & State University, researching realized states of a Circular Economy and the implications for product users, particularly with regards to their well-being. Her background is in law and business.

Nina Szczygiel is a professor of management at Aveiro University and a full researcher at the Research Unit on Governance, Competitiveness and Public Policies (GOVCOPP). With her background in economics, management and psychology, she has focused her research on intersectoral partnerships, social networks and integration, and well-being.

Alessandro Tavoni is professor at the Economics Department, University of Bologna. Before that, he was associate professor at the London School of Economics Grantham Research Institute on Climate Change and the Environment. He investigates the drivers of cooperative behavior in the global commons through a combination of non-cooperative and evolutionary game theory models, as well as experiments, surveys, and simulations.

Birgit Teufer is the head of the Department of Business & Psychology at the FERNFH University of Applied Sciences in Austria. She carried out her doctoral research at the University of Klagenfurt (Austria), where she investigated the dynamics of consumer cooperatives and related social innovations. Her research is particularly focused on the impact of these cooperatives on sustainable development, emphasizing not only ecological and economic factors but also the critical social dimensions.

John Thøgersen is a professor of economic psychology at the Department of Management, Aarhus University. He is the coordinator of the Marketing and Sustainability Research Group at the Department of Management. He is also an editor of the *Journal of Consumer Policy*, published by Springer-Nature.

Vanessa Timmer is executive director of OneEarth Living, a nonprofit "think and do tank" advancing sustainable everyday living around the world. Vanessa is also a senior research fellow at Utrecht University with the Copernicus Institute of Sustainable Development, the Urban Futures Studio, and Pathways to Sustainability.

Viivi Toivio is a senior data analyst at the Hot or Cool Institute. She develops data tools and insights on the environmental impacts of lifestyles, products, and services, helping inform policy and guiding organizations and individuals toward a sustainable future. Her recent work focuses on promoting 1.5-degree lifestyles to meet climate targets.

Arnold Tukker is distinguished professor for inter- and transdisciplinary sustainability research and professor of industrial ecology at Leiden University. He served a maximum term as Scientific Director of Leiden's Institute of Environmental Sciences (CML) from 2013 to 2022, overseeing a growth of the institute from some 50 to 170 staff, including PhDs. He led the EU Sustainable Consumption Research Exchanges (SCORE!) network project, which formed the inspiration for starting the SCORAI Global and European networks.

Edina Vadovics has a background in education (MEd) and environmental sciences and policy (MSc and MPhil). She is research director at GreenDependent Institute, a non-profit research and action organization in Hungary. Edina's work focuses on sustainable lifestyles, active (energy) citizenship, social innovation, and the role of communities, often in a multi-actor and transdisciplinary setting.

Karine Van Doorsselaer is an industrial chemical engineer with a specialization in plastics. She holds a master's degree in ecology and a PhD in human ecology (Free University of Brussels). She is a lecturer in materials science and ecodesign at the Product Development Curriculum of the University of Antwerp.

Rens van Tilburg is a researcher and advisor on sustainable economy and finance at Sustinnova. He is also a member of the Dutch academic think tank Sustainable Finance Lab, where he has served as executive director between 2011 and 2024. Before that, Rens worked as an economic advisor to the Greens in the European and Dutch Parliament and to the Dutch government on innovation policies.

Stephan Wallaschkowski is a researcher at Leuphana University, Lüneburg, and a lecturer and the study program coordinator for the bachelor's and master's degrees in sustainable development at Bochum University of Applied Sciences. He is also a co-chair of the joint SCORAI and Future Earth working group on sustainable consumption communication.

Thomas Webler is a senior research fellow at the Social and Environmental Research Institute in Massachusetts. His research focuses on community engagement in decision-making about risky and controversial technologies. He was a Fulbright Teaching Scholar at Pondicherry University in India on biodiversity protection, a Breuninger Fellow at the University of Stuttgart on the topic of cooperation, and a senior fellow at the Institute for Applied Sustainability Studies in Potsdam Germany.

Lorraine Whitmarsh, MBE, is an environmental psychologist specializing in perceptions and behavior in relation to climate change, based in the Department of Psychology, University of Bath, UK. She is the director of the UK Centre for Climate Change and Social Transformations (CAST) and co-director of Bath's Institute of Sustainability and Climate Change. She was lead author for IPCC's Working Group II Sixth Assessment Report.

Dominik Wiedenhofer works as senior research scientist in the field of industrial ecology at BOKU University, Austria. He investigates pathways towards a sustainable circular economy and futures with high well-being with lower energy and material demand. For this purpose, he develops biophysical systems models to quantify assess supply- and demand-side measures aimed at mitigating resource use and greenhouse gas emissions.

Nannan Xi is an assistant professor of gamification at the Faculty of Management and Business at Tampere University. Her research focuses mainly on game-based approaches (gamification, gameful systems, AR/VR/MR, metaverse, multisensory modalities, wearables, and streaming technologies) in information systems, management, consumer psychology, metaverse business, and organization studies toward sustainable decision-making and digital business transformation.

Lei Zhang is a lecturer at the Environmental Policy Group at Wageningen University. She joined the School of Ecology and Environment at Renmin University of China as an associate professor since 2009. As an environmental sociologist, her research covers topics ranging from environmental policy analysis and governance of environmental flows, to sustainable consumption.

Yulin Zhu is a PhD student at the School of Economics, Beijing Institute of Technology in China. Her research interests are sustainable consumption behavior and environmental policy analysis. She is currently researching the impact of the digital economy on consumption and its carbon footprints in China.

Caroline Zimm is a research scholar at IIASA (International Institute of Applied Systems Analysis) in Laxenburg, Austria. Caroline is working on inequality and justice aspects related to energy demand, technology diffusion, and climate change. Her research has also focused on identifying transformative pathways within safe and just Earth System Boundaries.

Index

For Product Safety Concerns and Information please contact our EU
representative GPSR@taylorandfrancis.com Taylor & Francis Verlag GmbH,
Kaufingerstraße 24, 80331 München, Germany

Printed and bound by CPI Group (UK) Ltd, Croydon, CR0 4YY
28/01/2026
02044214-0012